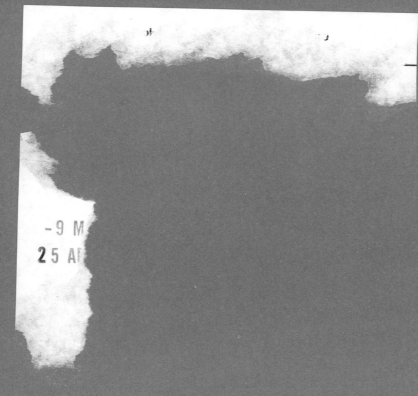

-9 M
25 A

REVISED EDITION

Pneumatic Control for Industrial Automation

Peter Rohner
Gordon Smith

John Wiley & Sons
Brisbane   New York   Singapore   Chichester   Toronto

First published 1987
by AE Press

Published 1989 by
JOHN WILEY & SONS
33 Park Road, Milton, Qld 4064
1 Thomas Holt Drive, North Ryde, N.S.W. 2113
90 Ormond Road, Elwood, Vic. 3184
236 Dominion Road, Mount Eden, Auckland 3, N.Z.

Typeset by Abb-typesetting, Collingwood, Victoria

Revised edition 1990

Printed in Singapore

National Library of Australia
Cataloguing-in-Publication entry

Rohner, Peter, 1939–
    Pneumatic control for industrial automation.
    Rev. ed.
    Includes index.
    ISBN 0471 33463 4.

    1. Pneumatic control. 2. Automatic control. I. Smith,
    Gordon. II. Title.

629.8'045

Cover photograph by courtesy of *Manutec*,
Gesellschaft für Automatisierungs-und
Handhabungssysteme

Design by Peter Yates

Illustrated by Peter Rohner

Foreword

'Automate or perish', the emotive buzz words of the '80s which turn away the indolent and excite the progressives dependent upon individual knowledge and experience of applied automation.

'When you begin to teach you begin to learn', words that go back in time but which are still very relevant in today's climate. This catch phrase particularly applies to the two authors who have condensed their combined industrial experience and knowledge into this manual on applied automation.

The text with its illustrations brings to the reader a clear concept and understanding of applied automation techniques. Moreover, the authors recognise the importance of combining pneumatic automation technology and electronic programmable controllers (PLC), together with descriptive language and graphic illustrations.

This manual is an invaluable tool to the practitioner, and essential reading for the student of fluid power and its applications throughout industry.

Fred Allen
Managing Director
Norgren Martonair Australia Pty Ltd

Acknowledgments

The authors wish to thank their respective wives, Heidi Rohner and Dilys Smith for the invaluable help and support they received in the preparation of this book for publication.

The cover photograph was obtained from Manutec, Gesellschaft für Automatisierungs-und Handhabungssysteme, D-8510 Fürth, West Germany.

The following materials have been reproduced by kind permission of the companies and organisations listed:

Appendix 4, Conversion table:	Applied Measurement Australia Pty Ltd
Figs. 5–3, 5–4, 5–5, 5–6, 5–7, 5–9, 5–11A, 5–11B, 5–16, 5–17, 13–4, 13–5, 13–6, 13–7, 13–9, 13–10, 13–14, 13–15, 13–17, 13–18, 13–22, 13–23, 13–24, 13–26, 13–27, 13–28, 14–7, 14–8, 14–9, 15–1, 15–3, 15–5, 15–6	Atlas Copco Australia Pty Ltd
Figs. 3–6, 3–12, 3–18, 3–19, 3–22, 3–26, 3–32, 6–8, 6–9, 6–15, 6–20, 6–21, 16–4	Ferrocast Williams Pty Ltd on behalf of Robert Bosch GMBH Stuttgart, West Germany
Fig. 3–17	Electro Hydraulic Industries Pty. Ltd., Mitcham, Victoria
Figs. 2–17, 2–18, 3–15, 3–30, 4–3, 4–4, 4–6, 12–5, 14–2	Festo Didactic GMBH, Esslingen, West Germany
Figs. 3–16, 8–41, 12–1, 12–2, 12–3, 12–6, 12–11, 12–13, 12–14, 12–15	Norgren Martonair Pty. Ltd., Mt Waverley, Victoria
Fig. 14–5	SMC Pneumatics Pty. Ltd., Mitcham, Victoria
Figs. 11–21, 11–26	Omron Tateisi Electronics Co, Bell Automation Pty Ltd, Preston, Victoria.

Please Note

Colour coding of fluid pressure used throughout this book:

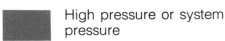 High pressure or system pressure

 Exhaust pressure, or background tint

 Reduced pressure

Contents

Foreword

Acknowledgments

Preface

1 Physical Principles in Pneumatics   1

2 Directional Control Valves   10

3 Linear actuators (cylinders)   36

4 Flow Controls for Pneumatics   60

5 Air Motors (Rotary Actuators)   68

6 Pneumatic Sensors   78

7 Circuit Presentation and Control Problem Analysis   89

8 Pneumatic Step-Counter Circuit Design   93

9 The Cascade System of Pneumatic Sequential Control   124

10 Combinational Circuit Design   137

11 Electronic programmable controllers for fluid power   158

12 Compressed Air Servicing   193

13 Air Compression   205

14 Compressed Air Drying   224

15 Compressed Air Distribution   232

16 Circuit Documentation and Circuit Construction   240

Appendices:

1 Industrial fluid power symbols   246

2 Fluid power formulae   251

3 Units of measurement and their symbols   252

4 Conversion table   254

Index

Preface

Automation of manufacturing processes in industry today is extremely important, not only for its economic benefits but also for its role in producing a consistent quality product, and whilst much is said about the negative aspects of automation in industry, automation has taken over many work duties which are dirty, monotonous and hazardous.

Automation has also created numerous job opportunities for properly trained and qualified people in all areas of fluid power: engineers, technicians, mechanics, sales and service personnel are in great demand, and there is also a shortage of trained and experienced teachers of fluid power subjects in post secondary, vocational and tertiary education.

For many years pneumatics has played an important role in automated manufacturing and handling processes ranging from the so-called "low cost automation" applications to the sophisticated and complex powering and control of entire transfer machines and robots. Pneumatics must therefore be seen and respected for its dual role: that of power transmission and that of logic control. The latter is often poorly understood and the widespread ignorance that pneumatics can effectively and cheaply play an important role in logic control may be attributed to one or both of the following factors:

- Superficially, the application of pneumatics to sequential and combinational control seems to be so simple that many assume there is little one needs to know about it to make use of the control components available. Such people will apply the controls and actuators without the mandatory understanding of their inherent characteristics; characteristics which are in fact very simple and based on the fundamental principles of logic control laid down by the rules and laws of Boolean Switching Algebra, Pascal's Law, and the Gas Laws.
- Teaching institutes and curriculum bodies make ample provision for subjects related to all facets of machine design and maintenance but pay little attention to fluid power control and power transmission, thus ignoring the fact that most production and manufacturing machines are up to 60% fluid power controlled. On the power transmission side, this percentage of fluid power involvement is even greater.

The purpose of this textbook is to facilitate a thorough understanding of the fundamental principles of industrial pneumatic control and power transmission as well as production and distribution of clean, compressed air. It is also intended to provide a practical working knowledge of the commonly encountered components for designing, installing and maintaining industrial pneumatic systems. As such, it is aimed at those students who are preparing themselves for industry, whether this be in design, maintenance or sales of pneumatics and related control systems.

The book is part of a comprehensive learning and teaching package which also includes more than 350 overhead projectorial masters to enchance effective teaching of this devise and involved subject. (These, and a series of workbooklets to match the text, are available directly from Peter Rohner, 14 Turner Street, Briar Hill, Vic 3088.)

Modern teaching method principles are applied throughout this book to stimulate and facilitate learning. Basic formulae, where introduced, are coloured with pink background tint, component functions are extensively described and simple circuits show their application and integration. Most components (actuators, valves, compressors, etc.) are depicted as sectioned illustrations together with their matching I.S.O. graphic symbol. The circuit design chapters show a variety of fully worked-out but typical industrial control circuits applying the widely used step-counter and cascade design methods. These are followed by a chapter dealing with the often least understood design of combinational logic circuits. Integrated into that chapter are such topics as truth tables, Boolean algebra concepts and Karnaugh-Veitch maps as used for the effective design of combinational controls. A complete chapter is dedicated to the programming of Electronic Logic Controllers (P.L.C.). The principles of circuit design set out in Chapters 2, 8 and 9 are also applicable to P.L.C. programming and are therefore repeated and integrated with electronic P.L.C. control in Chapter 12.

The book is divided into two sections: Chapters 1–12, written by Peter Rohner, explain the use of compressed air for control and power transmission, while Chapters 13–16, written by Gordon Smith, explain the production and distribution of compressed air. Also included in this section is a chapter on circuit construction and documentation.

The appendix contains a concise formulae collection, applicable I.S.O. fluid power symbols, units of measurement, prefixes for fractions and multiples of base units, an extenisve conversion table and, last but not least, a comprehensive index with over 800 listings.

Peter Rohner   Gordon Smith

1 Physical Principles in Pneumatics

Personnel who operate, service, or design fluid power systems should have a thorough understanding of the physics and properties of fluids and their behaviour under different circumstances.

Liquids and gases flow freely, and for that reason both are called fluids (from the Latin word *fluidus*, meaning flow).

A fluid is defined as a substance which changes its shape easily and adapts to the shape of its container. This applies to both liquids and gases. Their characteristics are discussed throughout this book.

Transmission of force by fluids

When one end of a bar of solid material is struck, for example with a hammer, the main force of the blow is transmitted straight through the bar to the opposite end. The direction of the blow determines the direction of the major force transmitted, and the more rigid the bar, the less force is either lost in it, or transmitted at angles different to the direction of the blow (fig. 1–1).

When a force is applied to the end of a column of a confined liquid (fig. 1-1), that force is transmitted straight through the column to its opposite end, but also—equally and undiminished—in every other direction, sideways, downwards, and upwards. This physical behaviour is defined by Pascal's Law (fig. 1–2). Pascal's discovery has opened the way to the use of confined fluids for power transmission and force multiplication.

Blaise Pascal (1623–1662) also discovered that pressure is equal to force per unit area, or the force divided by the area on which it acts. Thus, we must now define these three important quantities: Force, Pressure and Area.

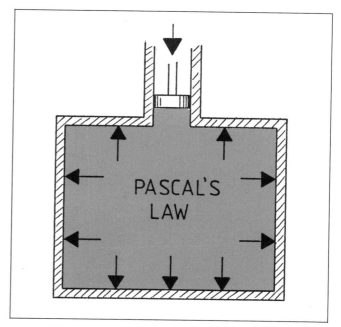

Fig. 1–2 Pascal's law states that pressure applied to a static and confined fluid is transmitted undiminished in all directions and acts with equal force on equal areas and at right angles to them.

Force

Forces cannot be seen but their effects can be. A force applied to an object has the effect of causing:

- a movement to begin or to change; or
- the shape of an object to change.

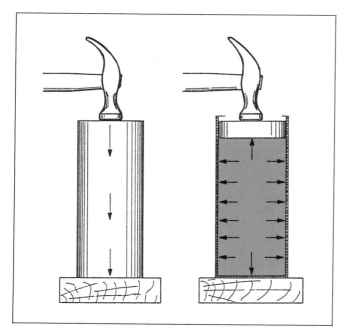

Fig. 1–1 Transmission of force through a solid material and through a static fluid.

In pneumatic applications force is produced by air pressure acting onto a surface (usually a piston, a valve or a diaphragm (fig. 1–3)).

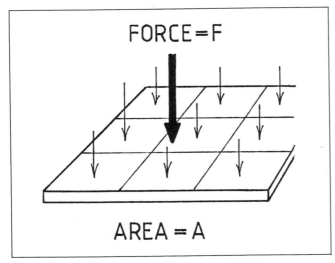

Fig. 1–3

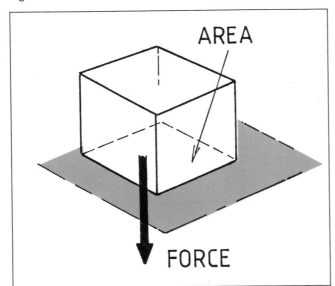

Fig. 1–4

Pressure

Pressure is defined as force per unit area (figs. 1–3, 1–4). This means that if we want to find out the pressure required for a certain pneumatic application, we have to measure or determine two quantities:

- the size of the force required; and
- the area over which the pressure is acting.

Area

The only area we are concerned with is the area or surface on which the pressure can act that will pro-

duce the force required for the job. This area is called *effective area* (fig. 1-5). By using these three quantities (Force, Pressure and Area) we can define their relationship as:

$$\text{Pressure} = \frac{\text{Force}}{\text{Area}} \qquad \text{Area} = \frac{\text{Force}}{\text{Pressure}}$$
$$\text{Force} = \text{Pressure} \times \text{Area}$$

The effective area "A" on which pneumatic pressure can act to produce the required pneumatic force (fig. 1–5) is either the:

- projection of the ball-seat area (check valve),
- projection of the pressure exposed piston area; or
- calculated effective piston area (spool valve).

Circular area

In pneumatic power transmission calculations the area of the circle is often used.

To calculate the area of a circle you may use one of the following formulae:

$$\text{Area} = \pi \times r^2$$
$$\text{Area} = \frac{\pi \times D^2}{4}$$
$$\text{Area} = D^2 \times 0.7854$$

Mass Compared to Force

The newton is the S.I. unit for force and is defined as: "That force which when applied to a mass of one kilogram produces in it an acceleration of one metre per second, per second in the direction of the force".

This definition gives no reference to gravitational force either on the earth or anywhere else and thus the newton is described as a non-gravitational or *absolute* unit of force. Newton found by experiments that on the surface of the earth acceleration (g) due to gravity is approximately 9.81m/s^2. Thus Newton's Law states:

$$\text{Force} = \text{Mass} \times \text{Acceleration} \qquad \text{or:}$$
$$F = m \times g$$

From the foregoing statements one can conclude that the mass of 660kg (fig. 1–6) exerts onto the platform on which it rests a force of 6474.4 newtons, calculated as follows:

$$660\text{kg} \times 9.81 \text{ N/kg} = 6474.4 \text{ newtons}$$

If one now assumes that the platform, the piston rod and the piston have a weight of an additional 140kg then the total weight to be lifted would amount to 800kg. Its force then is:

$$800\text{kg} \times 9.81 \text{ N/kg} = 7848 \text{ newtons}$$

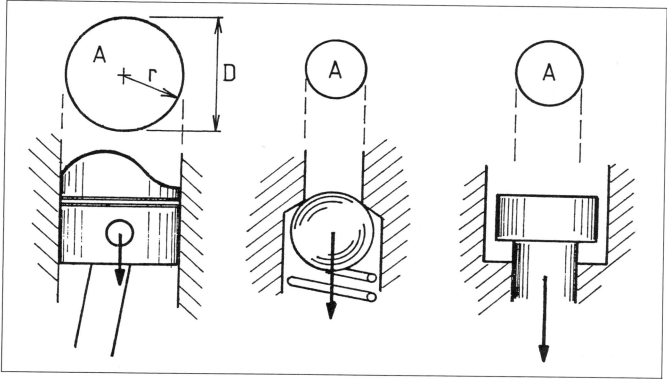

Fig. 1-5 Effective Areas.

The pressure in pascals (and bar) for this load to be lifted can now be calculated if one assumes the piston area to be $0.01m^2$ and actuator friction is neglected.

$$\text{Pressure} = \frac{\text{Force}}{\text{Area}} = \frac{7848\,N}{0.01m^2} = 784\,800\,Pa$$
$$= 784.80\,kPa = 7.848\,bar$$

Pressure in the S.I. system is expressed in *pascals*. Since the pascal is a very small unit, the prefixes kilo and mega are used to express pressure in industrial pneumatics. European countries predominantly use the unit "bar" to express pressure, whereby one bar equals 100 kPa or 100 000 Pa.

Flow and Pressure Drop

Whenever a gas is flowing a condition of unbalanced force must prevail to cause fluid motion. Hence, when a gas (and compressed air is a gas) flows through a pipe with a constant diameter, the pressure will always be lower downstream than at any point upstream (fig. 1-7).
The pressure differential or pressure drop is caused by the friction of the gas molecules amongst each other and friction of gas molecules on the walls of the pipe. Thus the inside surface

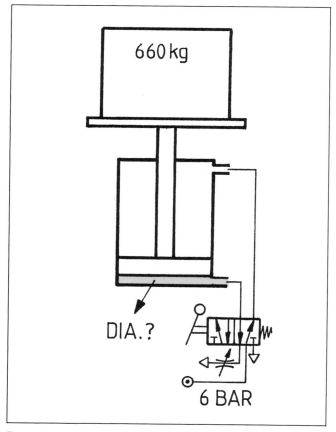

Fig. 1-6 Force-pressure-area relationship.

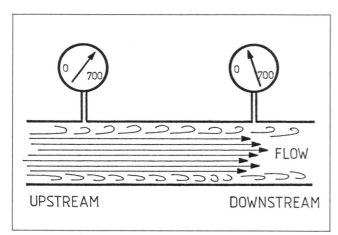

Fig. 1–7 The flow causes friction and thus a permanent pressure drop.

condition does influence the pressure drop in the piping system. This friction creates heat, and so the pressure drop (or loss) is a permanent loss, transferred into heat energy which cannot be regained (fig. 1–7).

As soon as the flow is stopped Pascal's Law must be applied for the now static condition, and pressures in all parts of the piping system (fig. 1–8).

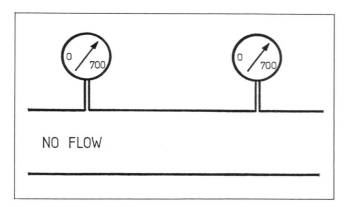

Fig. 1–8 Without fluid flow there is no pressure drop.

Flow through an Orifice to Atmosphere

An orifice is a hole with less cross sectional area than the pipes or cavities to which it is fitted. The orifice is generally used to control flow (speed control of actuators) or to create a pressure differential (pressure reducing valve).

When compressed air is discharged through an orifice to atmosphere, the speed at which it flows through the discharge orifice is either sonic (the speed of sound), or subsonic (slower than the speed of sound). This speed depends on two factors:

- the shape or type of orifice; and
- the pressure differential across the orifice.

If the pressure upstream of the orifice is 140 kPa (1.4 bar) higher than the downstream pressure then it can be surmised that the speed at which the compressed air flows is sonic. No matter how much the pressure upstream is increased now, the flow will remain sonic. This phenomenon is an important characteristic of compressed air on which the effective speed control of double acting linear actuators depends, and for that reason any pneumatic actuator or pneumatic motor (fig. 1–9).

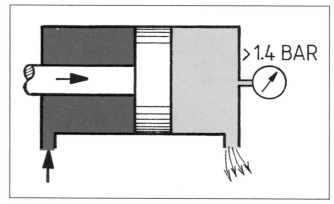

Fig. 1–9 Flow through an orifice to atmospheric pressure.

Air (Free Air) and Compressed Air

Air is a colourless, odourless and tasteless gas. In fact it is a mixture of individual gases, mainly nitrogen and oxygen.

Expressed in volumetric proportion the nitrogen makes up about 78%, the oxygen 20%, argon approx. 1% and the remaining 1% consists of other trace elements such as helium, hydrogen and neon.

Furthermore, atmospheric air has a great affinity with water, and unless dried for use in compressed air systems, may contain water in the form of vapour which sometimes can be as much as 6% by weight!

The air in the atmosphere contains not only gases and water-vapour, but also solid particles like dust, sand, soot and crystals of salt. In large cities and industrial areas, the number of such particles may reach 500 000 per cubic metre of air.

The force exerted on one square metre by an air column which reaches from sea level to the outermost layer of the atmosphere is approximately 101 300 newtons. Thus at sea level the absolute atmospheric pressure is about 101.3 kPa (1.013 bar).

At higher altitudes, the column of air to the outermost layer of the atmosphere is shorter, hence the mass of the gas (air) resting on each square metre is also less. For this reason the atmospheric pressure on top of Mont Blanc, the highest mountain in the European alps (4 810 metres above sea level) is only about 54 kPa. On the other hand, at the bottom of a deep mine shaft of about 800 metres below sea level, the atmospheric pressure is approximately 111 kPa (1.11 bar).

100 kPa = 1 bar
98.0665 kPa = 1 kg/cm^2
100 kPa = 14.5 p.s.i.
100 kPa = 1.02 atm
100 kPa = 750 torr
98.1 kPa = 10 m WG
1 mm Hg = 1 torr
101.3 kPa = 14.7 p.s.i.
1 atm = 14.7 p.s.i.
1 Pa = N/m^2 = 0.00001 bar

Atmospheric Pressure

The weight of the air (every gas has a mass) causes *atmospheric pressure*. Atmospheric pressure at sea level is approximately 101.3 kPa. Atmospheric pressure can be measured with a barom-eter. Torricelli's mercury barometer is shown in fig.1-10. The atmospheric pressure acts upon the open surface of the mercury container and thus supports the mercury column in the vacuum tube. The distance "h" is proportional to the atmospheric pressure and varies with altitude.

Vacuum Pressure

Any pressure below normal atmospheric pressure is termed *vacuum pressure*. It can be measured with the same barometer, depicted in fig. 1-10. Pressure readings below the calibration mark "A" denote vacuum pressure. At absolute zero pressure, the mercury column in the vacuum tube will disappear entirely, since no pressure will be present to support a mercury column. Therefore, it can be said that any pressure reading on the scale is *absolute pressure*, or an absolute pressure measurement.

Gauge pressure versus absolute pressure

Most pressure gauges used in pneumatic systems, however, have a pressure calibration based on *atmospheric pressure*. This means that atmos-

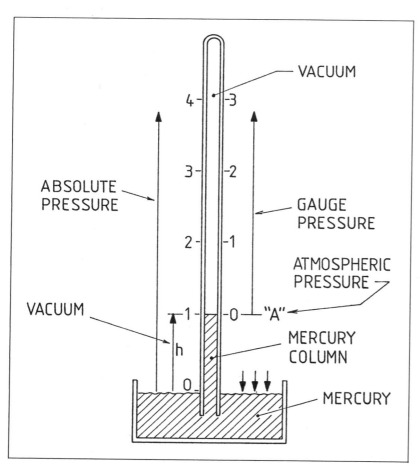

Fig. 1-10 Mercury barometer. Atmospheric pressure acting onto the mercury in the receptacle supports the mercury in the vacuum tube.

phere pressure is regarded as zero pressure, and any pressure above atmospheric pressure is thus a positive pressure reading. Most pressure gauges of this nature do not show pressure readings below atmospheric pressure. Pressure readings on such gauges are called *gauge pressure*, and pressure readings on gauges which start from absolute zero pressure are called absolute pressure (fig. 1–10).

Pressure in liquids

Under static conditions and without any external forces, the pressure at any point within a fluid system is proportional to the height of the fluid column above that point (fig. 1–11). Torricelli called the pressure at the bottom of the fluid column (tank) the *head pressure*. All that is required to work this pressure in Pa is the specific weight of fluid in the tank and the force that is produced by its weight. For example, a water column of 10 m height with a base area of 1 m^2 weighs 10 000 kg (specific weight for water is 1000 kg per 1 m^3).

Gas laws

Since gases are used in pneumatic systems it is important to mention at least some of the basic behavioural characteristics of such gases.

Boyle's Law

Robert Boyle (1627–1691) discovered by experimentation and direct measurement, that when the

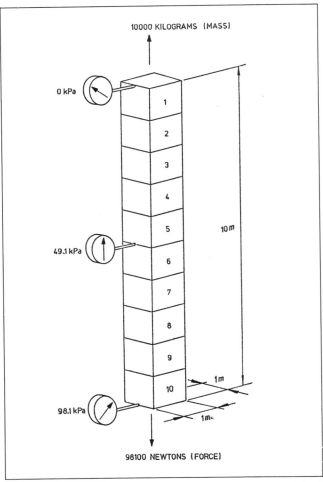

Fig. 1–11 A water column of 10 m with a base area of 1 m^2 produces a force of 98 100 Newtons, but its mass is 9.81 times less.

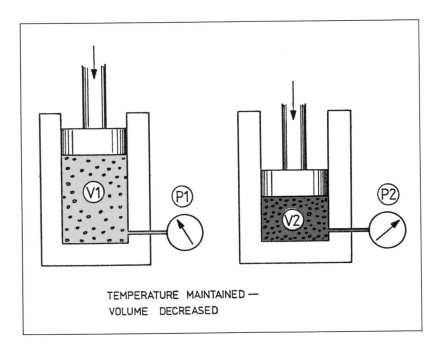

Fig. 1–12 Temperature maintained, volume decreased.

temperature of an enclosed sample of gas was kept constant and the pressure doubled by means of a piston (fig. 1–12) the volume was reduced to half the previous volume. As the piston retracted the volume increased again and the pressure decreased. Hence he concluded that the product of volume and pressure of an enclosed gas at a constant temperature remains the same (constant). Thus Boyle's Law states:

$$V_1 \times p_1 = V_2 \times p_2 = \text{constant, or } \frac{V_1}{V_2} = \frac{p_2}{p_1}.$$

It must be remembered that for calculation purposes one has to convert all given gauge pressures to absolute pressures.

Example

An air receiver with a volume of 6 m^3 must be filled with compressed air to a maximum pressure of 900 kPa (9 bar). Calculate the volume of free air (F.A.) to be pumped by the compressor.

$V_2 = ?$ m^3 $p_2 = $ Atmospheric Pressure
$\qquad\qquad = 101.3$ kPa (1.013 bar)

$V_1 = 6$ m^3 $p_2 = 9$ bar $+ 101.3$ kPa

$V_2 = \dfrac{V_1 \times p_1}{p_2} \quad \dfrac{6 \text{ m}^3 \times 1001.3 \text{ kPa}}{101.3 \text{ kPa}} = 69.3 \text{ m}^3$

Since the receiver already contained 6 m^3 of free air before the compressor started to pump, the pumped volume must be

69.3 m^3 − 6 m^3 = 63.3 m^3

Theoretically this is what would occur with an

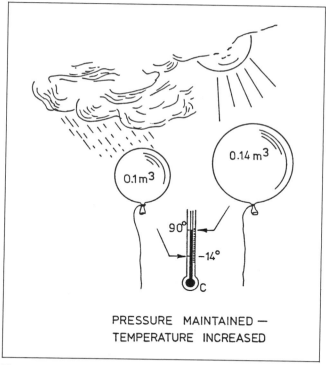

PRESSURE MAINTAINED —
TEMPERATURE INCREASED

Fig. 1–14 Pressure maintained, temperature increased.

ideal gas. But because density changes with temperature (that is, gases expand when heated and contract when cooled (see fig. 1–14), there will be some minor changes in practice with regard to pressure and volume. (A more detailed explanation is given in Chapter 13.)

Charles' Law

Whereas Boyle's Law describes the change of state of a gas at a constant temperature, experimental work by Charles (in which pressure was maintained constant while temperature varied, and the change in volume was measured) enabled him to formulate another gas law. (Fig. 1–14.)

Charles' Law states: at constant pressure the volume of a gas varies in direct proportion to a change in temperature. Expressed in a formula this is:

$$\frac{V_1}{V_2} = \frac{T_1}{T_2}, \text{ or, transposed } V_2 = \frac{V_1 \times T_2}{T_1}$$

To solve problems of this nature absolute values of temperature and pressure must be used.

Absolute zero on the Kelvin temperature scale is equivalent to −273 degrees Centigrade. One Kelvin temperature unit is equal to 1 degree Centigrade.

Thus 100°C = 373 Kelvin.

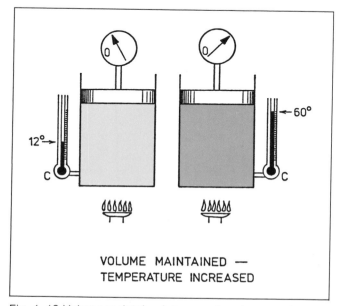

VOLUME MAINTAINED —
TEMPERATURE INCREASED

Fig. 1–13 Volume maintained, temperature increased.

Example

A balloon with a gas volume of 0.1 m^3 at a temperature of $-14°C$ is heated to a temperature of 90°C (fig. 1–14). What is its increased gas volume if the pressure remains constant?

$$V_2 = \frac{V_1 \times T_2}{T_1} = \frac{0.1 \text{ m}^3 \times (90°C + 273°C)}{(-14°C + 273°C)}$$

$$= 0.14 \text{ m}^3$$

Gay-Lussac's Law

Gay-Lussac's Law supplements the gas laws of Charles and Boyle. He observed that if a volume of a gas is kept constant, the pressure exerted by the confined gas is directly proportional to the absolute temperature of the gas (fig. 1–13). Expressed in a formula this is:

$$\frac{p_1}{p_2} = \frac{T_1}{T_2}, \text{ or, transposed } p_2 = \frac{p_1 \times T_2}{T_1}$$

The combination of the three gas laws of Boyle, Charles, and Gay-Lussac results in the *general gas law* reads:

$$\frac{p_1 V_1}{T_1} = \frac{p_2 V_2}{T_2} \text{ or } \frac{pV}{T} = \text{constant}$$

Advantages of pneumatic power transmission

- Freedom of location of pneumatic power converters such as compressors and actuators.
- Air for compression is available in almost unlimited quantities and can be exhausted again to atmosphere.
- Safety and overload protection by means of relief valves and the compressibility of the air.
- Compressed air can be stored and released from the air receiver (reservoir) which contains potential energy.
- Pneumatic systems work safely in explosion risk areas and are practically insensitive to high temperatures below 120°C.
- Pneumatic systems are clean, and should compressed air leak from a piping system, then there exists no contamination risk.
- Linear actuators operate with high speed (up to 2m/sec.) and their speed is infinitely variable.
- The operating elements (actuators, valve) are of simple construction and thus inexpensive.
- Pneumatic valves are ideally suited for logic functions and thus are used to control complex sequential and combinational machinery.

Transmission of power

There are four basic methods of transmitting the power of a prime mover to a machine. These methods are electric, mechanical, hydraulic and pneumatic. Each is used to transmit power (force over distance in time) and modify motion (accelerate, decelerate), and each method has its special capabilities and limitations. Pneumatic systems integrate all of these methods to accomplish a most effective and efficient form of power transmission (electrically operated valves, swivel actuators, hydro-check actuators, stored compressed air as potential energy).

The concept of power transmission

Power is the measure of a defined force moving through a given distance at a given speed. To understand this fundamental concept, the term force must now be explained.

Force may be defined as any cause which tends to produce or modify motion. Due to inertia, a body at rest tends to stay at rest, and a body in motion tends to maintain that motion until acted upon by an external force. Force is measured in newtons.

The concept of *pressure* must also be explained. Pressure is force per unit area and is expressed in pascals (Pa). Both force and pressure are primarily measures of effort. A force or pressure may be acting upon a motionless object without moving the object, if the force or the pressure is insufficient to overcome the inertia of the object.

Work is a measure of accomplishment. For example, the piston of a pneumatic actuator exerts a force on an object and moves the object over a given distance. Thus, work has been accomplished. The concept of work, however, makes no allowance for time. The work performed by the hydraulic actuator while moving the object from point A to point B in 8 seconds is precisely the same as when it moves the object from point A to point B in only 2 seconds, but the task performed in 2 seconds is obviously much greater. To explain the difference in performance one must resort to the definition of power.

Power is work performed per unit of time. Thus it can be said that power is the rate at which energy is transferred or converted into work. Power is expressed in watts. Mathematical relationships of force, pressure, area, work and power are shown in fig. 1–15.

$$\text{Pressure (p)} = \frac{\text{Force}}{\text{Area}}$$

$$\text{Force (F)} = \text{Pressure} \times \text{Area}$$

$$\text{Area (A)} = \frac{\text{Force}}{\text{Pressure}}$$

$$\text{Power (P)} = \text{Force} \times \text{Velocity}$$

$$\text{Force (F)} = \frac{\text{Power}}{\text{Velocity}}$$

$$\text{Velocity (v)} = \frac{\text{Power}}{\text{Force}}$$

$$\text{Power (P)} = \text{Pressure} \times \text{Flowrate}$$

$$\text{Pressure (p)} = \frac{\text{Power}}{\text{Flowrate}}$$

$$\text{Flowrate (Q)} = \frac{\text{Power}}{\text{Pressure}}$$

Fig. 1–15 Relationships of force, pressure, area, work, and power.

The basic concept of a pneumatic system

Pneumatics is the engineering science of pneumatic pressure, pneumatic potential energy production, its use in linear and rotary actuators and finally the use of compressed air as a signal carrier for the achievement of logic sequential and combinational control. Such pneumatic systems may include:

- Compressors which convert available power from an electric motor or internal combustion engine to pneumatic power at the actuator.
- Valves which control the direction of fluid flow to the actuators or process logic signalling functions to achieve sequential control of pneumatic actuators.
- Pressure and flow control valves which control the level of power produced on the actuators and pneumatic motors.
- Actuators which convert the potential energy of the compressed air into usable mechanical power output at the point required.
- Connectors which link the various system components provide power conductors for the compressed air in motion or in static condition.
- Sufficient quality, conditioning and storage equipment which ensure quantity and cleanliness as well as lubrication of the compressed air to provide a troublefree working life for its actuators and valves.

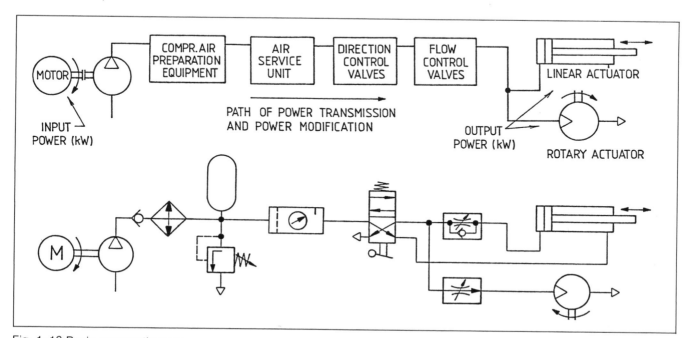

Fig. 1–16 Basic pneumatic system.

2 Directional Control Valves

A valve is a device which receives an external command (either mechanical, fluid pilot signal or electrical) to release, stop or redirect the fluid that flows through it.

Directional control valves in particular, as their name implies, control fluid flow direction. They are applied in pneumatic circuits to provide control functions which:

- control direction of actuator motion (power valve);
- select an alternative flow path (shunt valve);
- perform logic control functions ("AND", "OR" functions);
- stop and start flow of fluid (on-off valve);
- sense machine and actuator positions (limit valves).

Directional control valves are classified according to their specific design characteristics:

1. Internal valve mechanism (internal control element) which directs the flow of the fluid. Such a mechanism can either be a poppet disc, a poppet ball, a sliding spool, a rotary plug, or a combination of poppet and spool.
2. Number of switching positions (usually two or three). Some valves may provide more than three, and in exceptional cases up to six switching positions (selector valves).
3. Number of connection ports (not called number of ways!). These ports connect the pneumatic pressure lines to the internal flow channels of the valve mechanism and often also determine the flow rate through it.
4. Method of valve actuation which causes the valve mechanism to move into an alternative switching position. Figs. 2–7, 2–9, 2–13 and 2–15 show an array of frequently used valve actuation methods.

Valve symbols

Symbols are an ideal way of drawing and explaining the function of fluid power components and of directional control valves in particular. Symbols exist for most of the more commonly used valves. They are standardised to provide and safeguard an internationally agreed form of circuit drawing. These standards should be closely adhered to, and only deviated from if they are not available for a particular valve, or if a valve consists of several standardised parts. Most fluid power valves are drawn according to I.S.O. standards (International Standards Organisation) or C.E.T.O.P. standards (European Fluid Power Standards Committee).

Valve switching positions

For each of the switching positions provided by a valve, the graphic symbol shows a square (sometimes called a box). This means the valve on the left in fig. 2–1 provides two switching positions and the valve on the right three. The *rest position* of a valve is assumed by its moving parts (the valve mechanism) when the valve is not connected, not pressurised and not actuated. In most pneumatic control circuits all valves are depicted in the rest position and by convention valves with two switching positions should normally be connected to the right hand square (figs. 2–1 and 2–3).

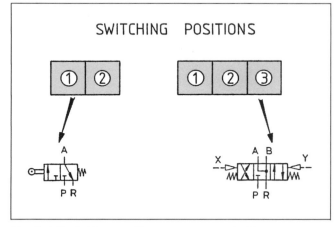

Fig. 2–1 Switching positions are represented by squares.

Valves with three switching positions should, however, be connected to their centre square (figs. 2–1 and 2–2), which again depicts the valve's rest position (or neutral position). However, it is sometimes the practice to draw valves in their actuated or *initial position*. This position is assumed when the valve is installed in the machine and a lever, a cam or a machine part is pressing onto the valve's operating element so the valve's mechanism is no longer in the rest position. Since this is not a standard practice, it is imperative to state on the circuit diagram whether all valves are

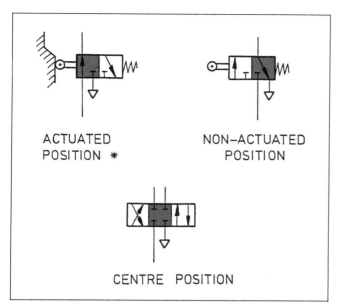

ACTUATED POSITION *

NON-ACTUATED POSITION

CENTRE POSITION

Fig. 2-2 Various valve connecting methods.

Fig. 2-4 Valve ports.

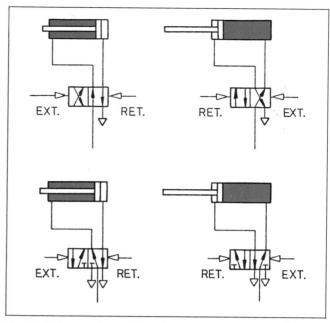

Fig. 2-3 Flow path configurations depend on initial actuator position.

shown in their rest position. Each of the valves being actuated must also be shown with a cam, and all pressure lines are to be connected to the left square (fig. 2-2).

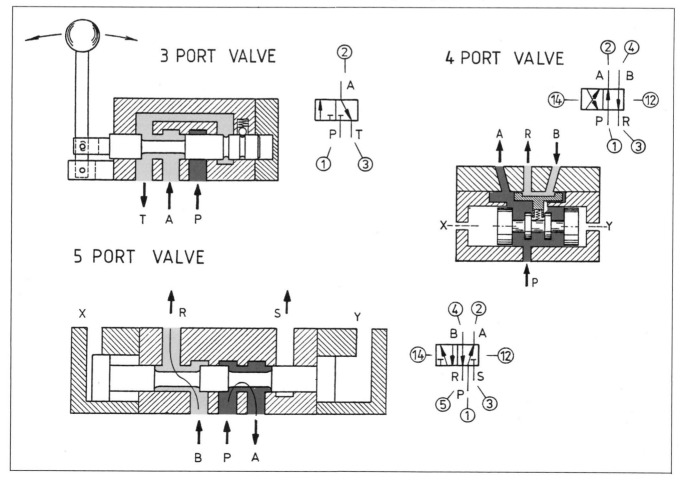

Two-position valves which are not spring biased and can freely assume both switching positions (bi-stable or memory valve) may be drawn in whatever flow-path configuration is demanded by the actuator position. But here again the pressure lines must be connected to the right hand square (fig. 2–3), and unnecessary crossing of pressure lines should be avoided (see also fig. 2–26).

Valve ports and port labelling

Directional control valves are frequently described by the number of their working ports (flow openings). Two-port valves are only used to turn airflow either on or off (fig. 2–28).

Three-port valves have three working ports (fig. 2–4). If a three-valve has two inlet ports and one outlet port, it is used as a selector valve ("OR" function valve) to guide two flow lines to one common outlet (fig. 2–38). Three-port valves are predominantly used to create pilot signals (fig. 2–35) or make logic functions (figs. 2–33, 2–38, 2–39 and 8-7).

Four- and five-port valves are used to control double acting linear actuators or reversible motors (fig. 2–3). The pressure port connects the valve to the pressure source, the two outlet ports are connected to the actuator and exhaust air escapes through the exhaust port(s).

The labelling of ports is presently undergoing some radical changes. The new port numbering system is based on a very logical concept. The pilot signal 12 for example, connects the pressure port 1 to outlet port 2, and the pilot signal 14 shifts the valve mechanism so that pressure port 1 is connected to outlet port 4. For three-port valves with only one exhaust, the exhaust port is labelled 3 (for old and new labelling comparison, see fig. 2–4). It must be noted that pilot signal ports do not count for the valve classification. Thus the five-port valve in fig. 2–4 is not a seven-port valve and is classified as a 5/2 valve (see also fig. 2–8).

Valve mechanism

Directional control valves consist of a valve body or valve housing and a valve mechanism. Some valves are mounted to a sub-plate. The ports in the sub-plate or on the valve body are threaded to hold the tube fittings which connect the valve to the fluid conductor lines (tubes). Some valves have the tube fittings moulded into the valve body. This is normally the case where the valve body is made from plastic. The valve mechanism directs the pressurised fluid (compressed air) through the valve body to the selected output ports or stops the fluid from passing through the valve.

External signal commands (electrical manual or air pilot pressure), and internal signal commands (air pilot pressure, spring force) may be applied to shift the valve mechanism (figs. 2–6, 2–7, 2–9 and 2–12).

Poppet, spool, flat slide, toggle disc and spool/poppet mechanism are predominantly used

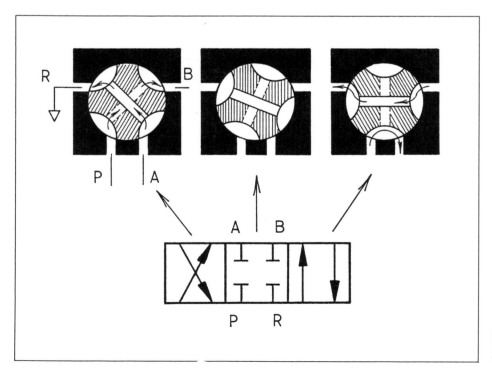

Fig. 2–5 Rotary plug valve as used in hand lever operated shunt valves.

in directional control valves whereas poppets and balls are the preferred mechanism in check valves and logic "OR" function and "AND" function valves. Rotary plugs or rotary plate valves are used only in hand lever operated selector or shunt valves (fig. 2–5).

Normally closed and normally open valves

These terms are generally applied to three-port valves. The valve shown in fig. 2–6 (top), a two-position valve with three ports, permits no flow from pressure port P to output port A in its normal (spring biased) position. In the case of a normally open valve (fig. 2–6, bottom), the supply is permitted to stream to output port A, while the valve is in its normal position. Both valves, when actuated, change their flow mode to the flow configuration depicted in the left-hand square of the valve symbol (fig. 2–6).

Methods of valve actuation

The term *actuation* in relation to pneumatic valves refers to the various methods of moving the valve mechnism into any of its alternative switching positions. Valves can be actuated by five basic methods: Manually, mechanically, electrically, hydraulically or pneumatically.

In complex machine applications any of these methods, or combinations of them, may be used to gain optimum control. Manual methods use hand or foot actuators such as levers, push buttons, knobs and foot pedals, electrical methods use A.C. or D.C. solenoids. For pneumatic/electronic control integration some manufacturers make low wattage coils for direct P.L.C. connection. Solenoids range from 12 to 240 volts. Most solenoid and air pilot valves are also equipped with a manual override button or plunger. Thus the valve mechanism can be operated even when the air pilot signal or the electrical signal should fail. In hydraulically and pneumatically operated valves the fluid pilot signal is made to act onto a piston or it may act directly onto the valve mechanism and thus moves it to its alternative position (figs. 2–4, 2–7, 2–9, 2–13 and 2–14).

Spring-actuated valves

The terms spring-biased or spring-offset, and in the case of three-position valves spring-centered, refer to the application of springs which return the valve mechanism to its *rest position* as soon as the pilot signal command is cancelled. Such commands, whether they be pneumatic, electric or manual, must be maintained as long as the valve is to remain in its non-normal or actuated position.

A valve without a spring bias is actuated entirely by its external controls, and can therefore "float" or "drift" between its two extreme positions wher-

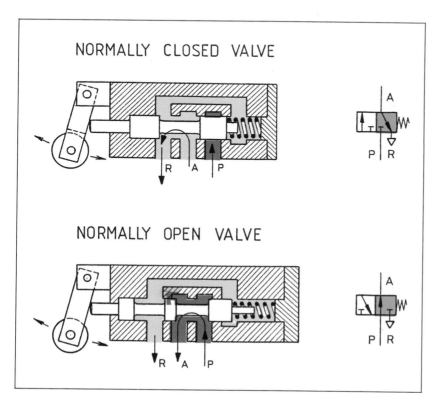

NORMALLY CLOSED VALVE

NORMALLY OPEN VALVE

Fig. 2–6 Valves with spring bias (normal position).

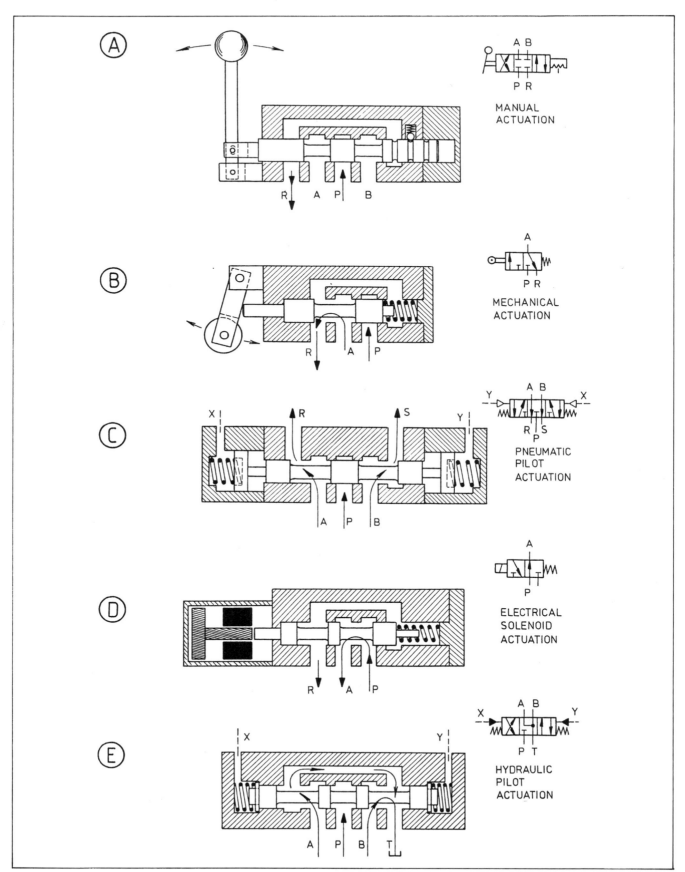

Fig. 2-7 Basic valve actuation methods applied to spool valves.

SYMBOL	PORTS	POSITIONS	COMMENTS
	2	2	N/C
	2	2	N/O
	3	2	N/C
	3	2	N/O
	4	2	
	5	2	
	3	3	FULLY CLOSED CENTRE
	4	3	FULLY CLOSED CENTRE
	5	3	PRESSURE TO A & B, EXHAUSTS CLOSED
	5	3	FULLY CLOSED CENTRE
	5	3	PRESSURE CLOSED, A & B TO EXHAUST

IF SPRING RESET

Fig. 2–8 Commonly used valves.

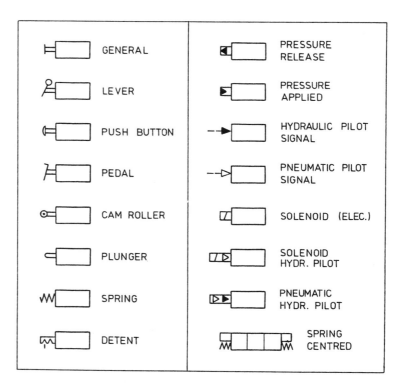

Fig. 2–9 Commonly used valve actuation methods.

GENERAL		PRESSURE RELEASE	
LEVER		PRESSURE APPLIED	
PUSH BUTTON		HYDRAULIC PILOT SIGNAL	
PEDAL		PNEUMATIC PILOT SIGNAL	
CAM ROLLER		SOLENOID (ELEC.)	
PLUNGER		SOLENOID HYDR. PILOT	
SPRING		PNEUMATIC HYDR. PILOT	
DETENT		SPRING CENTRED	

ever the external control is cancelled. To avoid this, a "detent" mechanism or friction-producing seal rings are used to give the valve a bi-stable (or memory) characteristic. Thus when the valve is switched (selected) into a specific flow condition, it will remain in that condition until switched into its alternative condition (figs. 2–7A, 2–12 and 2–17).

Spool valves

In modern valve designs, the spool valve (fig. 2–10) is most frequently used. Spool valves offer a number of advantages over poppet and flat-slide valves (fig. 2–12).

- Spool valves may be used for flow reversal. This is only possible with one other valve type, the spool/poppet valve (See also: spool/poppet valves, valve conversion).
- Since flow forces acting onto the lands are equal (equal pressure on equal area) the valve is said to be balanced and uses therefore only small actuating forces.
- Spool valves may operate without elastomeric seals if the spool to body tolerance is kept small (below 0.003 mm) and fluid pressure does not exeed 900 kPa (9 bar).
- Spool valves may be built for more than two switching positions, but then the spool is normally spring centered (fig. 2–10) or hand lever operated with detented positions (fig. 2–7A)
- Spool valves may be built for negative or positive overlap to achieve a specific cross-over condition or, if the centre is selectable, a specific centre condition (figs. 2–8, 2–10 and 2–21).

With spool valves one must be careful to mount the valve onto a reasonably flat (preferably machined) surface, since only slight body distortion may warp the valve body sufficiently to make the closely toleranced and lapped spool seize, or hinder it from moving freely when actuated. The same may also be said for bolting the valve too firmly against its mounting surface.

Spool valves are susceptible to contamination of the compressed air and therefore require a close monitoring of the air quality—its filtration and lubrication. The required filtration rate must not exceed 40 micrometres (40 μm).

Flat slide valves

Flat slide valves are a modern adaption of the old steam valve used on steam locomotives. The flat slide is carried to and fro by a grooved piston. The piston may either be air-pilot signal or spring or solenoid/air pilot operated (figs. 2–4 and 2–12). The valve can not be flow reversed (P cannot be used as R and vice versa) but it offers the advantage of a combined exhaust port which may be useful for actuator speed control (see speed control fig. 4-2). Since compressed air forces the flat slide against its bearing surface, the flat slide provides absolute leakfree (pressure assisted) sealing and does automatically adjust for wear on its gliding surface. Flat slide valves are also made with a double slide to gain two exhaust ports and some valves have also snap action toggle discs instead of carrier spools.

Poppet valves

Poppet valves are still used and offer at least two significant advantages over spool and flat slide valves (figs. 2–14):

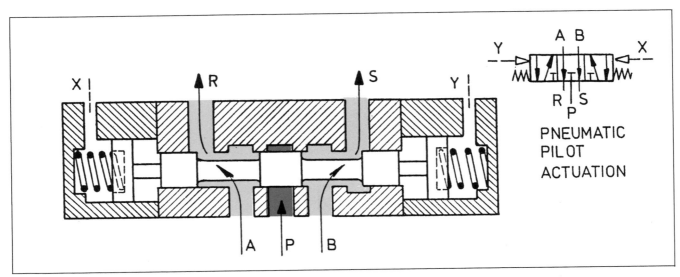

Fig. 2–10 Spool valve spring centred with specific centre condition and positive overlap.

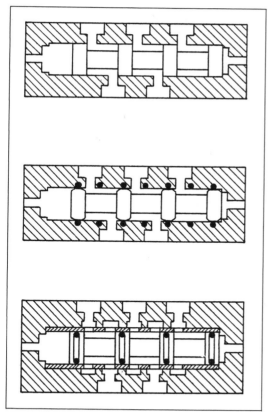

Fig. 2-11 Various spool valve designs.

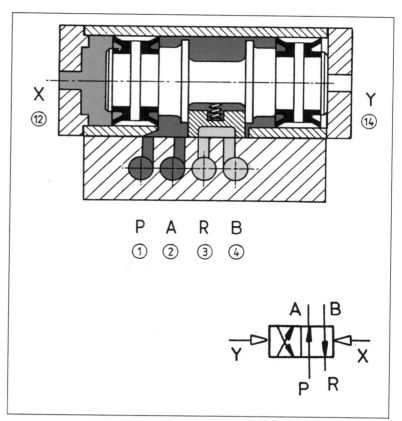

Fig. 2-12 Air-pilot operated flat slide valve.

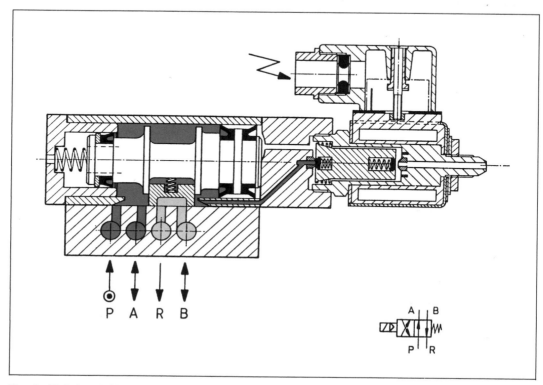

Fig. 2-13 Solenoid/Air pilot operated flat slide valve with reset spring.

- Poppet valves are robust and do not require lubrication. They have no close tolerance gaps on moving parts since they provide head-on instead of dynamic sealing.
- Poppet valves provide positive and air pressure assisted sealing (air pressure acts onto the underside of the poppet, figs. 2–14 and 2–15).

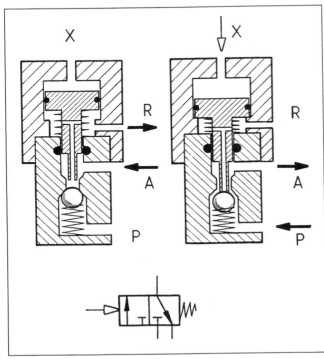

Fig. 2–14 Ball type poppet valve.

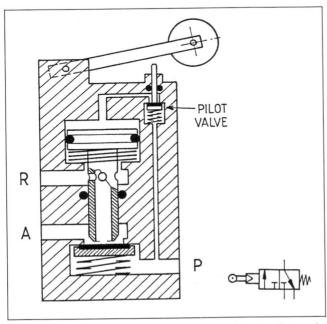

Fig. 2–15 Roller operated poppet valve with air pilot assistance.

Because of the sealing design mentioned earlier, poppet valves require a much higher operating force than any other valve design type and thus are often built with air–pilot operating assistance. The operating mechanism (roller, plunger, lever) opens a small pilot valve. This pilot valve directs pilot air to a large piston or diaphragm which operates the poppet mechanism.

True poppet valves (figs. 2–14 and 2–15) are not flow reversible (port P can not be interchanged for port R). Hence some component manufacturers offer a poppet valve for normally open as well as for normally closed flow configuration (see: valve conversion).

Spool/poppet valves

Being a combination of spool and poppet, these valves offer advantages assigned to both design mechanisms (fig. 2–16). These valves may be connected as normally closed or normally open (they are flow reversible) and therefore they find frequent application as *logic valves* ("YES", "NOT", "AND", "OR" function. See also logic functions). When used as logic valves they are air pilot operated but with plunger or cam roller operation they are often used as position sensing valves (limit valves). Their obvious advantages are:

- Low operating force. Light springs suffice to provide leak free static sealing.
- Spool/poppet valves are flow reversible (port P may be exchanged with port R, port A remains the same).
- Spool/poppet valves provide positive overlap (pasage A to R is closed before passage P to A opens).

Friction free valves (toggle disc valves)

A new generation of friction free valves has emerged on the valve market. These valves are specifically designed to combine all the advantages of spool and poppet valves and at the same time eliminate all the disadvantages of spool and poppet valves. Friction free valves, as their name implies, work totally friction free and offer also the following advantages:

- High air flow characteristics. Equal or better than spool valves. Compare figures 2–10, 2–11 and 2–17.
- Static, long life sealing. The seal ring moves head-on against the sealing lips (same as in poppet valves).
- The seal carrier (spool) is suspended between the two pre-tensioned discs and moves friction-free to and fro.

Fig. 2–16 Spool/poppet valve.

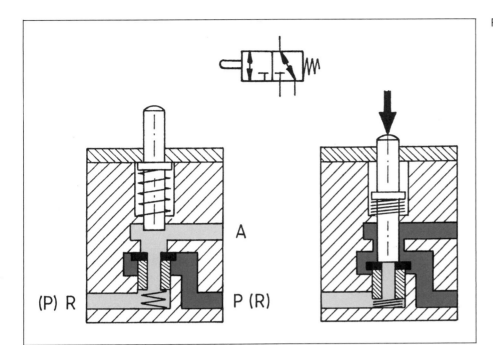

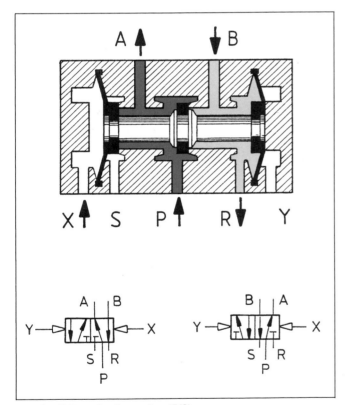

Fig. 2–17 Toggle disc valve (5/2).

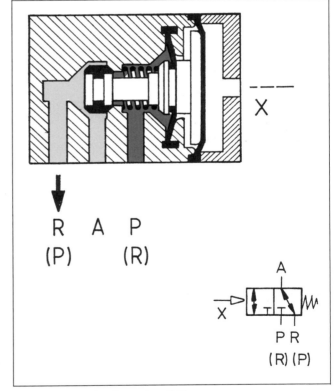

Fig. 2–18 Toggle disc valve (3/2).

- Actuation of the disc moves it over the toggle point and tension forces snap the disc rapidly into its alternate position, thus giving rapid flow switching and memory function!
- Air contamination cannot impair valve movement

and lubrication of the compressed air is not required.
- The valve is flow reversible which means port P may become port R and vice versa. Port A remains (fig. 2–18).

It shows that these valves provide an impressive number of advantages and therefore will become the valve of the future. It must, however, be noted that these valves have negative overlap, but due to their extreme fast switching action this may not provide any problem for most industrial applications.

Cross-over condition

Directional control valves with only two selectable valve postions are in reality three-position valves with a special centre condition. This centre condition is known as the transition or cross-over condition. For pneumatic valves this cross-over condition should be "CLOSED" as otherwise compressed air would momentarily stream to all ports. This could produce faulty control, particularly where valves are used to provide logic switching, or it would mean an unnecessary loss of compressed air during the valve transition (fig. 2–19).

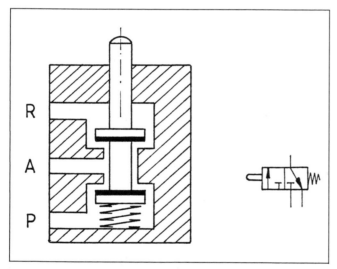

Fig. 2–19 Cross-over condition for a poppet valve with negative overlap. Compressed air streams momentarily to both outlet ports (ports A and B).

For spool valves the closed condition cross-over is achieved with a positive spool overlap (fig. 2–21). For poppet and spool/poppet valves positive overlap (or closed cross-over) is achieved by the moving plunger. Here the plunger moves first onto the valve disc before the disc is being forced off its sealing seat. Thus in a normally closed valve, the air passage A to R is closed off before the passage P to A is opened (figs. 2–14, 2–15, 2–16 and 2–20).

For flat slide valves the groove in the flat slide is designed in such a way that the valve provides also positive overlap (fig. 2–12).

Toggle disc (or friction free) valves do not provide positive overlap, but their extremely fast switch-over compensates for this, and air loss during valve transition is minimal and tolerable (fig. 2–16)

Special centre condition

For three-position valves the centre position is a specially selected (switched) position; either by hand lever or centering springs. In such a case the centre condition has significance and is deliberately chosen for this application within the machine. Valve manufacturers offer a variety of valve centre flow configurations (fig. 2–8). These flow configurations are mainly determined by the land width of the spool (figs. 2–7C, 2–7E and 2–21).

Valve conversion

In order to achieve logic switching functions it is often essential and very practical to convert one valve to perform the function of another (figs. 2–22 and 2–23). The logic "YES" function for example requires a normally closed valve, and so does the logic "AND" function (figs. 10-3 and 2–33). The logic "NOT" function and the logic "INHIBITION" function, however, use a normally open valve (figs. 8–7 and 10–8). The logic "OR" function may be achieved using a double-sided check or shuttle valve but equally simple and often more practical is the use of a normally closed valve (figs. 2–36 and 2–38). Similarly the logic "AND" function may be achieved with a twin pressure valve, but (as mentioned before) the normally closed valve, air pilot operated, may be used instead (figs. 2–32 and 2–33).

In industry, for economy and stock-keeping, it is therefore practical to store only a small number of air-pilot operated normally closed valves which can be converted to normally open configuration. Should the need arise (sudden machine breakdown; replacement valves and spares not readily available) these valves can then be converted to perform any of the four previously listed logic functions.

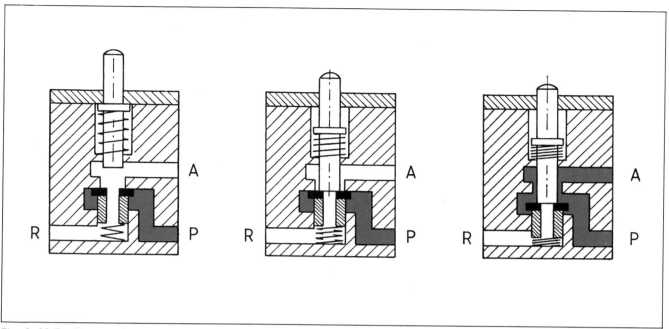

Fig. 2–20 Positive overlap cross-over for spool/poppet valve.

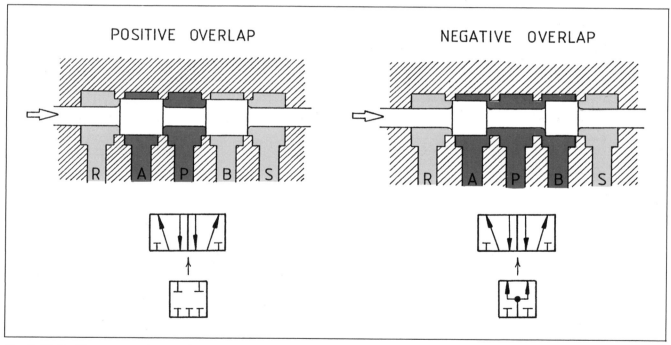

Fig. 2–21 Positive or negative overlap for spool valves.

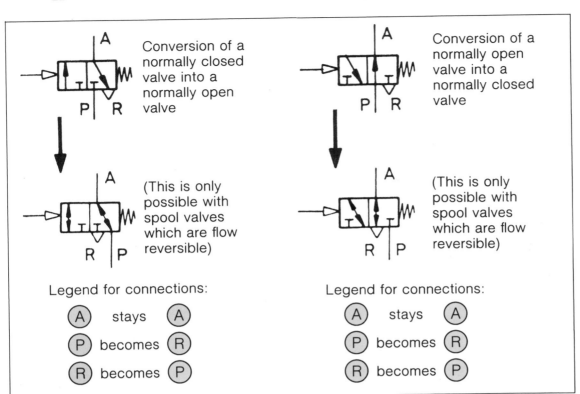

Conversion of a normally closed valve into a normally open valve

(This is only possible with spool valves which are flow reversible)

Legend for connections:

(A) stays (A)

(P) becomes (R)

(R) becomes (P)

Conversion of a normally open valve into a normally closed valve

(This is only possible with spool valves which are flow reversible)

Legend for connections:

(A) stays (A)

(P) becomes (R)

(R) becomes (P)

Fig. 2–22 Valve conversion.

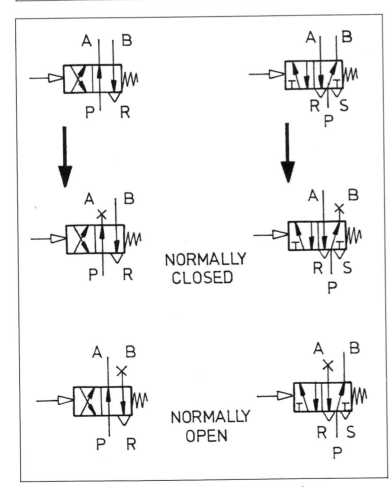

NORMALLY CLOSED

NORMALLY OPEN

Fig. 2-23 Conversion of four- or five-port valves.

Actuator Control

Directional control valves, when used for the controlling of linear or rotary actuators, are called power valves. Such power valves may have memory characteristics (*bi-stable*) or may reset to their normal position (*mono-stable*) when the pilot signal or any other valve actuation is cancelled.

Figure 2–24 shows the application of direct actuator control; e.g. the operator actuates the power valve directly. The following circuits show the principle of pilot or remote control: figs. 2–25 to 2–30. On machines pilot control is much more frequently used than direct control.

For the control circuits in fig. 2–25 and 2–26, when the pilot valve resets the power valve also resets (spring actuation) and the actuator will move into the retracted position. This form of control is called single pilot control. The control circuits in figs. 2–27 to 2–30 show double pilot control. Here the actuator is controlled by a power memory; e.g. the power valve has memory characteristics (bi-stable valve).

The actuator in fig. 2–27 (right) extends when the push-button pilot valve is actuated and retracts automatically when the position sensing valve a_1 (roller valve), being in the fully outstroked position, is actuated. The real position of the roller valve in

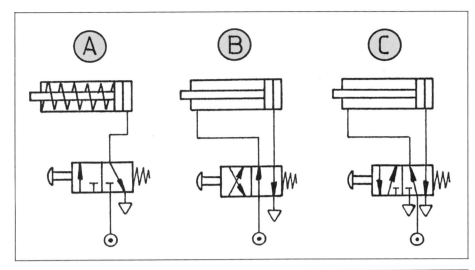

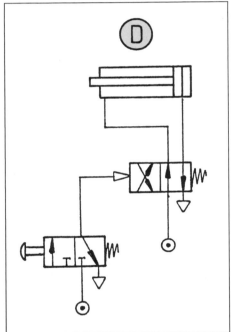

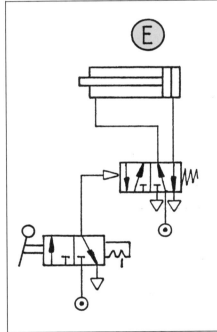

Fig. 2–24 Various direct control methods. Direct actuator control, however, is seldom used on machinery.

Fig. 2–25 *Far left:* Pilot control with push-button valve.

Fig. 2–26 *Left:* Pilot control with detented lever valve.

Fig. 2–27 *Below:* Double pilot control.

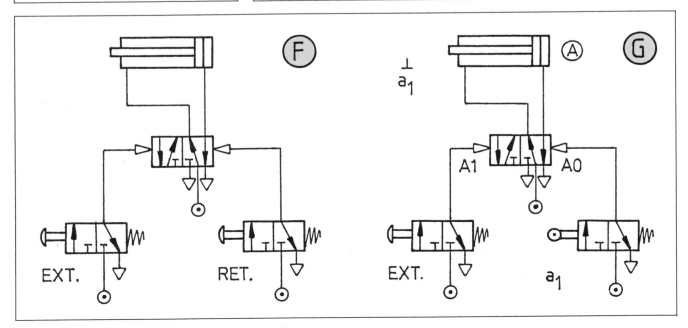

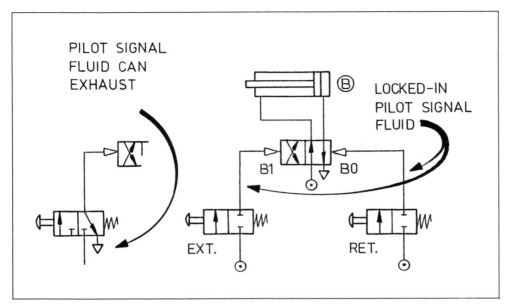

Fig. 2-28 Prevention of locked-in pilot signals.

relation to the actuator is shown with an inverted T ($\perp$) and is labelled a_1 (see also circuit labelling).

Pilot signals must be able to exhaust when the opposite signal on the power valve is applied. Two-position valves as depicted in fig. 2-28 do not allow the trapped pilot signal to exhaust and are therefore not suitable for pilot control.

The control circuit in fig. 2-29 shows an actuator being extended in the rest position. The push-button pilot valve causes the actuator to retract (pilot signal AO). When fully retracted, the actuator will operate the position sensing valve a_0 (roller valve) and thus signals the actuator to extend (pilot signal A1) automatically. Compare this circuit with the circuit shown by fig. 2-27. Compare also the labelling of the two circuits and note the difference in the flow configuration of their power memory valves. The control circuit in fig. 2-30 is an extension of the circuit depicted in fig. 2-29. The full extension position of the second actuator is not controlled. Multiple actuator control of a common power valve is frequently used, but in order to achieve sufficient actuator speed, the power valve must be large enough to provide the air flow required. Uniform actuator movement, however, may only be achieved if both actuators are of equal size and equal load requirement.

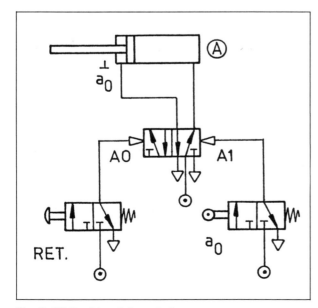

Fig. 2-29 Double pilot control.

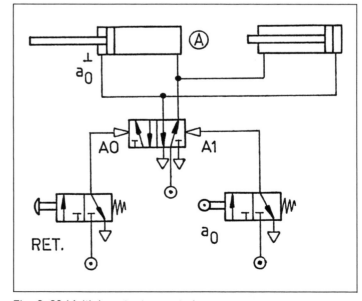

Fig. 2-30 Multiple actuator control.

Circuit labelling

In order to direct the reader to logical control design, this book uses only the binary movement expression and labelling method. Thus actuator extension signals (pilot signals) at the power valves are denoted with the binary figure 1 and retraction pilot signals with the binary figure 0. Power valves carry the letter of the actuator they control. In figure 2–31, the actuator for example is labelled A, hence its extension pilot command is A1 and its retraction pilot command is AO. In fig. 2–48 the second actuator is labelled B.

Position sensing valves also carry the letter of the actuator by which they are actuated. The valve placed in the fully outstroked position carries in addition also the binary figure 1 and the valve in the retracted position the binary figure 0. These letters, however, are lower case with subscripted figures (see figs. 2–31, 2–48) whereas the pilot signals on the power valves are capital letters with the figure on the same line (see also figs. 2–43 and 2–44).

Position sensed control
(Limit valve control)

Cam roller operated pilot valves located at the stroke limits of an actuator are called position sensing valves or limit valves. Position dependent

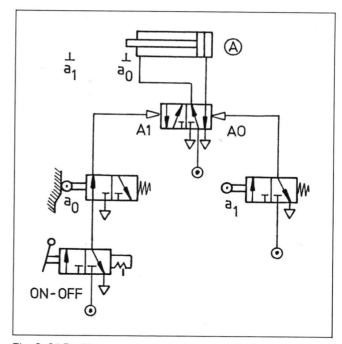

Fig. 2–31 Position sensed oscillation control. The two position sensing valves (a_1 and a_0) check full stroke movements. The on-off valve terminates these oscillations and causes the actuator to stop in position a_0.

sequencing of an actuator (or other actuators) is controlled and signalled by these valves; hence their name (figs. 2–31, 2–29, 2–30 and 2–35). Figure 2–31 depicts an oscillation control using

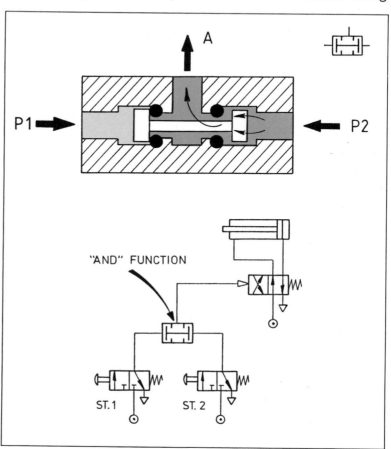

Fig. 2–32 ''AND'' function valve. Both input signals must be present. Then the weaker or second arriving passes through to the outlet port A.

two position sensing valves, one at each stroke limit. As long as the compressed air supply is switched on (on-off valve), the actuator will oscillate, but when the air supply is switched off, it stops in retracted position, resting on position sensing valve a_0.

"AND" Function valve (two pressure valve)

Some component manufacturers make a special purpose "AND" function valve (two pressure valve). This valve ensures that *both* expected input signals *must be present* on the valve before either of them is permitted to pass through to outlet port A (see fig. 2–32). The first arriving signal shifts the valve element (twin poppet), and the weaker or second arriving signal finds its way to the outlet port; thus it may be said that both input signals are required. The valve is extremely simple in its construction and costs less than an air pilot operated spool or spool/poppet valve (see figs. 2–14, 2–16 and 2–18). As previously mentioned, the "AND" function valve (two pressure valve) may be replaced by a two position, normally closed and air pilot operated spool or spool/poppet valve (figs. 2–14, 2–16, 2–18 and 2–33). This type of valve is widely available and has the advantage that the weaker signal may be used as the pilot signal and the stronger signal passes through the valve.

The "AND" function may also be achieved by arranging the pilot signal valves contributing to the

"AND" function in series order (series arrangement) as shown in fig. 2–34. Here the actuator is signalled to extend if the start valve is actuated and the actuator is fully retracted, thus actuating roller valve a_0. Pilot valves contributing to a series function may with the exception of time delay and impulse valves be arranged in any order. That means the start valve in fig. 2–34 could also be above the roller valve. Up to eight valves may be connected in series, providing the signal leaving the series chain is strong enough to switch the power valve at the end (pressure drop).

"OR" Function valve (shuttle valve)

Most component manufacturers make a special purpose "OR" function valve (shuttle valve). This valve connects either input 1 or input 2 to the outlet port A, but prevents any cross flow from input 1 to input 2 or vice versa. It is to be noted that a *T-connector* must not be used to merge two pilot

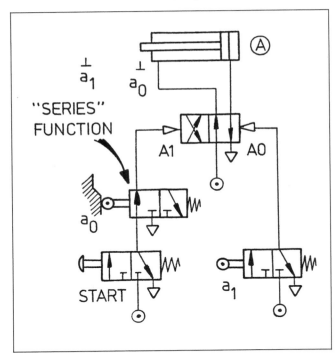

Fig. 2–34 "AND" function with pilot valves in series order.

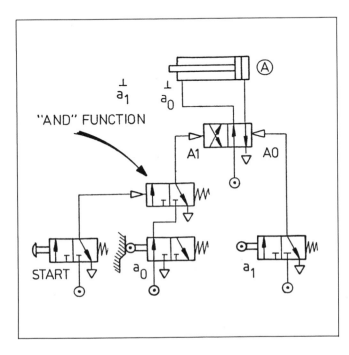

Fig. 2–33 "AND" function with 3/2 pilot operated normally closed valve.

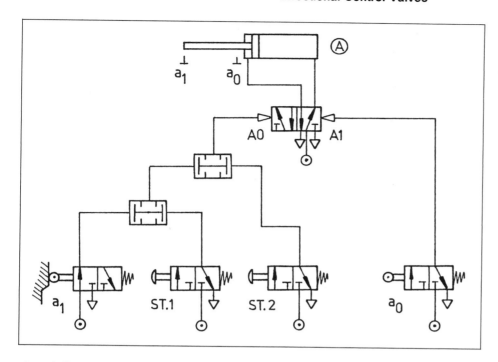

Fig. 2–35 "AND" function with three input signals. A multiple "AND" function of several inputs may be formed (up to 12 inputs without a significant pressure loss).

signal lines (compare figures 2–36 and 2–37). Using a T-connector as a substitute for an "OR" function valve would not suffice, since the pilot signal would immediately exhaust through the alternate push-button pilot valve (fig. 2–37). The "OR" function may also be achieved by using a

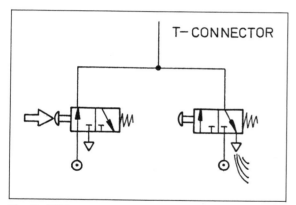

Fig. 2–37 T-connector does not work for the gathering of signals.

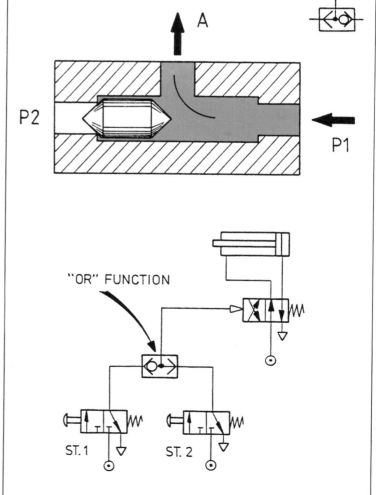

Fig. 2–36 "OR" Function valve.

two position, normally open pilot valve (fig. 2–38) or by arranging the pilot valves contributing to the "OR" function as shown in fig. 2–39. All three circuit versions are equally effective, but the version depicted in figure 2–39 is obviously less expensive and thus should be given preference providing the second pilot valve is flow reversible (may be connected as normally open, with port R becoming port P).

The control circuit in fig. 2–40 shows how three pilot signals can be "OR" connected (three input "OR" function). This pattern may be extended for several more "OR" connected pilot signals, and it is obvious that the number of "OR" function valves required is one less than the number of "OR" connected input signals (fig. 2–40).

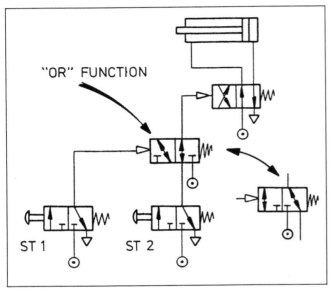

Fig. 2–38 "OR" Function valve.

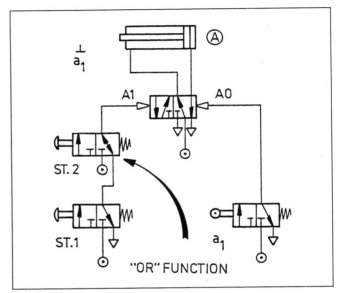

Fig. 2–39 Two start valves in "OR" function arrangement. Start valve 2 must either be flow reversible or a normally open valve.

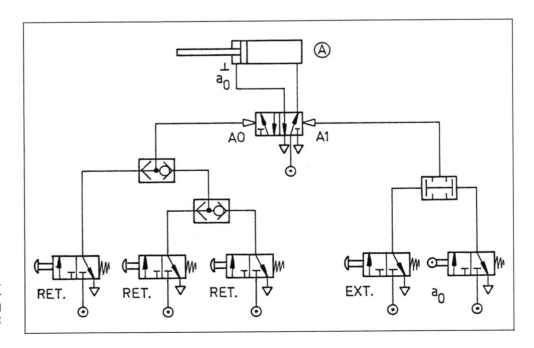

Fig. 2–40 Multiple "OR" connection permitting a cylinder retraction to be signalled from three different start locations.

Time delay valves

Pneumatic time delay valves (timers) are used to delay a pilot signal. The time delay valve is normally placed between the valve causing the pilot signal and the destination of the pilot signal (power valve, see fig. 2–41). The valve causing the pilot signal in the illustration (fig. 2–41) is a push-button valve; the valve receiving the delayed pilot signal is a spring reset type power valve. Figure 2–41A shows a pilot signal without delay. Figure 2–41B shows the pilot signal being delayed with a passive timer. Figure 2–41C shows the pilot signal being delayed with an "active" timer. Active timer integration has the advantage of "restoring" a weak signal to full system pressure.

As shown in fig. 2–41, a timer may be built from two separate valves; namely a flow control valve with free reverse flow and an air pilot operated normally closed 3/2 valve. For longer time delay functions, one must also connect a small air reservoir into the pilot line connecting the flow control to the 3/2 valve. However, most manufacturers of pneumatic components make a complete timer consisting of all three aforementioned components (fig. 2–42). Such time delay valves are predominantly of the spool/poppet or spool design which

permit normally closed or normally open flow configuration.

Valve operation

Compressed air is connected to port P (figs. 2–41 and 2–42). Port A is connected to the pilot signal port of the power valve. The pilot signal valve (push-button valve) is connected to port X of the timer. Once the push-button valve is actuated, compressed air flows via the flow control valve orifice into the air reservoir and pressure inside the reservoir starts to build up. The spool moves to the right when the air pressure force from the reservoir exceeds the spring force holding the valve closed; thus the flow passage from P to A opens and the pilot signal is transferred to the power valve (fig. 2–43). The time required for the pressure build-up to reach spool shift force is equal to the required time delay and is adjusted on the flow control valve (flow rate = pressure build-up time).

For the time delay valve to resume its initial or spring reset position, the pilot signal connected to port X must cease and the compressed air inside the reservoir must exhaust. In order to make the timer quickly available again for the next time delay function, exhaust air can flow through the inbuilt

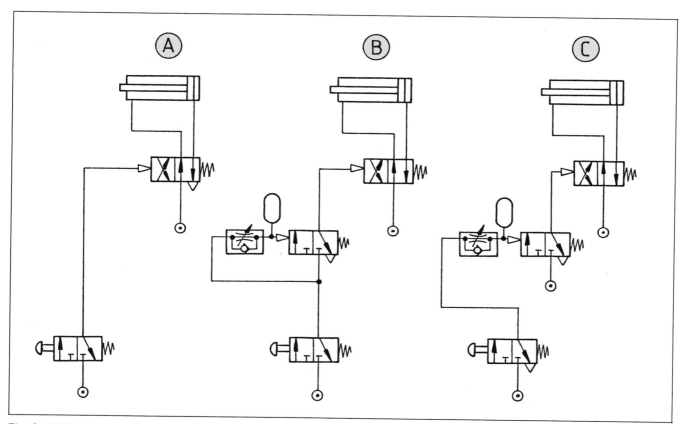

Fig. 2–41 Timer integration.

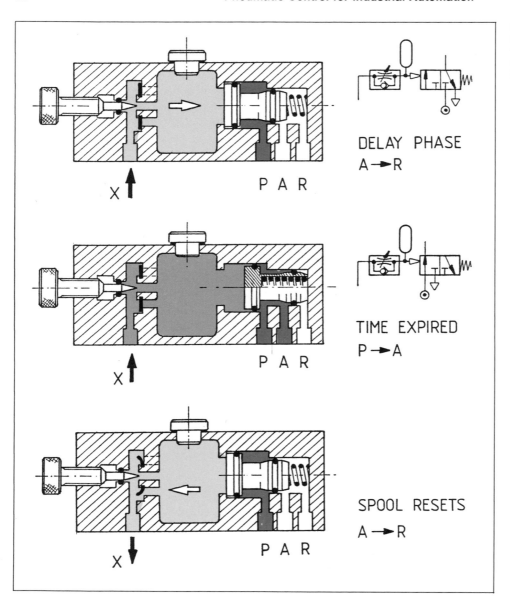

Fig. 2-42 Time delay valve consisting of a 3/2 spool valve, a flow control valve with free reverse flow, and a reservoir in between.

DELAY PHASE
A → R

TIME EXPIRED
P → A

SPOOL RESETS
A → R

check valve (rubber disc) and thus the reservoir is rapidly emptied.

However, any pressure fluctuation on the pilot line will cause inconsistent time delay. It is therefore recommended to keep pressure in the pilot line as stable as possible or use reduced pilot pressure; set well below the minimum pressure fluctuation level.

Figure 2-43 shows a typical timer (time delay valve) application. The actuator is permitted to extend when the start valve is actuated and the position sensing valve a_0 is actuated. The actuator then must fully extend to valve a_1 and remain in that position for 15 seconds before it retracts automatically. Fig. 2-44 depicts a timer application to make the actuator extend automatically after

having been in the fully retracted position for 7 seconds.

Note: The position sensing valve a_1 is shown in its actuated position; hence the valve is attached to the left hand square and the cam is indicated.

Impulse valve (pulse valve)

Impulse valves are used to shape a long signal into a short pulse. Their most frequent application is in sequential control circuits where persisting signals which could otherwise cause an opposing signal on a power valve have to be removed (fig. 2-45). The valve is placed into the pilot signal line between the pilot valve and the power memory valve (fig. 2-45).

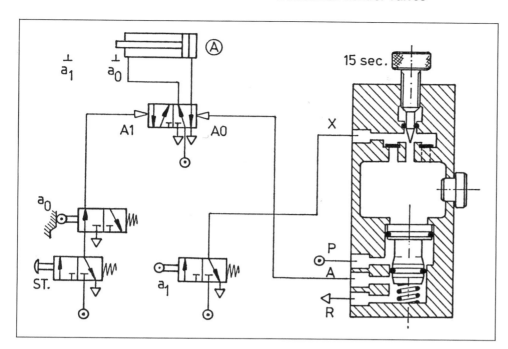

Fig. 2-43 Typical timer application.

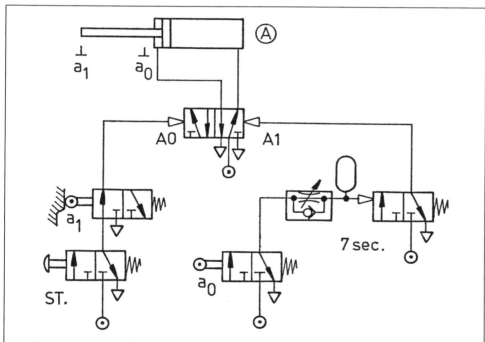

Fig. 2-44 Timer application.

Valve operation

The pilot signal is connected to port X and to port P. Port A is connected to the pilot signal line leading to the signal port of the power valve. Once the push-button pilot valve is actuated, the emerging pilot signal will flow immediately through the open pulse valve into the power valve signal port, where it actuates the power valve. Simultaneously the air pressure in the reservoir starts to build up. The spool moves to the right (fig. 2-46)

when the air pressure force from the reservoir exceeds the spring force holding the valve open; thus the flow passage from P to A closes and the pulse expires. The time required for the pressure build-up to reach spool shift force is equal to the required pulse length and is adjusted on the flow control valve (flow rate is equal to pressure build-up time).

For the pulse valve to resume its initial or spring reset position, the pilot signal connected to ports X

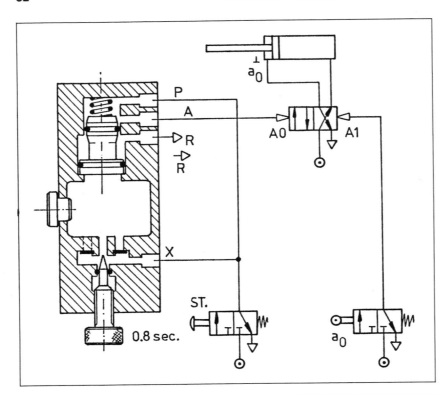

Fig. 2-45 Impulse valve integration.

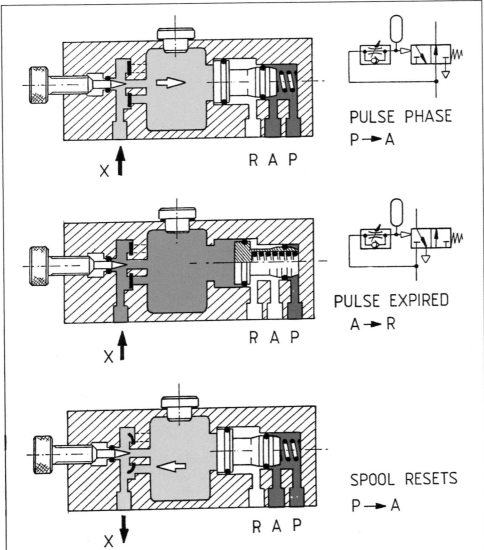

Fig. 2-46 Impulse valve (pulse valve).

and P must cease and compressed air inside the reservoir must exhaust. In order to make the pulse valve quickly available for the next function, exhaust air can flow through the inbuilt check valve (rubber disc) and thus the reservoir is rapidly emptied.

Figure 2–47 shows a typical impulse valve application. The actuator must retract immediately after having reached the fully outstroked position a_1. This could be foiled if an opposing signal on the power valve (caused by the operator remaining on the start valve) prevented the power valve from reacting to the AO pilot signal (fig. 2–47). The impulse valve shapes a persisting start signal into a pulse.

Impulse valves may also be used as normally open timers; e.g. to turn an actuation or machine function off after a predetermined time. A typical example of this is shown in fig. 2–48 (refer to: timer N/O).

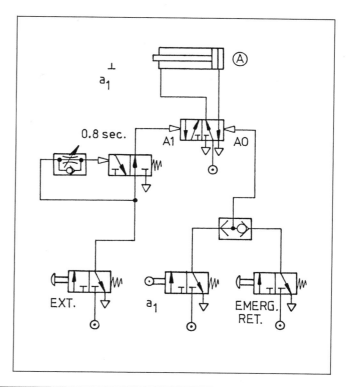

Fig. 2–47 Impulse valve application. The impulse valve shapes the start signal into a short pulse of 0.8 seconds.

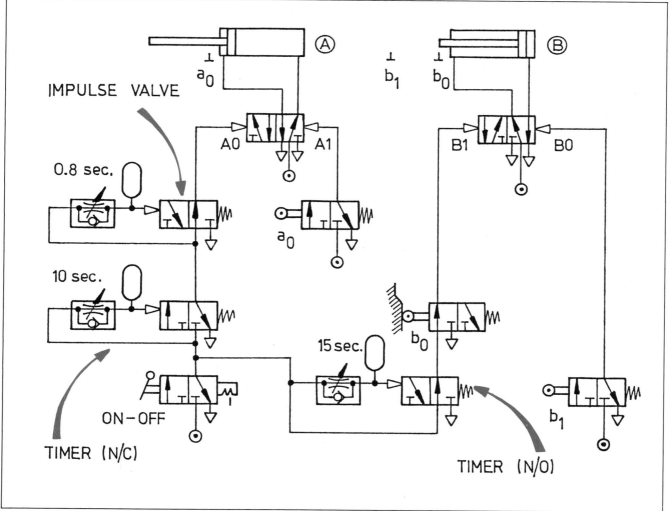

Fig. 2–48 Typical circuit application for a time delay valve and impulse valve in series connection.

Timer and impulse valve in series arrangement

Timers and impulse valves may be placed in series arrangement but the order of placement is important. Impulse valves are usually set to a pulse length of about 0.5 to 0.8 seconds. After this time the pulse expires. Timers need a constant input which exceeds the time setting. It is therefore imperative to place the timer upstream of the impulse valve (fig. 2–48).

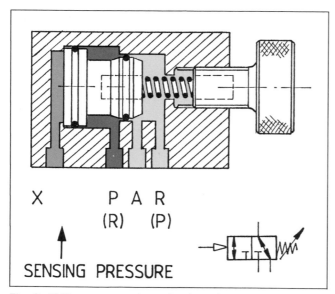

X   P A R
             (R) (P)

↑

SENSING PRESSURE

Fig. 2–49 Pressure sensing valve (spool type).

Pressure sensing valve (Pressure sequence valve)

The pressure sensing valve is basically a 3/2 spool valve which is predominantly used as a normally closed valve (fig. 2–49). The spool-shift spring can be adjusted to different spool-shift forces, thus the valve opens at different but selectable sensing pressures; normally from 100 kPa to 1000 kPa (1 to 10 bar). Pressure sensing valves are used where a specific pressure is required for a machine function (a clamping function for example). Figure 2–50 shows a control circuit for an actuator that is only permitted to extend if the system pressure is at least 600 kPa (6 bar) and the start valve is actuated. With a minimal pressure of 600 kPa the spool will move upwards and the start signal can pass through the valve (from P to A) and actuate the power valve at signal port A1. After being fully outstroked, the actuator retracts automatically.

The actuator in fig. 2–51 extends when either start 1 or start 2 is actuated ("OR" function) and retracts automatically when full stroke extension is reached and the extension pressure sensed on the actuator is at least 600 kPa ("AND" function with series arrangement).

Time and pressure sensed control

Very few controls use pressure sensing or time sensing in its pure form. Most controls use a combination of position sensing, time sensing and pressure sensing and in most cases manual sequencing in one form or another is added (fig. 2–52).

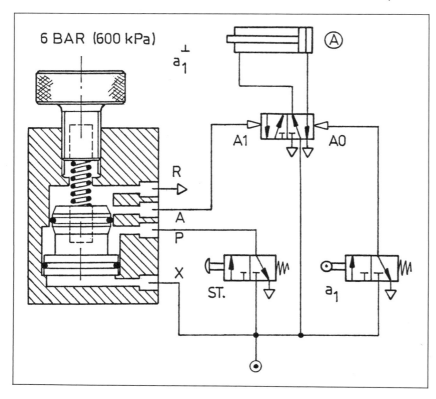

6 BAR (600 kPa)

a_1 ⊥

Ⓐ

A1   A0

R

A

P

X

ST.

a_1

Fig. 2–50 Typical pressure sensing valve application.

Fig. 2–51 The actuator retracts automatically when position a_1 and an extension pressure of 600 kPa (6 bar) is reached.

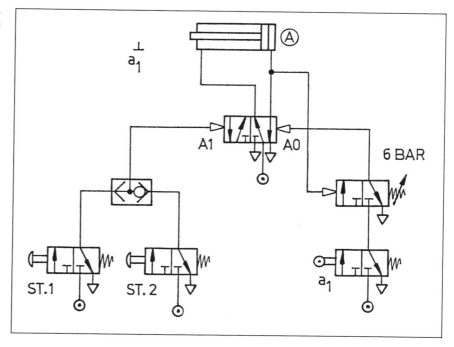

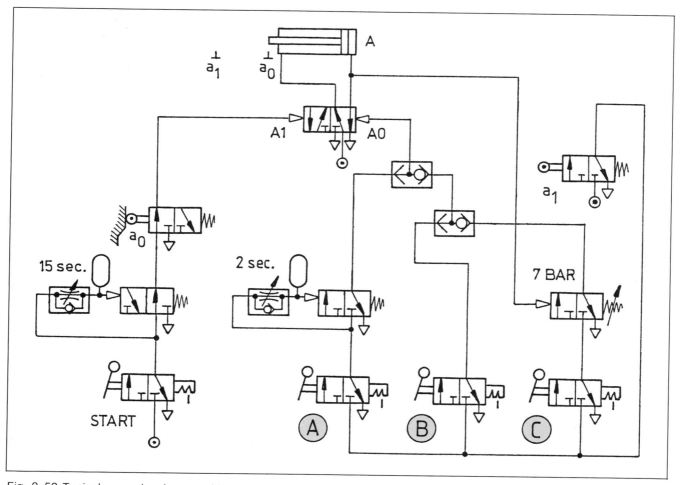

Fig. 2–52 Typical example where position sensing, time sensing and pressure sensing are applied in the same control circuit.

3 Linear actuators (cylinders)

Pneumatic linear actuators are used to convert the stored (static) energy of compressed air into linear mechanical force or motion. Although the actuator itself produces linear motion, a variety of mechanical linkages and devices may be attached to it to produce a final output force which is rotary, semi-rotary or a combination of linear and rotary. Levers and linkages may also be attached to achieve force multiplication or force reduction as well as an increase or reduction of motion speed (fig. 3–1).

The main parts of a pneumatic linear actuator are shown in figs. 3–2 and 3–5. A variety of refinements, additions and options can be added to this basic actuator to achieve a diversity of engineering features (figs. 3–6, 3–8, 3–10, 3–12 and 3–14). The generation of linear thrust force with a pneumatic actuator is very simple and direct. The compressed air when delivered to one end of the actuator, acts against the piston area and produces a force against the piston (force = pressure × area). The piston with the attached piston rod starts to move in linear direction as long as the reacting force is smaller. The developed force is used to move a load which may be attached either to the protruding piston rod or to the actuator housing (fig. 3–1). The distance through which the piston travels is known as the stroke.

Single acting actuators

Single acting actuators produce pneumatic force in one direction only. These actuators may be mounted in vertical direction, thus permitting the previously moved or lifted load to return the piston to its initial position. When the actuator must be mounted horizontally or when no external force can be used to return the piston, then an inbuilt spring is used to cause retraction or return of the piston (fig. 3–2).

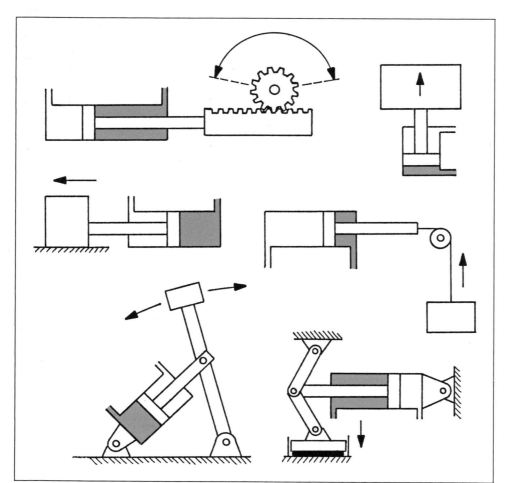

Fig. 3–1 Linkages and loads attached to linear actuators.

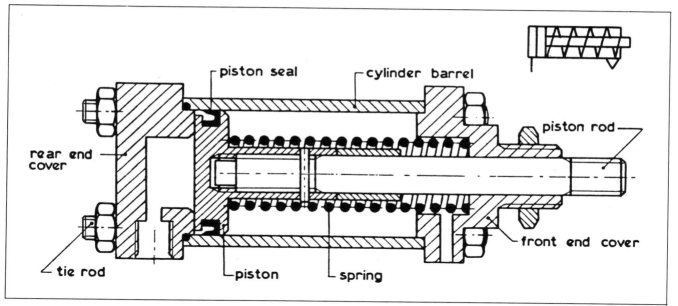

Fig. 3–2 Single acting actuator.

When using a single acting actuator the force available from the air operated stroke is reduced by the opposing spring force. It must also be borne in mind, that the force due to the spring compression is progressively decreasing along its stroke. Single acting actuators are usually built with a stroke length up to 100 mm.

Some manufacturers of pneumatic actuators make a special type of single acting actuator, called a short stroke clamping cylinder or diaphragm actuator. These actuators have extremely short strokes ranging from 1 to 10 mm and piston areas ranging from 100 to 3000 mm^2, whereby the large piston (diaphragm) actuators normally have the extremely short strokes of 1 to 2 mm.

Such clamping actuators are normally retracted by either an inbuilt spring or by the pre-tensioned diaphragm (fig. 3–3).

A very special type of single acting actuator is the air bag. The air bag actuator consists of a flexible ''bellow'' (or bellows) made from plies of nylon reinforced cord, encased by neoprene rubber. The cord is meant to provide a restraining effect as pressure is applied and helps to maintain the effective area of the internal air column, which provides the thrust when filled with compressed air. The bellows may be made in single, double or triple convolutions. ''Girdle hoops'' of metal or metal wire covered with rubber, provide additional restraint on multiple bellow types (fig. 3–4).

Metal end pieces are attached to the top and bottom of the bellow(s) and the air enters and exhausts through a port in the end piece. Manufacturers of air bags provide performance charts that contain such information as: useable stroke length, load-carrying capacity related to stroke and air pressure, air volume at various stages of inflation, and mounting methods. The main applications are heavy duty lifting, logging industry, air presses, isolation of vibration, cause of vibration, shock absorption and clamping.

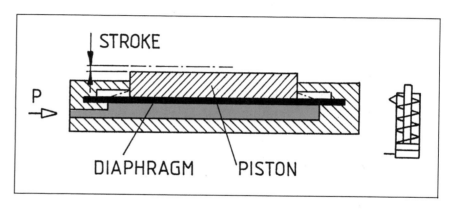

Fig. 3–3 Diaphragm actuator.

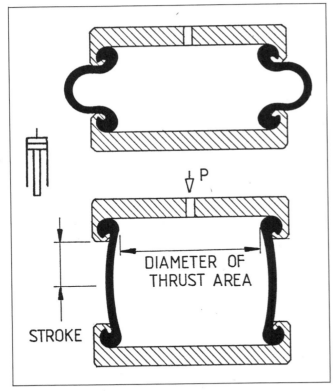

Fig. 3–4 Air bag.

Double acting actuators

Double acting actuators permit the application of pneumatic force in both stroke directions. However, the retraction stroke develops a much smaller force than the extension stroke, since the compressed air acts on a smaller area known as the annular area (fig. 3–5).

Double-ended actuators with rods protruding on both actuator ends are used where the developed force must be equal for both piston-rod movement directions (extension and retraction). Since the voids to be filled with compressed air are equal for extension and retraction, the resulting piston speeds are also equal for both strokes. Double ended actuators provide also better piston rod guidance since loads are borne over two bearing bushes; one on each actuator end (fig. 3–6). Double ended actuators may also prove advantageous where the limit valves (position sequence valves) or electrical limit switches cannot be placed near the reciprocating piston rod. This may be the case where the transported load moves in a hostile environment, or in an area of extreme heat, or in a liquid. In such a case, the rear piston rod is used to actuate the limit valves and the front piston rod moves the load (fig. 3–7).

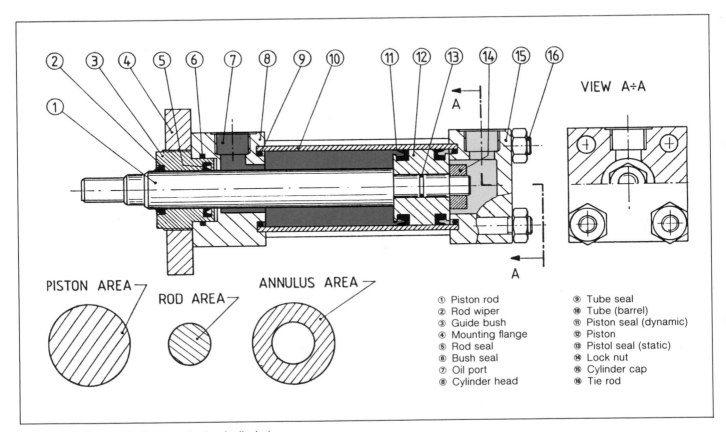

① Piston rod
② Rod wiper
③ Guide bush
④ Mounting flange
⑤ Rod seal
⑥ Bush seal
⑦ Oil port
⑧ Cylinder head
⑨ Tube seal
⑩ Tube (barrel)
⑪ Piston seal (dynamic)
⑫ Piston
⑬ Pistol seal (static)
⑭ Lock nut
⑮ Cylinder cap
⑯ Tie rod

Fig. 3–5 Pneumatic linear actuator (cylinder).

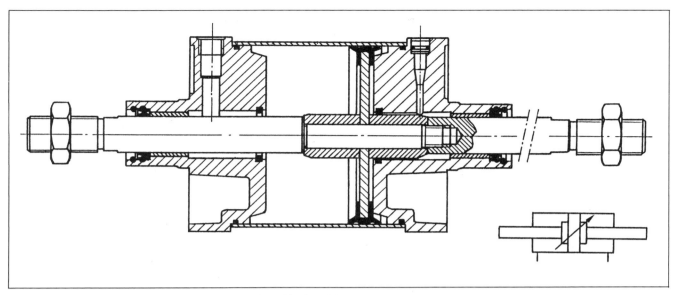

Fig. 3-6 Double-ended linear actuator with end cushioning. Note the two piston rods.

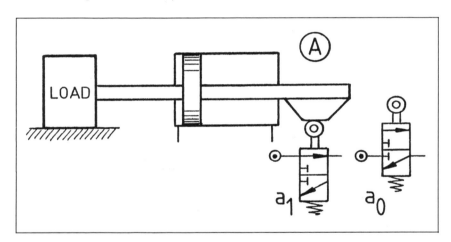

Fig. 3-7 Actuator with rear rod actuating the limit valves.

Pneumatic end-position cushioning

Pneumatic end-position cushioning (simply called cushioning) refers to the controlled deceleration of the mass in motion during its final part of the actuator stroke. The kinetic energy released on impact at the stroke end must be absorbed by the internal stroke limit stops. These stops are usually a protruding bush which is an extended part of the end caps. Their capacity to absorb shocks depends entirely on the elasticity of their material. On some actuator types the rubber seal shoulder of the piston seal is used to absorb the shock (fig. 3-6); other designs use a rubber or plastic ring embedded into the endcap as a shock or impact damper. For large and fast moving masses or for actuators with piston diameters above 25 mm a pneumatic end-cushioning is recommended. Figure 3-8 shows a cross-section of a pneumatic cushioning mechanism built into the end caps of

pneumatic actuators. The piston is fitted with a tapered cushioning bush (boss). When this bush enters into the air exit bore of the end cap (during the final part of the stroke), the main air exit begins to shut until it finally closes off completely. This first stage of exit flow throttling may cause an initial slight speed reduction. The remaining air in the annular cavity around the cushioning boss is now trapped and the momentum of the mass carries the piston on against the trapped air; thus compressing it to a high pressure. As the pressure due to Boyle's Law rises, the resulting resisting force forms a cushion and begins to decelerate the piston and its mass (load).

As the kinetic energy is dispelled, the mass is slowed down to a much reduced speed which matches the rate of air flow permitted and adjusted by the cushioning valve. A check valve is also built in as part of the cushioning mechanism. This check

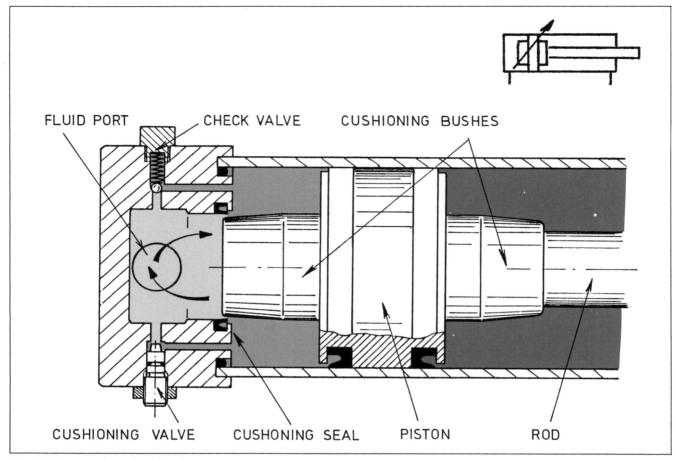

FLUID PORT   CHECK VALVE   CUSHIONING BUSHES

CUSHIONING VALVE   CUSHONING SEAL   PISTON   ROD

Fig. 3–8 Pneumatic end-cushioning.

valve allows full air flow in reverse direction to achieve fast and full force break-away from the end position (fig. 3–8).

In order to utilize the full potential of a cushioning provision, the complete stroke length of the actuator must be used. An effective stroke under-utilization of as small as 5 mm would already diminish the cushioning effect to up to 50% and the machine stop had to bear the brunt of the impact, making the machine a noisy and self-destructive implement. If on the other hand the kinetic energy is gently dispelled, the machine can look forward to a long and trouble free life (see also cushioning curve, fig. 3–25).

Seals in linear actuators

Seals are grouped into static and dynamic applications. Static seals are fitted between two rigidly connected components. The seals between the piston rod and the piston in fig. 3–9G are static seals and so are the seals that prevent external leakage past the joint where the tube is mounted onto the endcap in fig. 3–9G and 3–9H.

Dynamic seals prevent leakage between components which move relative to each other. Therefore, dynamic seals are subject to wear, whereas static seals are essentially wear free. The seals between the moving piston and the stationary tube (barrel) are typical dynamic seals, and so are the seals that stop compressed air from escaping past the clearance gap between the moving rod and the end cap (figs. 3–9E and 3–9H). The rod wiper ring may also be regarded as a dynamic seal. It prevents contaminants from being drawn into the actuator when the rod retracts (figs. 3–9D and 3–9H). Rod wiper rings are made from extremely hard rubber to make them more wear-proof.

O-ring seals are generally used as static seals, whereas dynamic seals range from simple square rings to complex groove rings (lip seals) and specially designed and custom built double cup piston seals (fig. 3–9 and 3–6). Lip ring seals provide excellent sealing even under stationary piston conditions. When a lip ring is fitted, the lip must be slightly pre-tensioned. This alone produces satisfactory sealing even under low pressure conditions. This tension is of course increased when

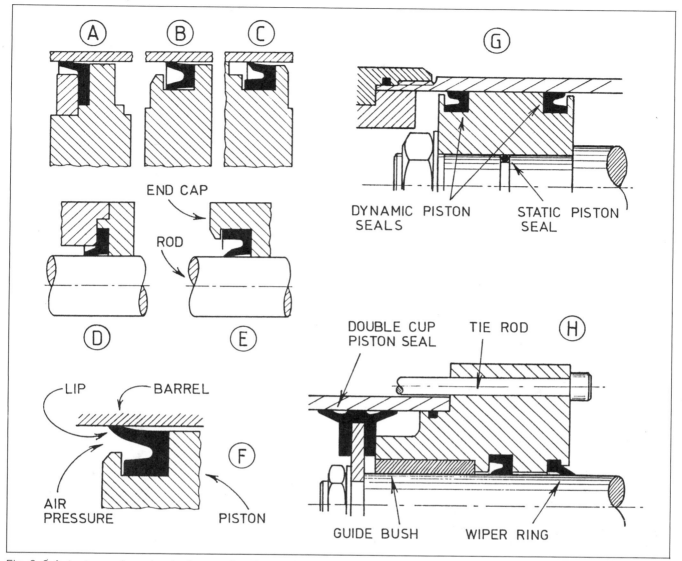

Fig. 3–9 Actuator seals and actuator construction methods.

pressure increases and thus it adjusts itself for any wear occurring during its service life. Piston seals may be made from a variety of materials to suit a range of environmental conditions. Perbunan seals are suitable for temperatures from −20°C to +80°C; Viton seals range from −20°C to 190°C; and Teflon seals may be used for a temperature range from −80°C to 200°C.

Actuator construction

Pneumatic actuators are often required to work under adverse conditions. On the outside they may come into contact with abrasive or aggressive chemicals, water, extreme hot or cold temperatures, and often also with mechanical abuse. On the inside they invariably have to withstand the ever-present moisture contamination of the com-

pressed air, and misalignment of the moving load takes its toll on the guide bush, the piston rod and the piston seals (figs. 3–5, 3–9).

Therefore, pistons, end caps, cylinder barrels and actuator mountings are either made from aluminium alloys, stainless steel, brass or plastics. Piston rods are fabricated from either stainless steel or hard chromium plated high-tensile steel. Guide bushes or linear-bearing bushes are made from brass, bronze, nylon, P.T.F.E., or self-lubricating (graphite impregnated) metals. More recent trends in material selection are yielding to the ever increasing demand for non-lubricated air, thus plastics are more frequently used, and on some linear actuators today the only metal part to be found is the stainless steel piston rod.

The actuator end caps (fig. 3–10G) are either

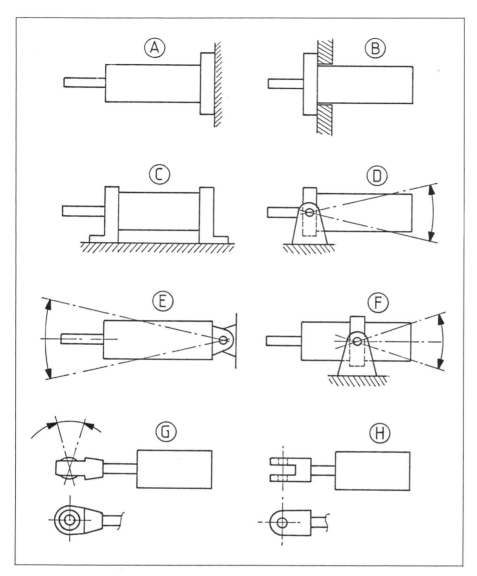

Fig. 3–10 Various kinds of actuator mounting methods.

screwed onto the cylinder barrel or, on cheaper actuators, the barrel may be crimped onto the end caps. With larger actuators tie-rods may be used to hold the end caps against the cylinder barrel. On smaller actuators stroke cushioning pads are embedded into the end caps and on larger actuators pneumatic end cushioning is standard equipment. For actuator mounting methods see figures 3–1 and 3–10.

Actuator failure

Standard type actuators are not designed to absorb piston rod side loading. Thus actuators must be mounted with care and accuracy, to ensure that the load moves precisely parallel and in alignment with the actuator centreline (fig. 3–11).

For many applications the piston rod is best fitted with a clevis or a spherical bearing rod-eye (fig. 3–10H and 3–10G) or the actuator must be allowed to swivel around a trunion mount, permitting it to swing as the direction of the load changes. A self aligning piston-rod coupling can also be used to compensate for both angular and radial misalignment, but the angular misalignment must not exceed 4°C in either direction and the radial alignment must not be more than 1 mm out (fig. 3–12).

Failure of the rod bearing (guide bush, fig. 3–9H) usually occurs when side load is not detected early enough or cannot be avoided. In such cases one may use a double ended piston rod type actuator (fig. 3–6), or an actuator with oversize piston rod which is less flexing than a standard size piston rod.

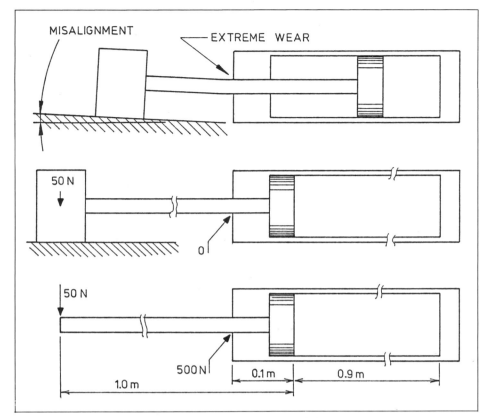

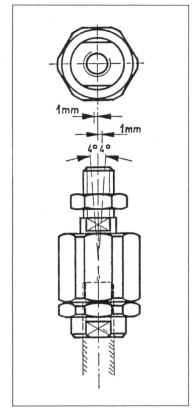

Fig. 3–11 Accurate actuator alignment and prevention of side-load.

Fig. 3–12 Self-aligning piston rod coupling.

Special actuators

In pneumatic control systems one may find many special types of actuators built to perform in a unique situation to gain optimal mechanical advantage, force output, speed and impact control. The symbols for these actuators are depicted in fig. 3–13. The last three symbols are not standardised but may be used until adopted into the catalogue of international standardisation.

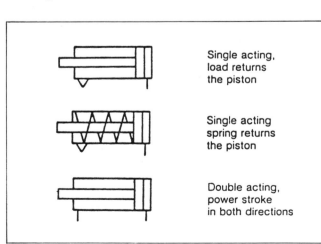

Single acting, load returns the piston

Single acting spring returns the piston

Double acting, power stroke in both directions

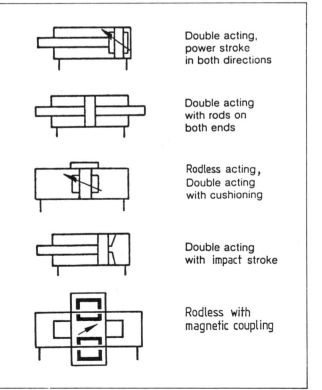

Double acting, power stroke in both directions

Double acting with rods on both ends

Rodless acting, Double acting with cushioning

Double acting with impact stroke

Rodless with magnetic coupling

Fig. 3–13

Impact cylinder (special actuator)

For work functions requiring a sharp and massive impact force similar to that of a hammer, the *impact cylinder* has been developed. Its high kinetic energy is only developed during the extension stroke and is based on the formula:

$$E = \frac{(m \times v^2)}{2}$$

The mass (m) is a heavy duty piston rod about twice as thick as a normal piston rod and a heavy duty piston (fig. 3–14). In the start position air pressure acts onto the underside of the piston and when the power valve is changed over, a pressure (p_2) starts to build up in chamber A, which acts onto piston area A1. Simultaneously the air pressure in chamber B is being exhausted. As long as

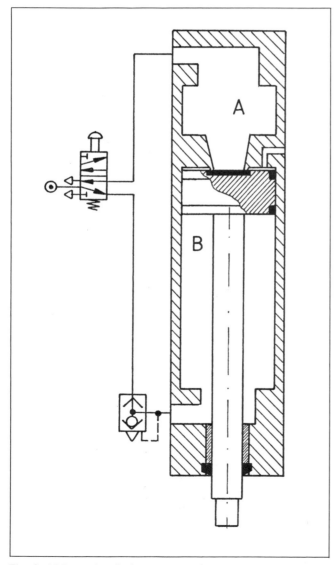

Fig. 3–14 Impact cylinder.

the piston is held against its upper end cap, the pressure exposed piston area (A1) is only about 1/9 of the annular area below. The piston will therefore not start to move down until the pressure in chamber B has fallen to below 1/9 of the pressure in chamber A. As soon as the piston starts moving downwards, the high system pressure in chamber A starts to act onto the now fully exposed upper piston area which is by now nine times larger than the previously, partially covered upper piston area. This causes an extremely sudden and great increase in force and results in a high acceleration of the mass of its piston and rod. The extent of this acceleration is governed by the pressure acting onto the full piston area and the resistance against the accelerating mass. After about 50 mm of piston travel, the piston will reach a speed of about 5 m/s. This is also the beginning of the work stroke. The speed will increase (if the moving mass is unopposed) and may reach a maximum speed of about 6–7 m/s at 75 mm of piston travel. The *kinetic energy* is governed by the velocity of the mass (see given formula); thus the first 50 to 75 mm should be a free and unopposed piston travel, and the work stroke should fall between 50 and 75 mm of the piston travel. Every impact cylinder has besides its acceleration and work-stroke travel also an overtravel distance, through which deceleration and impact cushioning takes place (see also impact cylinder calculation example).

Rodless actuators

Few pneumatic devices have enjoyed such immediate success and widespread application as the rodless actuator. An ingenious design eliminates the piston rod and cuts installation length almost to half compared to conventional double acting linear actuators. Presently, there are four design types of rodless actuators available.

With the "FESTO" rodless actuator, a strong magnetic coupling transmits the force from the moving piston assembly inside the stainless steel tube to the guided carriage sleeve on the outside of the tube. Permanent magnets are embedded into both the piston assembly and the outer carriage sleeve. Together these magnets in the piston and in the carriage create a strong force field which acts through the fully closed actuator tube and holds them together. Thus piston and carriage move in unity along the guiding actuator tube when compressed air is applied to either end of the piston (fig. 3–15). To reduce friction the carriage sleeve slides along on nylon bearing liners, and bearing liners also guide the piston assembly which slides on the inside of the tube. Thus the magnets never "touch" the tube.

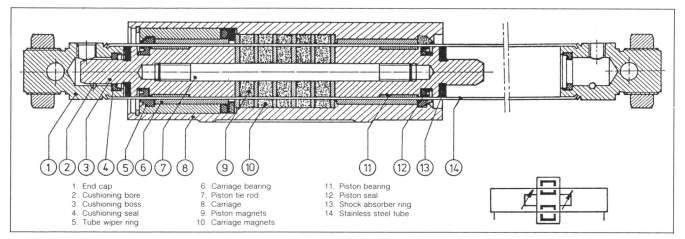

1. End cap
2. Cushioning bore
3. Cushioning boss
4. Cushioning seal
5. Tube wiper ring
6. Carriage bearing
7. Piston tie rod
8. Carriage
9. Piston magnets
10. Carriage magnets
11. Piston bearing
12. Piston seal
13. Shock absorber ring
14. Stainless steel tube

Fig. 3–15 ''FESTO'' magnetic coupling type rodless actuator.

''FESTO'' rodless actuators are built for piston diameters ranging from 16 to 40 mm and stroke lengths from 50 to 4000 mm. These actuators travel with extremely high speed (3 m/s) and thus come with standard built-in pneumatic cushioning. The magnetic rodless actuator offers a range of impressive advantages:

- Absolutely no leakage and hence the system is cheaper to operate and can be used with non-lubricated air.
- The actuator tube is hermetically sealed (no slot) and thus no dirt can enter the system from out-

side. This ensures trouble-free operation.
- The carriage can rotate around the stainless steel tube. This may prove advantageous for certain machine applications.

However, it must be noted that excessive pressure or force beyond 1.6 times the stated maximum output force may cause the carriage to break away from its magnetic coupling. But, should the coupling part, it can easily be united without dismantling the actuator.

The ''LINTRA'' rodless actuator consists of a slotted aluminium tube (barrel) with its longitudinal

Fig. 3–16 ''LINTRA'' rodless actuator.

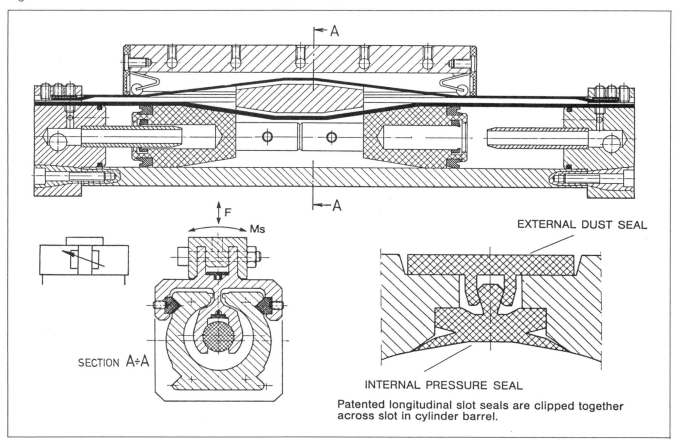

SECTION A÷A

EXTERNAL DUST SEAL

INTERNAL PRESSURE SEAL

Patented longitudinal slot seals are clipped together across slot in cylinder barrel.

pressure and dirt slot seal bands, a piston assembly that is connected through the slot to the outside carriage, and end covers for sealing the tube (fig. 3–16). The rodless piston with its two lip seals at each end runs inside the slotted aluminium extrusion tube and is sealed with two patented seal strips that prevent leakage and dirt ingression. The carriage onto which the load is attached runs externally and is guided on "V" slots extruded in the barrel.

The "LINTRA" rodless actuator sold by "MARTONAIR" offers an impressive range of advantages over standard double acting actuators and other brands of rodless actuators:

- The extrusion construction of the aluminium barrel allows high bending moments and transverse forces.
- The integrated adjustable guide system offers maximum simplicity as separate carriage guidance is not required.
- The rigidity of the aluminium extrusion barrel allows stroke lengths up to 10 m (10000 mm)!
- The unique sealing strip closure concept guarantees leak proof, long life sealing performance.
- The actuator can safely operate with air pressures up to 1000 kPa (10 bar), giving thrust forces up to 2650 newtons with a 63 mm piston.

With the "ORIGA" rodless actuator principle, force from the moving piston is transferred to the carriage by means of a piston yoke that slides along the barrel slot (similar to the "LINTRA" actuator principle). Its slot seal, also claiming to be absolutely free of air leaks, consists of a thin steel strip on the inside of the tube and a dust or dirt seal strip on the outside of the tube. Both steel strips pass through the piston yoke, which separates them as the piston slides along, and closes them again and forces them against a holding magnet attached to the barrel slot. The air inlet holes are located in the end-caps and so are the end-cushioning bushes.

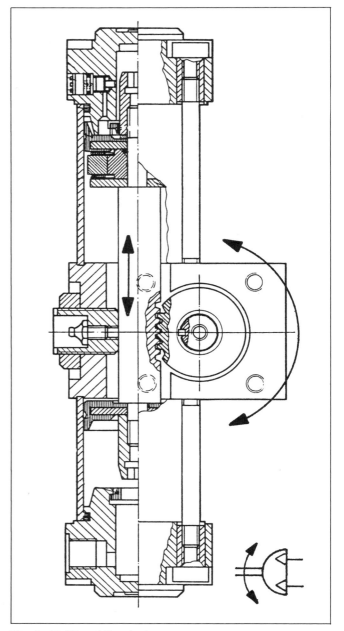

Fig. 3–18 "Bosch" swivel actuator.

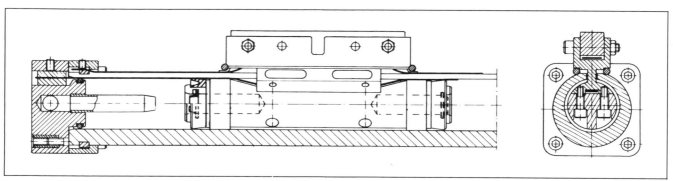

Fig. 3–17 "ORIGA" rodless actuator.

The "ORIGA" rodless actuator is extremely rigid due to its extruded type barrel construction and provides a maximum force output of 2600 newtons with a piston diameter of 80 mm at a maximum air pressure of 800 kPa (8 bar). "ORIGA" actuators are made with 80, 63 and 40 mm piston diameter and a maximum travel distance of 7000 mm (fig. 3–17).

Swivel Actuator

The swivel actuator is a linear actuator with a rack and pinion drive mechanism sandwiched between its two pistons. The angle of rotation can be chosen from 90° to 360°. These actuators are excellently suited for limited swivel action and are built with piston diameters from 32 to 100 mm and a torque range of 0.9 Nm to 23.5 Nm per 100 kPa (1 bar) pressure. Thus a 100 mm piston type actuator provides 235 Nm at 10 bar minimal system pressure. These actuators may also be ordered with or without pneumatic endcushioning (fig. 3–18).

Locking units for pneumatic actuators

Due to the compressibility of air, the locking of an actuator in any desired position between the two stroke limits has proved to be very unsatisfactory.

Mechanical locking units can hold any rod shaped mechanical part, especially piston rods of pneumatic actuators in any required stroke position. The locking unit can hold the piston rod absolutely creep free even when full system pressure is applied onto the piston. This means that it is also able to support a load of corresponding size to the full actuator force absolutely firm, regardless of direction. If when a locking unit is gripping a piston rod, pressure is also applied to the underside of the piston, it is possible to support a load which is nearly twice the normal load force. The locking unit described here may either be used with spring force locking and pneumatic pilot signal release or with pneumatic pilot signal locking and pneumatic pilot signal release (fig. 3–19).

Actuator sizing

The main criteria on which the size of a linear pneumatic actuator is based are:

- Force output for extension and retraction;
- Piston speed for extension and retraction;
- Impact cushioning at the end of the piston stroke;
- Mechanical stability of the piston rod.

The first three listed items are very complex to calculate, since they depend on a number of basic

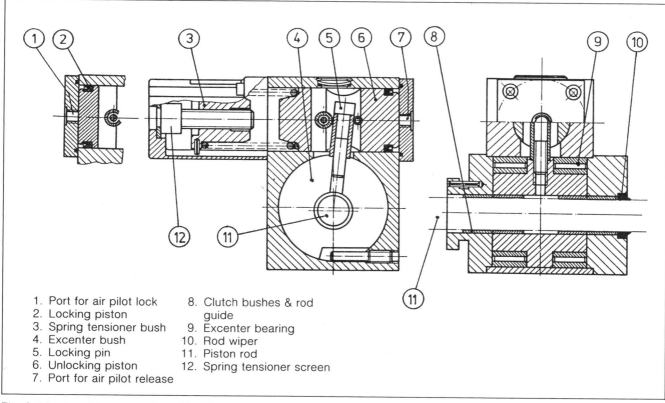

1. Port for air pilot lock
2. Locking piston
3. Spring tensioner bush
4. Excenter bush
5. Locking pin
6. Unlocking piston
7. Port for air pilot release
8. Clutch bushes & rod guide
9. Excenter bearing
10. Rod wiper
11. Piston rod
12. Spring tensioner screen

Fig. 3–19 "Bosch" pneumatic rod locking unit.

concepts such as Boyle's Law, Pascal's Law, Amonton's Law, Flow of compressed air through a nozzle discharging to atmosphere, and finally friction and mass of the load as well as its direction of motion. Therefore each of these items is developed separately and explained in detail.

Calculation of external forces (load)

The load which the pneumatic actuator has to move can be separated into the lifting force, the friction force of the load, and the force required to accelerate the mass.

The lifting force depends on the mass (m) to be moved and the inclination angle (α) on which the mass is moved (fig. 3–20). Thus the lifting force (F_L) is calculated as follows:

$$F_L = m \times g \times \sin \alpha$$

If the mass is to be lifted vertically then the formula is:

$$F_L = m \times g \quad \text{(for g use 9.81)}$$

The friction force (F_F) depends on the force developed by the mass and the contact surface condition of the mass where it rests against the bearing surface, and the bearing surface below that mass (fig. 3–20).

If the two contacting surfaces produce low friction (due to finely machined surfaces and good lubrication), then the friction coefficient μ will be low. Conversely, if the two contacting surfaces produce high friction then the friction coefficient will also be high. The frictional force for a mass movement in the horizontal direction is calculated as follows:

$$F_F = m \times g \times \mu \quad \text{(for g use 9.81)}$$

The friction force for a mass movement in an inclined direction (other than horizontal) is calculated as follows:

$$F_F = m \times g \times \mu \times \cos \alpha \quad \text{(for g use 9.81)}$$

The force required to accelerate the mass (m) is also important for actuator calculations. This force, called inertia force (Fm), depends on the mass (m) to be accelerated, the stationary speed (v) which is obtained after acceleration stops (fig. 3–25), and the distance (s) over which the acceleration takes place. Assuming a constant acceleration (a) the formula to be used is:

$$F_m = m \times a \quad \left(a = \frac{v^2}{2 \times s} = \text{m/s}^2\right)$$

Thus the total force of all forces encountered by

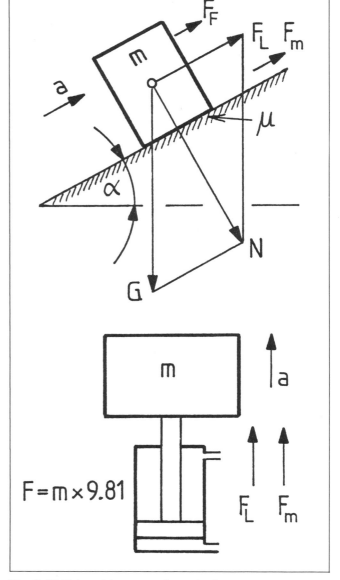

Fig. 3–20 External forces on the actuator.

the actuator piston, excluding frictional forces inside the actuator, is:

$$F_{TOT} = F_L + F_F + F_m$$

For an actuator which has to move a load supported or suspended by its piston rod, the calculation of the total force is simply: $F_{TOT} = m \times g$.

For an actuator moving a load in perfect horizontal direction the total force is: $F_{TOT} = m \times g \times \mu$. But both of these cases did not consider the inertia force for mass acceleration. Thus if only moderate acceleration is required, the inertia force (F_m) should be included when calculation of the total force is made.

Static thrust force calculations

The static thrust or pull can only be used for clamping actuators which usually have a very short stroke and produce no work until the component to be clamped is contacted. For such calculations the force produced is the product of the minimal system pressure times the effective piston area (fig. 3–21). The basic formula to calculate theoretical actuator force output (N), required system pressure (Pa), and effective piston area (m²), can be presented in a simple triangular illustration (fig. 3–21). The letter F stands for force, the letter p for pressure, and the letter A for the effective piston area.

- Force = Pressure × Area → $N = Pa \times m^2$

- Pressure = $\dfrac{Force}{Area}$ → $Pa = \dfrac{N}{m^2}$

- Area = $\dfrac{Force}{Pressure}$ → $m^2 = \dfrac{N}{Pa}$

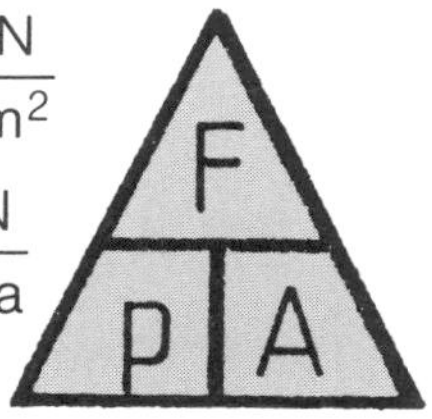

Fig. 3–21 Force-pressure-area relationship.

Three different formulae may be used to calculate piston area, these are:

$$\text{Area} = d^2 \times 0.7854 = \frac{\pi \times d^2}{4} = \pi \times r^2$$

If friction is to be included, the basic formula for force output is:

$$\text{Force} = \text{Pressure} \times \text{Effective Area} \times \frac{\text{Efficiency}}{100}$$

$$N = \frac{Pa \times m^2 \times \eta}{100}$$

Therefore:

$$\text{Force}_{\text{(EXTENSION)}} = \frac{Pa \times d^2 \times 0.7854 \times \eta}{100}$$

$$\text{Force}_{\text{(RETRACTION)}} = \frac{Pa \times (d_p^2 - d_r^2) \times 0.7854 \times \eta}{100}$$

where

d_p = diameter of the piston (m)

d_r = diameter of the rod (m), and

η = mechanical efficiency (%)

Efficiency (%) = 100% force − friction loss (%)

The chart in fig. 3–22 shows an array of effective piston forces for "BOSCH" actuators. These fig-

dia. Cylinder	dia. Piston rod	Piston area A (cm²)		Effective force (N) at p (bar)															
				3		4		5		6		7		8		9		10	
		●	○	●	○	●	○	●	○	●	○	●	○	●	○	●	○	●	○
10	4	0,8	0,66	21	17	28	21	35	29	42	35	49	41	56	46	63	52	70	58
12	6	1,1	0,8	29	22	39	30	48	37	58	45	68	52	77	60	87	67	97	75
16	6	2,0	1,73	53	46	70	61	88	76	106	91	123	107	141	122	158	137	176	152
20	8	3,1	2,6	82	69	109	92	136	114	164	137	191	160	218	183	246	206	273	229
25	10	4,9	4,1	129	108	172	144	216	180	259	216	302	253	345	289	388	325	431	361
32	12	8,0	6,9	212	182	282	243	352	304	422	364	493	425	563	486	634	546	704	607
40	16	12,6	10,6	333	280	444	373	554	466	665	560	776	653	887	746	998	840	1109	933
50	20	19,6	16,5	517	436	690	581	862	726	1035	871	1207	1016	1380	1162	1552	1307	1725	1452
63	20	31,1	28,0	824	739	1098	986	1373	1232	1647	1478	1923	1725	2196	1971	2471	2218	2746	2464
80	25	50,0	45,3	1328	1199	1771	1598	2213	1998	2656	2397	3098	2797	3541	3196	3984	3596	4426	3995
100	25	78,5	73,6	2072	1943	2763	2591	3454	3238	4145	3886	4836	4534	5526	5181	6217	5829	6908	6477
125	32	122,7	114,6	3239	3028	4319	4037	5399	5047	6479	6056	7558	7066	8638	8075	9718	9084	10798	10094
160	40	201,0	188,4	5309	4976	7079	6635	8848	8294	10618	9953	12388	11612	14157	13270	15927	14929	17697	16588
200	40	314,1	301,4	8295	7962	11060	10616	13825	13270	16590	15924	19355	18579	22120	21233	24885	23887	27650	26541
250	50	490,6	471,0	12960	12442	17280	16590	21600	20737	25920	24885	30239	29032	34559	33180	38879	37327	43199	41474

● FULL PISTON AREA
○ ANNULAR AREA

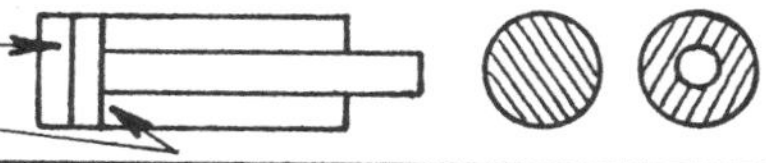

Fig. 3–22 Effective piston forces for "BOSCH" actuators.

ures include a 12% friction loss for internal actuator friction and are based on the formulae:

$$N = \frac{Pa \times m^2 \times 88}{100} \quad Pa = \frac{N \times 100}{m^2 \times 88} \quad m^2 = \frac{N \times 100}{Pa \times 88}$$

Example 1 (static force calculation)

A pneumatic double acting actuator is used to clamp work pieces in a machine tool (fig. 3–23). The actuator available has a piston diameter of 125 mm. The required clamping force is calculated to be 6000 newtons. Select the appropriate minimal system pressure to achieve this force. Assume a friction loss of 5% (near static thrust).

Solution:

$$\text{Pressure} = \frac{\text{Force} \times 100}{\text{Area} \times \eta}$$

$$p = \frac{6000 \times 100}{0.125^2 \times 0.7854 \times 95 \times 10^5}$$

$$p = 510 \text{ kPa (5.1 bar)}$$

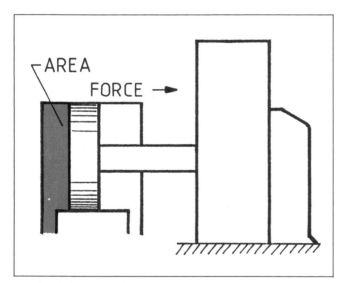

Fig. 3–23 Static force calculation applied to clamping.

Example 2 (static force calculation)

Calculate the force output for a double acting linear actuator with a piston diameter of 100 mm and a rod diameter of 25 mm. The minimal system pressure is 500 kPa (5 bar) and the internal friction loss of the actuator is 12% of its theoretical force. Calculate both the extension and the retraction force.

Solution:

$$\text{Force}_{EXT} = \frac{500 \times 10^3 \times 0.1^2 \times 0.7854 \times 88}{100}$$

$$= 3455 \text{ N}$$

$$\text{Force}_{RET} = \frac{500 \times 10^3 \times (0.1^2 - 0.025^2) \times 0.7854 \times 88}{100}$$

$$= 3240 \text{ N}$$

Example 3 (near static force calculation)

A pneumatic linear actuator must slowly raise a platform with a load on it (fig.3–24). The platform and the load have a mass of 660 kg. The actuator internal friction is assumed to be 12% and the minimal system pressure adjusted on the pressure

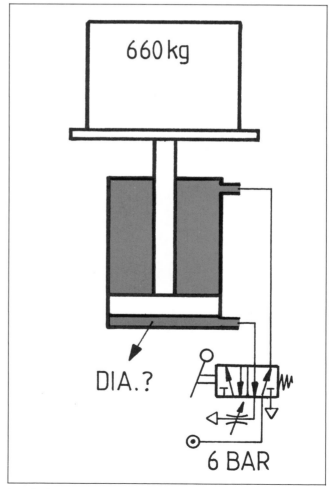

Fig. 3–24 Near static force calculation.

The South East Essex
College of Arts & Technology
Carnarvon Road Southend on Sea Essex SS2 6LS
Tel: Southend (0702) 220400   Fax: Southend (0702) 432320

reducing valve of the air service unit is 600 kPa (6 bar). Inertia force can be neglected. Calculate the minimal required piston diameter for this load and from the chart in fig. 3–22 select the smallest possible standard piston diameter. Compare the calculated result with the listed force output figure in the chart.

Solution:

$$d = \sqrt{\frac{F \times 100}{p \times 0.7854 \times \eta}}$$

$$d = \sqrt{\frac{660 \times 9.81 \times 100}{600 \times 10^3 \times 0.7854 \times 88}} = 0.12495 \text{ m}$$

The chart shows for the force of 6474 newtons a piston diameter of 125 mm which renders an effective force of 6479 newtons. This is slightly more than the calculated force, so the piston of 125 mm is sufficient to raise the load if no back pressure is encountered.

Relationship of pressure to load

The diagram shown in fig. 3–25 explains the complex relationship of pressure and load to the speed of a double acting pneumatic actuator. It shows that the minimal system pressure should be about 10% above the breakaway pressure and that the back pressure is about 40% of the break away pressure if meter out speed control is used; thus the back pressure load which must be added to the maximum load force is also 40%. These figures are also in agreement with load curve Ⓒ in fig. 3–26. For this reason some circuit designers suggest that 50% extra load force should be added to the maximum load force calculated before, and the actuator should then be sized for this new load which automatically takes care of back pressure

Fig. 3–25 Relationship between pressure and load and speed and back pressure and cushioning.

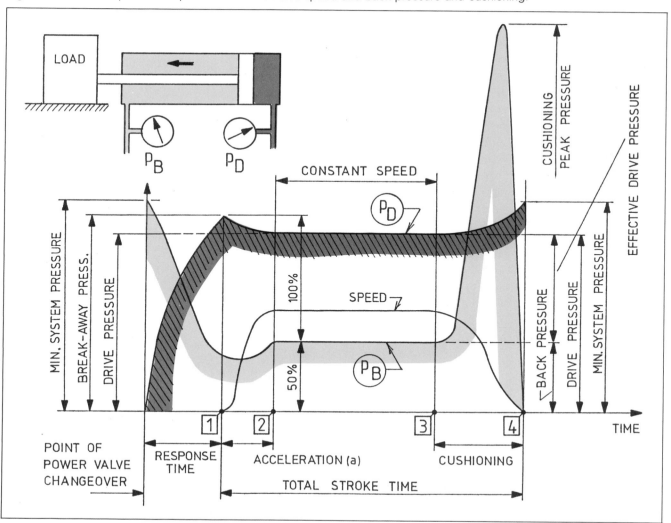

and internal actuator friction losses. The diagram (fig. 3–25) shows that after the valve has been switched over, the back pressure (p_B) drops rapidly and the drive pressure (p_D) increases to a point where the force imbalance causes the piston with its attached load to break away (see fig. 3–26, point [1]).

As the load starts to move and accelerate, the back pressure continues to fall until the acceleration of the load and the piston catches up with the exhausting air [point 2] and forces it out of the actuator exhaust port. This also reestablishes the near balance of forces and with the load and back pressure remaining constant, the speed must also remain constant since the exhausting air exists at sonic speed (the speed of sound).

When the pneumatic cushioning starts to become effective [point 3], the back pressure starts to rise rapidly, thus providing the necessary counterforce to transfer the kinetic energy into heat (Amonton's Law). As the speed is reduced to a constant governed by the flow rate past the cushioning valve, the load comes eventually to a standstill at the end of the piston stroke at point [4] in fig. 3–25 (see also pneumatic end cushioning fig. 3–8 and Boyle's Law fig 1–12).

Dynamic thrust force calculations

For the sizing of double acting actuators which are to exert force while reciprocating or have to accelerate the load rapidly, the following practical aspects must be taken into consideration:

- Applied air pressure to the force side of the piston and the effective piston area must be large enough to cope with all calculated external forces (F_L, F_F and F_m; see fig. 3–20).
- There must be sufficient back pressure on the exhaust side of the moving piston to ensure that the exhausting air has a sonic flow rate; thus giving constant flow rate. This provides effective and constant speed control whilst the load is being moved (meter-out speed control). See also flow through an orifice to atmosphere (Chapter 1) and fig. 3–25.
- The exhausting air must have the minimal required back pressure for effective pneumatic cushioning. Should the exhaust air pressure drop to atmospheric pressure, then the pneumatic cushioning would no longer be effective and the piston with its attached load would have to be braked by hard impact on the endcaps.

The actuator sizing nomogram in fig. 3–26 makes this seemingly complex task very simple. Curve Ⓐ is to be used for static thrust force calculations. It includes a 12% friction loss for internal actuator

friction. Figures obtained from this curve must only be used for clamping applications or slow moving loads with minimal acceleration and no back pressure!

For reciprocating actuators which must impart

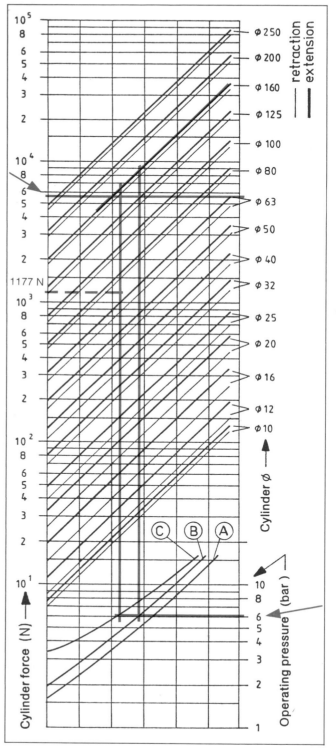

Fig. 3–26 Actuator sizing nomogram.

acceleration to the load attached to them, curve Ⓑ must be used (see fig. 3–25).

Now use Curve Ⓒ to check whether the actuator diameter selected with curve Ⓑ is also capable of moving the load at constant speed (stationary speed). The stationary speed distance spans from point [2] to point [3] in fig. 3–25. This speed must be kept constant despite the now higher back pressure. From the fig. 3–25 and 3–26 one can easily see that the back pressure increases for the constant speed phase. This causes the force output now to be lower than during the acceleration phase. Should the readings on the sizing nomogram (fig. 3–26) for curve Ⓑ and Ⓒ give different actuator diameters, then one must use the larger of the two piston diameters (see also effects of undersizing an actuator).

Example 4 (dynamic force calculation)

An actuator must reciprocate a load in horizontal direction (fig. 3–27). The load mass is 1000 kg, the friction coefficient is 0.12, the acceleration travel distance is 40 mm and the stationary speed after acceleration is 600 mm/s. The system pressure is 600 kPa (6 bar) calculate the piston diameter.

Solution:
The lifting force (F_L) is zero since the inclination angle α is also zero

$$\longrightarrow (m \times 9.81 \times \alpha\,0° = 0)$$

The friction force (F_F) is:

$$F_F = m \times 9.81 \times \mu$$
$$= 1000 \times 9.81 \times 0.12 = 1177 \text{ N}$$

The inertia force (F_m) is:

$$F_m = \frac{m \times v^2}{2 \times s} = \text{N}$$

$$F_m = \frac{1000 \times 0.6^2}{2 \times 0.04} = 4500 \text{ N}$$

The total external force (F_{TOT}) is:

$$F_{TOT} = F_F + F_m$$
$$= 1177 + 4500 = 5677 \text{ N}$$

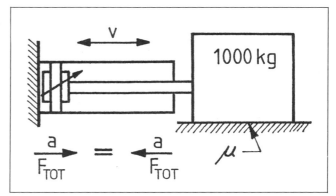

Fig. 3–27 Dynamic force application.

In fig. 3–26 of the actuator sizing nomogram one finds with a minimal system pressure of 600 kPa (6 bar) on curve Ⓑ and a total force output (F_{TOT}) of 5677 newtons a piston diameter (for the annular area) of 160 mm. The actuator is therefore suitable for imparting to the mass the envisaged acceleration to the stationary speed v (600 mm/s within the acceleration distance s (40 mm).

With curve Ⓒ in fig. 3–26 one now checks the actuator's suitability to hold that speed constant against the back pressure which is necessary to brake the mass with the inbuilt pneumatic cushioning. The so obtained reading of 5500 newtons is well above the necessary friction force of 1177 newtons, so the actuator is suitable.

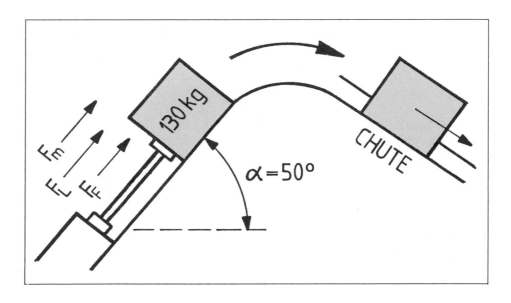

Fig. 3–28 (For example 5.)

Example 5

A pneumatic linear actuator must push parcels up a ramp onto a sorting shute. The ramp inclination is 50°. The friction coefficient is 0.10 and the mass of a parcel is 130 kg (fig. 3–28). A pneumatically cushioned actuator is used with a cushioning travel distance (l_c) of 0.028 m, the acceleration (a) of the load must occur within the cushioning travel distance (l_c) and must reach a stationary speed (v) of 0.5 m/s. Minimal system pressure is 600 kPa (6 bar) . Calculate:

1. The total force (F_{TOT}) of all external forces including the lifting force (F_L), the frictional force (F_F), and the inertia force (F_m) (also called mass force).
2. The piston diameter for the extension stroke. Assume an internal actuator friction loss of 12%.
3. Compare the calculated figure with the effective force figure given in the chart (fig. 3–22).
4. Compare the figures obtained from calculation and chart (fig. 3–22) with the result obtained in the actuator sizing nomogram of fig. 3–26.

Solution:

The lifting force (F_L) is:

$$Force\ F_L = m \times 9.81 \times \sin\alpha = N$$
$$= 130 \times 9.81 \times 0.766 = 977\ N$$

The friction force (F_F) is:

$$Force\ F_F = m \times 9.81 \times \cos\alpha \times \mu$$
$$= 130 \times 9.81 \times 0.643 \times 0.1 = 82\ N$$

The inertia force (F_m) is:

$$Force\ F_m = \frac{m \times v^2}{2 \times s} = N$$
$$= \frac{130 \times 0.5^2}{2 \times 0.028} = 580\ N$$

The total external force (F_{TOT}) is:

$$Force\ F_{TOT} = F_L + F_F + F_m = N$$
$$= 977 + 82 + 580 = 1639\ N$$

The piston diameter for extension is:

$$d = \sqrt{\frac{F \times 100}{p \times 0.7854 \times \eta}}$$

$$d = \sqrt{\frac{1639 \times 100}{600 \times 10^3 \times 0.7854 \times 88}} = 0.063\ m$$

The next standard size actuator is 63 mm with an effective force output of 1647 newtons. Its stan-dard size piston rod has a diameter of 0.02 m or 2 cm or 20 mm.

By using the actuator sizing nomogram (fig. 3–26) one finds that a piston diameter of 63 mm (0.063 m) would not suffice and thus the next larger piston diameter of 80 mm must be chosen.

Effects of undersizing an actuator

If an actuator is undersized it may either not cope with the given load force or it may just cope but the response and acceleration times become extremely long, such that the back pressure falls below the level required to maintain sonic speed. Thus the necessary back pressure needed to provide constant speed is no longer there and the speed becomes erratic. In addition, the pneumatic cushioning is totally ineffective since the initial back pressure used for cushioning is too low.

Speed control and impact control

Speed control is based on the flow rate either into or out of the actuator and is dealt with in the next chapter. Apart from speed control, pneumatic end cushioning must also be regarded as a form of speed control. Impact energy at the end of the piston stroke must be kept very small and if the load speed on impact should exceed 50 mm/s one should consider external impact cushioning or pneumatic internal end cushioning.

If the inbuilt pneumatic cushioning proves to be inadequate to break the load sufficiently, then one has only two possibilities available to remedy the problem.

1. The minimal system pressure (fig. 3–25) and the back pressure which depends on the minimal system pressure can be increased. If the back pressure is increased, the cushioning potential is also increased and thus the load may be sufficiently cushioned.
2. One may select an actuator with a larger piston diameter. This means a smaller pressure difference between drive pressure and back pressure is required to move the load. The back pressure throttling can then be increased, causing a higher back pressure, which in turn increases the cushioning.

Conclusion to thrust force calculations

An actuator can only be properly sized if all external forces are calculated and included in the total force. These forces include the load force, the friction force, and most importantly also the inertia force. Many actuators fail to perform as anticipated because the inertial force was not considered to be important and hence was not included in the total anticipated load force.

The load plus 50% rule of thumb can be used providing the inertia force is again included in the total force, and the back pressure force is not more than 40% of the drive pressure force (stationary speed force). However, it is advisable to use a precise actuator sizing nomogram similar to the one given in figure 3–26 which is based on calculation as well as on laboratory test figures (For friction figures see Appendix).

For downward moving loads the lifting force becomes negative and the friction force and inertia force remain the same.

Piston rod buckling

Column failure, or buckling of the piston rod, will occur if the actuator stroke is out of safe proportion in relation to the piston rod diameter, at a required force output (fig. 3–29). Piston rod buckling is calculated according to the *Euler formula*, where the piston rod is regarded as the buckling member. In pneumatics, Euler cases 3 and 4 are less frequent.

For pneumatic actuators the free buckling length (l) is calculated as follows (see figs. 3–10 and 3–29):

Euler case 1; rod end mounting: l = stroke length
Euler case 1; rear end mounting: l = stroke length*
Euler case 2; rod end swivel: l = stroke length
Euler case 2; rear end swivel: l = stroke length × 2**

Note: * The actuator barrel is assumed to be a stiff machine component which does not pivot. ** The actuator barrel can pivot and thus permits rod side swing away from the straight force line.

Euler formula: $K = \dfrac{\pi^2 \times E \times J}{{}^sK^2}$

(At this load rod buckles!)

The maximum safe operating load (in newtons) therefore is:

$F = \dfrac{K}{S} = \dfrac{\text{critical load}}{\text{safety factor}} = \text{permissible force}$

K = critical load (N)

l = free buckling length (m) with actuator fully extended
S = safety factor (usually 2.5–5)
E = elasticity modulus (Pa) → for steel use: $2.1 \times 10^{10} \times 9.80665$
F = force (N)

$J = \text{moment of inertia (m}^4) \rightarrow \dfrac{\pi \times d^4}{64}$

d = piston rod diameter

${}^sK = $ equivalent buckling length (m)

Example 6

Piston diameter   = 40 mm (0.04 m)
Piston rod diameter (d) = 16 mm (0.016 m)
Stroke length   = 550 mm (1.10 m)
Euler case (fig. 3–26)   = situation 2
Safety factor (S)   = 5 (max. safety)
Mounting type   = rear end swivel

Find safe operating pressure:

Solution:

$K = \dfrac{\pi^2 \times 2.1 \times 10^{10} \times 9.80665 \times \pi \times 0.016^4}{1.10^2 \times 64}$

K = 5403.86 newtons → Rod buckles at this load!

Safe loading: $F = \dfrac{K}{5} = 1081$ N (with safety factor 5)

Safe pressure for this actuator is:

$Pa = \dfrac{N}{m^2} = Pa = \dfrac{1081}{0.016^2 \times 0.7854} = 5375306$
$= 540$ kPa (5.40 bar)

For ease of calculation the formula for safe operating load can be simplified by grouping all constants in the formula together:

$F = \dfrac{K}{S} = \dfrac{\pi^2 \times 2.1 \times 10^{10} \times 9.80665 \times \pi \times d^4}{S \times {}^sK^2 \times 64}$

$= \dfrac{10^{11} \times d^4}{S \times {}^sK^2}$

For ease of piston rod sizing, most manufacturers of pneumatic linear actuators provide sizing charts or nomograms (fig. 3–30). Such graphic aids are based on calculations as well as data obtained by experience in the field. The two parameters required are the system pressure and the required force. When plotted into the nomogram one can then derive the piston rod which *would not buckle* as long as the plotted parameters are not exceeded towards an unsafe condition.

Air consumption calculations

Air consumption of the various pneumatic actuators represents an important part of the operating costs of a pneumatic system. It also provides a guide for the proper sizing of the compressor plant, the air distribution system, and air preparation system with its air dryers, air receiver, after cooler and air service units.

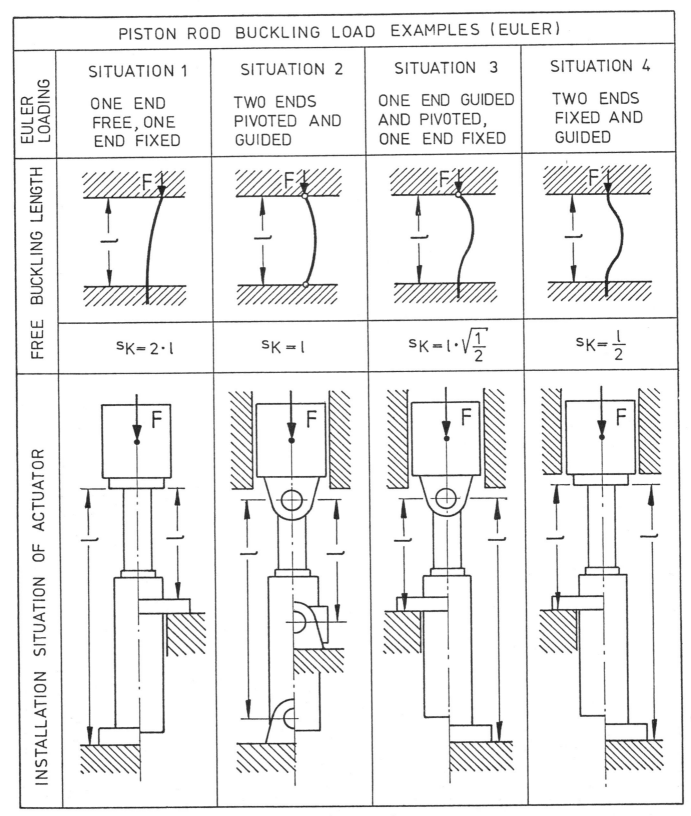

Fig. 3–29 Piston rod buckling load examples (for pneumatic actuators).

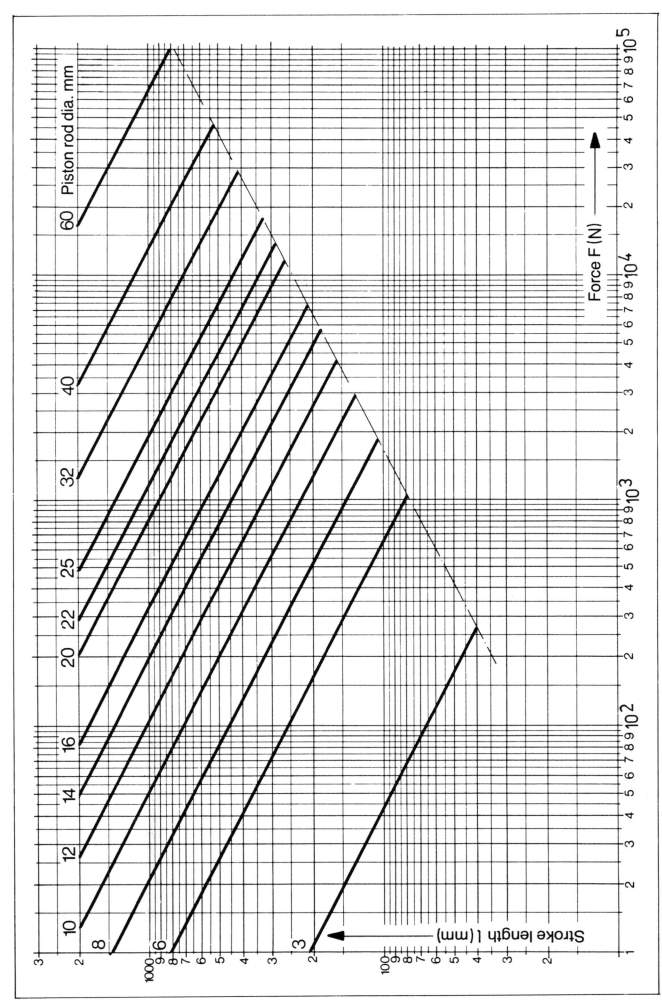

Fig. 3-30 Buckling load nomogram.

For a given minimal system pressure (operating pressure), piston area and stroke length, the air consumption is calculated with the formula:

> Compression ratio $(\varepsilon) \times$ geometrical volume (V)

The compression ratio is determined by the quotient of the absolute minimal system pressure and the absolute ambient pressure which is normally about 101.3 kPa (1.013 bar) at sea level.

$$\varepsilon = \frac{101.3 \text{ kPa} + \text{minimal system pressure (kPa)}}{101.3 \text{ kPa}}$$

The geometrical volume (V) for an actuator is calculated with the following formulae:

> $V = d^2 \times 0.7854 \times \text{stroke length}$
> (extension stroke)
> $V = (d^2_{PISTON} - d^2_{ROD}) \times 0.7854 \times \text{stroke length}$
> (retraction stroke)

For double acting linear actuators with a piston rod on one end only, the annular area geometrical volume is therefore less than the full piston area geometrical volume. For precise air volume calculations, one could therefore use this lesser volume for the retraction stroke.

However, it must be noted that the end cap "cavities" with their "dead" volume must also be filled with compressed air before the piston can move and the same applies for the air lines between the power valves and the actuator. As a rule of thumb one can therefore use for both strokes the full piston area geometrical volume, and assume that the oversizing for the piston rod volume makes up for the neglected end cap cavity volumes and air line volumes (fig. 3–5).

For single acting linear actuators one calculates only one stroke per one full cycle (extension and retraction) since the return stroke is not air powered and thus needs not to be included in the calculation.

Some typical end cavity volumes for "FESTO" actuators:

Piston Dia. mm	Rear End Cap cm^3	Rod End Cap cm^3
25	6	5
35	13	10
50	19	16
70	31	27
100	88	80
140	150	128
200	448	425

Fig. 3–31 End cap cavity volumes.

Example 7

Minimal system pressure	= 600 kPa (6 bar)
Piston diameter	= 80 mm (0.08 m)
Stroke length	= 400 mm (0.4 m)
Cycles per minute	= 5, (which equals 10 strokes)

Calculate flow rate per minute (Q)

Solution:

$$Q = \frac{101.3 + 600}{101.3} \times 0.08^2 \times 0.7854 \times 0.4 \times 10$$
$$= 0.14 \text{ m}^3/\text{min}$$

By using the air consumption nomogram in fig. 3–32 one can find approximate values quicker and more easily. The actual piston stroke is plotted in the nomogram on the left and a line is drawn up to the appropriate piston diameter then reflected horizontally to the right hand graph where it intersects with the minimal system pressure (gauge pressure) chosen. The line drops now vertically to the bottom end of the nomogram where one can read off the air consumption in litres per cycle (double stroke). The air consumption is always calculated in free air!

To compare the calculated value from the previous example with the obtained free air volume in the nomogram one calculates:

> 27 litres $\times$ 5 cycles/min = 135 litres/min

This value equates to 0.135 m^3/min, which is very close to the previously calculated result.

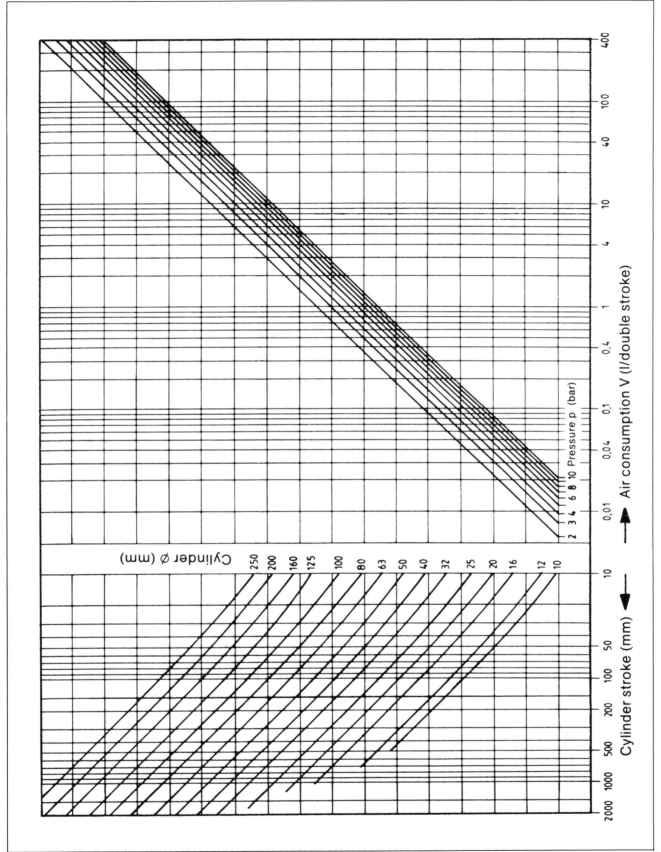

Fig. 3–32 Air consumption calculation nomogram.

4 Flow Controls for Pneumatics

Flow control valves are frequently used in pneumatic systems. Their use is limited to the control of compressed air to and from the actuator and their main functions are;

- to control the maximum speed of the actuator piston and its attached load,
- to provide constant speed for the main part of the actuator piston stroke when acceleration is completed,
- to cushion the load impact at the end of the actuator stroke or decelerate a load to a lower speed,
- to prevent free falling of a heavy load on the downward movement.

Simple restrictor type flow control

This valve, being variable as far as the cross section of the orifice is concerned, is therefore also called variable restrictor flow control valve or variable orifice flow control valve. It is mostly used on the outlet of the power valves (fig. 4–2) and restricts the flow of air in both directions (fig. 4–1).

These valves consist of a valve body and a throttling screw which adjusts the orifice cross-section to achieve the desired speed on the actuator.

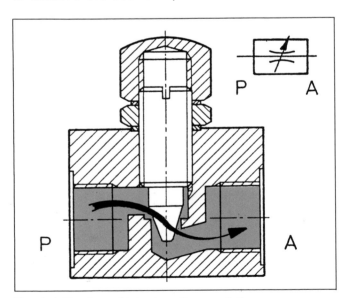

Fig. 4–1 Simple restrictor type flow control.

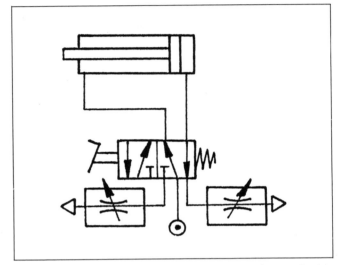

Fig. 4–2 Two simple restrictor type flow controls attached to the power valve exhaust ports.

Variable restrictor with free reverse flow

Where speed control (flow control) in only one valve direction is required, a check valve is built into the flow control. This permits unrestricted flow of air in the reverse direction (fig. 4–3).

The construction of this type of valve apart from the inbuilt check valve is identical to the simple restrictor type. The check valve consists of a rubber ring, which can flex upward if reverse flow is required, and a ring hub to hold the disc onto the check valve orifice during the throttling function.

Some pneumatic component manufacturers also make miniature type flow control valves which are directly built into the exhaust ports of the power valve. Others manufacture an actuator swivel-fitting which contains a small flow control valve with free reverse flow. Such inbuilt valves save space, extra connectors and installation costs.

Cam roller operated flow control

Cam roller operated flow control valves perform the same function as the previously explained flow controls but here a cam roller is used to alter the flow-rate (orifice area). Thus variable speed control during the stroking of the actuator can be achieved (figs. 4–4 and 4–5).

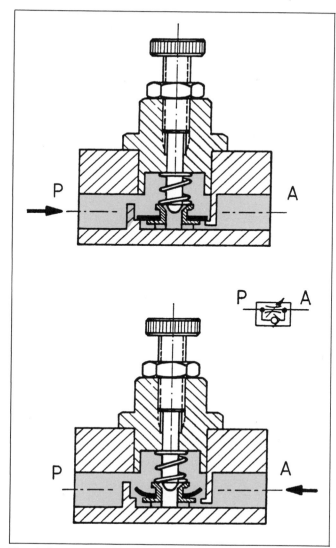

Fig. 4–3 Variable restrictor with free reverse flow.

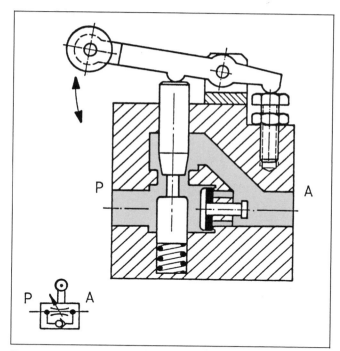

Fig. 4–4 Cam roller operated flow control valve.

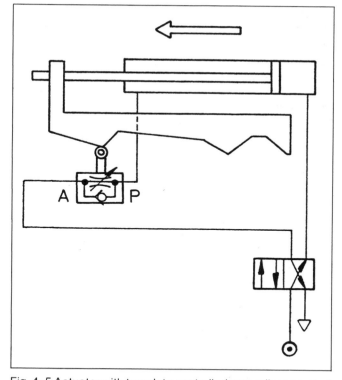

Fig. 4–5 Actuator with template controlled cam roller operated speed control (meter-out).

The minimal speed (orifice opening) is limited on the rear end of the roller lever by means of a screw (fig. 4–4). A template is attached to the moving piston rod or to the moving machine part driven by the piston rod (fig. 4–5).

A downward movement of the roller reduces the cross section of the orifice and thus reduces the speed of the actuator under control.

The inbuilt check valve permits unrestricted air flow in the reverse direction. The check valve poppet moves to the left when unrestricted flow is required and resets itself when the flow is in the opposite direction (P → A).

Quick exhaust speed control

The quick exhaust valve, although used as a speed controlling device, is not a flow control valve in the proper sense, but rather a self steering direc-

tional control valve similar to the "OR" function valve (fig. 4–6).

It allows air to flow from the power valve (port P) to the actuator to which it is attached (port A).

When the actuator is exhausting (flow in reverse direction), the valve element is shuttled to the left where it blocks port P and the exhausting air can rapidly flow through a large exhaust port (R) to atmosphere. Thus the exhaust air must not be forced all the way back through the power valve (fig. 4–7). This valve often permits a speed increase up to 2 m/s, but care must be taken to control load impact. The fitted exhaust silencer does not affect exhaust efficiency.

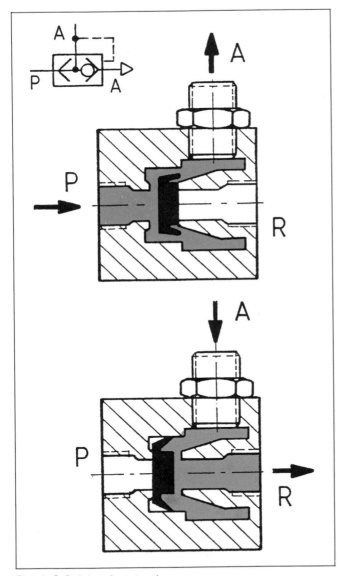

Fig. 4–6 Quick exhaust valve.

Hydraulic check speed control

Any form of pneumatic speed control no matter how effective, can never provide absolute speed consistency. Speed consistency is imperative for machining operations or applications in manufac-

turing where the quality of the product depends on absolute uniform speed over the entire stroke length.

For such cases an hydraulic check cylinder may be used. The hydraulic oil, which is virtually incompressible, is used to provide precise and uniform speed control which is independent and unaffected by load variations of the pneumatic actuator. The pneumatic actuator is linked to the hydraulic check cylinder. The force needed to move the hydraulic piston and its piston rod comes from the pneumatic actuator which literally drags the hydraulic piston along. The displaced oil is forced via a hydraulic flow control valve which provides the required speed control. An oil reservoir with a spring loaded piston compensates for the differing oil volumes displaced by the hydraulic piston and makes up for oil losses due to piston rod lubrication (fig. 4–8).

Installation methods for restrictor type flow controls

Two basic methods are commonly used to install restrictor type flow controls such as shown in figs. 4–1, 4–3 and 4–4. These methods are widely known as *meter-in* flow control and *meter-out* flow control (figs. 4–9 and 4–13). Both methods have their distinct advantages and disadvantages, but where possible meter-out should be given preference.

Meter-in speed control

Where an actuator has insufficient air volume on the exhaust side of the moving piston, so that the exhausting air can not be sufficiently compressed by the flow control to cause an air-cushion, then meter-in speed control must be used. This is mainly the case with small diameter or short stroke actuators. Meter-in must also be used for single acting actuators (fig. 4–17).

Meter-out speed control

Meter-out provides the necessary back-pressure which is imperative for smooth and constant piston speed (see Chapter 3, fig. 3–25).

The back pressure brought about by meter-out speed control reduces the force output of the actuator, which demands that a larger piston must be selected to move the given load. The advantages gained from meter-out speed control are numerous and may be listed as:

- A free falling, suspended or supported load can be controlled with ease (figs. 4–10 and 4–11).

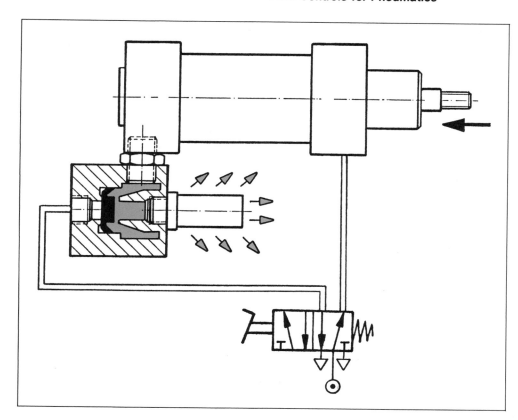

Fig. 4–7 Installation of quick exhaust valve on the actuator.

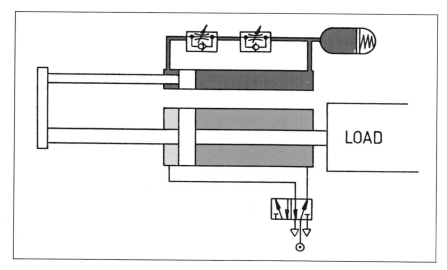

Fig. 4–8 Hydraulic check linked to a pneumatic double acting actuator.

- The back pressure caused by a meter-out speed control restrictor (flow control valve or an orifice) resists the advancing piston which is driven forward by the pressure on the opposite side (note pressure readings on the pressure gauges in fig. 4–12). This form of speed control provides excellent and constant control of the stationary actuator speed (see fig. 3–25).
- Pneumatic end-cushioning is greatly enhanced with the air pressure cushion caused by the meter-out speed control restrictor valve (fig. 4–9,

see also effects of undersizing an actuator in Chapter 3).

Power valve exhaust port speed control

To achieve optimal speed control it is advisable to mount the flow restrictors (speed control valves) directly onto the actuator (fig. 4–9).

One frequently finds these valves mounted onto the exhaust ports of the power valves (fig. 4–13). With this installation, however, one runs the risk of damaging the pneumatic seals inside the power

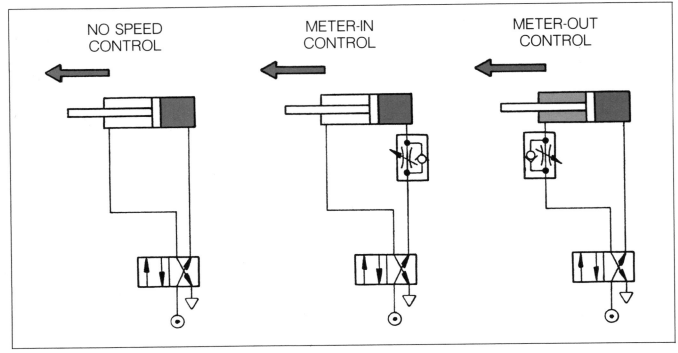

Fig. 4–9 Speed control installation.

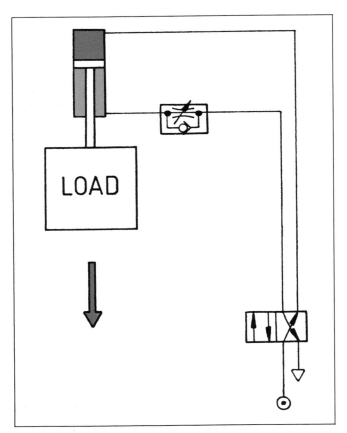

Fig. 4–10 Suspended load.

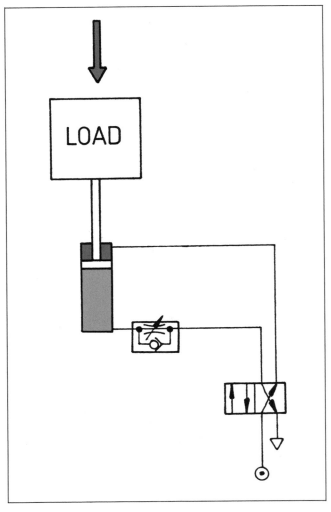

Fig. 4–11 *Right:* Supported load.

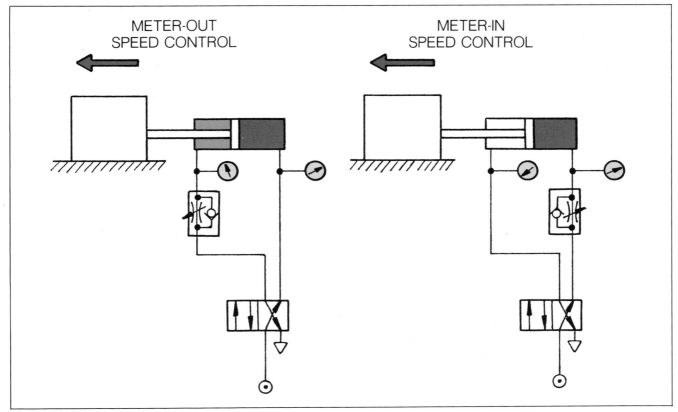

Fig. 4–12 Pressure comparison for meter-out and meter-in speed control.

valve or rupturing the air hose between the actuator and the power valve. This may occur because of pressure intensification induced by an oversize piston rod (see calculation fig. 4–14).

Example:
A custom built linear actuator has a piston diameter of 200 mm and an oversized piston rod diameter of 180 mm. The system air pressure is 1000 kPa (10 bar). Calculate the intensified pressure on the rod end air outlet, if the meter-out air outlet of the flow control valve is by mistake almost closed off.

$$\text{Pressure} = \frac{\text{Force}}{\text{Area}} \rightarrow Pa = \frac{N}{m^2} \rightarrow \text{Pressure}$$

$$= \frac{\text{Force to left}}{\text{Annular Area}}$$

$$\text{Pressure} = \frac{10 \times 10^5 \times 0.2^2 \times 0.7854}{(0.2^2 - 0.18^2) \times 0.7854 \times 10^5}$$

$$= \frac{0.4}{0.0076} = 5{,}263 \text{ kPa } (52.63 \text{ bar})$$

System sizing
System sizing is the simplest and least costly speed control method. Even if it does not provide adjustable speed control, it has a proper place especially in large series of identical controls. It is mostly established by trial and error but saves costly control valves. The fittings screwed into the actuator ports act as natural restrictors and where necessary, reduction fittings may be used to cause more flow restriction (fig. 4–15).

Pneumatic end-cushioning
Pneumatic end-cushioning is also a form of speed control. However, it only affects the very last portion of the piston stroke and, does, although adjustable, only count as an auxiliary to the other speed control methods discussed so far. Pneumatic end-cushioning is often used in addition to meter-out, quick exhaust control and system sizing (fig. 4–16).

Summary of speed control types
The speed control methods used to control the speed of pneumatic actuators are many and varied, but may be essentially summarized by the seven main methods discussed previously. They are listed in order of importance and frequency of use.

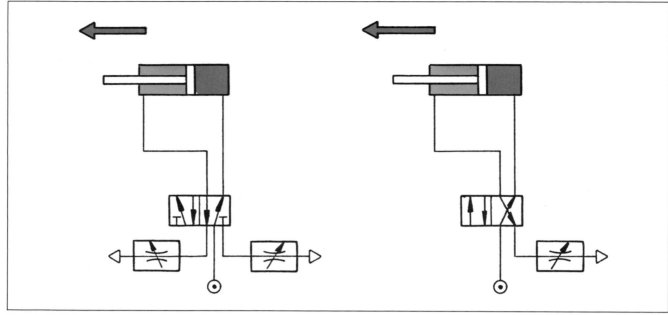

Fig. 4–13 Exhaust port speed control.

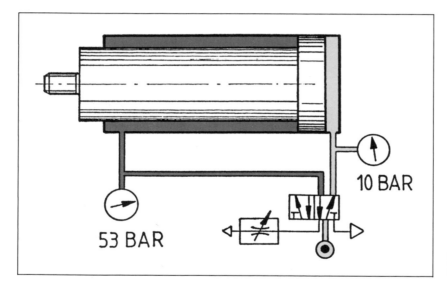

Fig. 4–14 Pressure intensification due to oversize piston rod.

- Variable restrictor with reverse flow installed as meter-out or meter-in (figs. 4–3 and 4–9).
- Simple restrictor type flow control installed mostly in power valve exhaust ports (figs. 4–1 and 4–13).
- Pneumatic end-cushioning in conjunction with other speed control types (fig. 3–8).
- System sizing used in conjunction with pneumatic-end cushioning (figs. 3–8 and fig. 4–15).
- Quick exhaust speed control in conjunction with other speed control types such as variable restrictors (fig. 4–18).
- Cam roller operated flow control in conjunction with other speed control types (fig. 4–5).

- Hydraulic check speed control usually installed as an ancillary unit atttached to an existing actuator.

Summary of installation methods

The most recommended method is meter-out, but power valve exhaust port installation, even if it is also a form of meter-out, is not as effective as having the speed control directly mounted onto the actuator. Meter-in speed control should be avoided where possible.

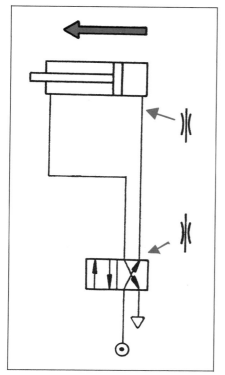

Fig. 4–15 System sizing.

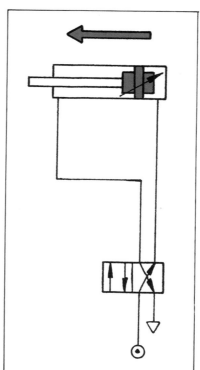

Fig. 4–16 Pneumatic end-cushioning.

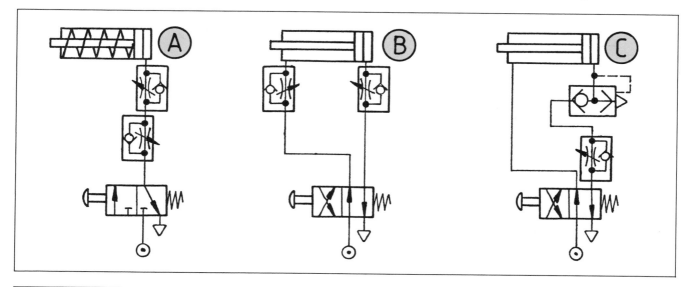

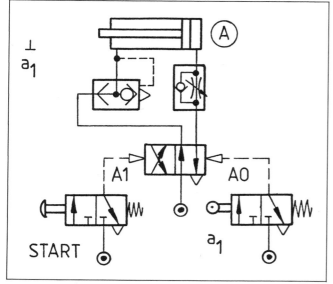

Fig. 4–17 *Above:* Shows a combination of quick exhaust and meter-in on the same port. Thus the actuator will retract rapidly but extend with adjustable speed.

Fig. 4–18 Shows a combination of quick exhaust and meter-out speed control. Thus the actuator will extend rapidly but retract with adjustable speed.

5   Air Motors (Rotary Actuators)

By the use of rack and pinion drives or other mechanical systems the linear actuator (or cylinder) can be used to provide a rotary movement and so provide a torque output such as shown in fig. 3–18. Although such swivel actuators are suitable for many applications, they are not to be mistaken for air motors. Swivel actuators or semi-rotary actuators are generally built and designed to provide a rotary torque output up to a 360° rotation.

Air motors, however, produce a continuous torque output—thus driving a revolving drive shaft—and are regarded as the pneumatic counterpart to an electric motor. Revolving air motors are put to extremely diverse uses: from the not too popular dentist's drill to the drive for a caterpillar tracked tunnel boring machine. The manufacturing industry in particular is a large user of a wide range of types and sizes of air motor driven tools such as nut runners, drills, screw drivers, grinders and sanders.

Air motors use the characteristics of compressed air power transmission to their best advantage, which means they are easily adjusted for varying speeds and torque, can be reversed, can safely be operated in an explosive environment, and can be stalled indefinitely (thus are overload safe). These operational characteristics, combined with their excellent power features, that make air motors so versatile and widely accepted.

Air motors can be classified into five basic types, each of which produces a specific range of power output and has an operating characteristic which

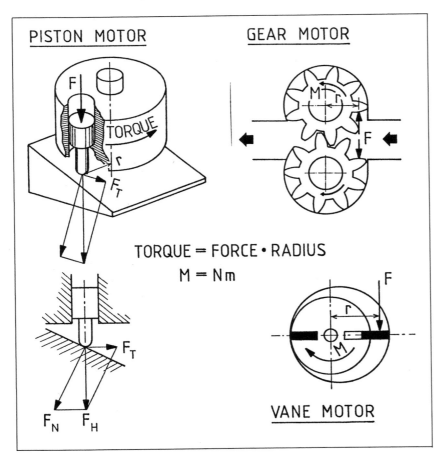

Fig. 5-1 Development of torque in pneumatic motors.

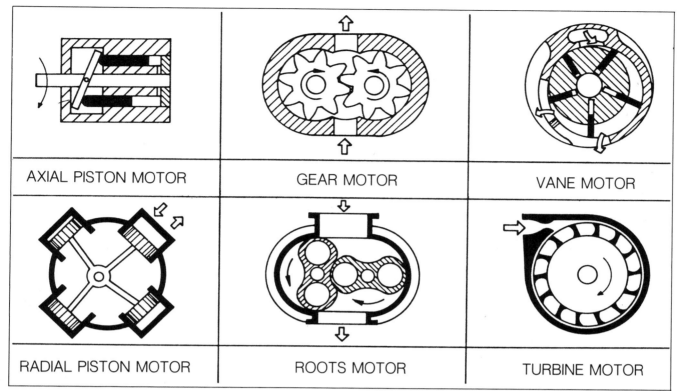

Fig. 5–2 Design concepts for air motors.

makes it ideally suited for a particular industrial application. These types are:

- vane motors (figs. 5–2 and 5–6)
- gear motors (figs. 5–2 and 5–10)
- piston motors (figs. 5–2 and 5–13)
- turbine motors (figs. 5–2 and 5–14)
- roots motors (fig. 5–2).

Each design type must have a driving surface area (A) subject to a pressure differential (Δ_p). For vane and gear motors, this surface is rectangular. For radial and axial piston motors the surface is circular (see fig. 5–1).

- In each design type the pressure exposed driving surface area (A) must be connected mechanically to the motor output shaft.
- Inlet and outlet air must have a timed porting arrangement to produce continuous rotation (this feature does not apply to turbine motors).

Maximum performance varies greatly between the various design types of air motors and is determined by:

1. the ability of its internal pressure exposed areas to withstand the forces acting upon them;
2. the internal leakage characteristics of the moving parts which seal the high pressure inlet from the low pressure outlet; and
3. the efficiency of the mechanisms which link the

moving pressure exposed parts to the motor drive shaft (output shaft).

Torque (turning moment)

Torque is defined as a twisting or turning moment and is expressed in newton metres (Nm). Torque is a function of air pressure and leverage whereby the leverage is measured from the centre of the drive shaft to the centre of the pressure exposed area. The torque output of a typical air motor (except turbine motors) is at its maximum when the motor is stalled, and theoretically also at the start. This, however, is not so in practice due to various design characteristics such as the starting position of the pressure exposed surfaces relative to the air inlet port (see fig. 5–5).

It must therefore be noted that *breakaway torque* is generally much higher than *running torque*. Breakaway torque is the turning moment required to accelerate a load from standstill into motion. Running torque is the amount of torque required to keep a load in rotation after it has been accelerated (see figs. 5–1 and 5–3).

Motor speed

Motor speed (n) is a function of motor displacement (V) which is its geometrical volume and the input flow delivered to its input port (flowrate through the motor). This flowrate is usually given in m³/min free air.

The unloaded maximum speed is called *free speed*, at which the motor's torque is zero (see fig. 5-3 for speed torque relationship). As the motor is loaded the speed drops and the motor torque increases linearly, thus showing an interdependent relationship of these two parameters.

Air motor driven machinery should, therefore, always be designed to permit the motor to rotate within its most optimal torque speed range. For this purpose air motors can be governed with a speed regulator which limits the operating speed to the point of maximum power and also conserves air consumption when the motor is not loaded (free wheeling). A typical speed–torque characteristics curve for a governed air motor is given in fig. 5-4.

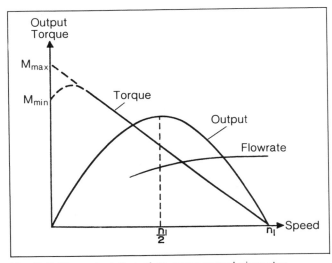

Fig. 5-3 Characteristics of an ungoverned air motor.

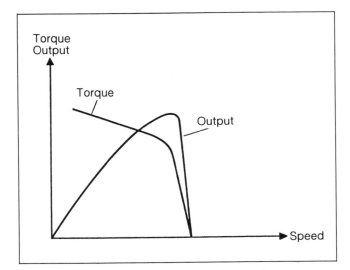

Fig. 5-4 Characteristics of a governed air motor.

Power output

The maximum power output for a typical ungoverned air motor is reached when the motor rotates at approximately half of its free speed (fig. 5-3). Power output for air motors is best applied with the formula

$$P = Q \times p \quad \text{OR:} \quad P = \frac{2 \times N \times n \times M}{60} = \text{Watts}$$

Selection of the correct design type and size of air motor for a particular application is most important, and the procedural steps for selection are given in a worked example at the end of this chapter.

Vane motors

The vane motor is the most common type of air motor used in the manufacturing industry. It covers a large range of power outputs from less than one kW to around 25 kW, which makes them versatile and widely accepted. Fig. 5-5 shows a cross section of a reversible vane motor. A rotor shaft which carries a number of sliding vanes is mounted eccentrically in a cylindrical casing or stator. The vanes are usually made of fibre which has excellent wear properties, and the more vanes used (up to a practical limit) the better the starting torque will be. However, increased friction from the vanes reduces efficiency.

In vane motors torque is produced by the compressed air acting onto the pressure exposed rectangular area of the vanes. These vanes in fact seal the low pressure exhaust port from the high pressure inlet port. According to Pascal's Law, which also applies to pneumatic systems, pressure acts undiminished in all directions and acts also with equal force on equal areas. It is therefore obvious that the motor depicted in fig. 5-5 must turn in a clockwise direction, since the sum of all vane areas being pressurised in a clockwise direction is much larger than the sum of vane areas being pressurised in an anticlockwise direction (see also fig. 5-2). As the rotor continues to rotate in this manner, any residual air remaining between the blades is vented via the anticlockwise supply port B. If the motor is to be operated to produce torque in the opposite direction, the anticlockwise port B is pressurized and port A is vented. The action is then exactly the reverse of that described for clockwise rotation.

As the motor turns the sliding vanes follow the contour of the cam ring (housing), thus forming sealed cavities. Since the vanes must seal and thus maintain cam ring contact at all times, and centrifugal force is absent during motor start, these vanes are usually spring loaded or air pressure loaded for smaller motors (see fig. 5–6).

During motor rotation the compressed air in the pockets between the vanes is permitted to expand, since the pockets also expand as they rotate towards the exhaust opening. This permits the potential energy stored in the compressed air to be used and thus the "technical work" of the motor is increased and its efficiency is therefore also increased (see fig. 5–8). Figure 5–8 clearly shows which part of the potential energy in the compressed air can additionally be utilized. It must be noted that this principle also applies to radial and axial piston motors, but does not apply to gear motors.

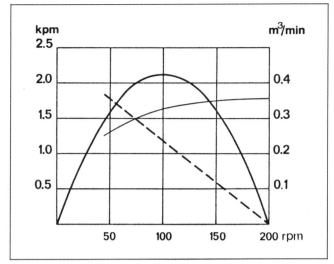

Fig. 5–7 Typical vane motor performance curves.

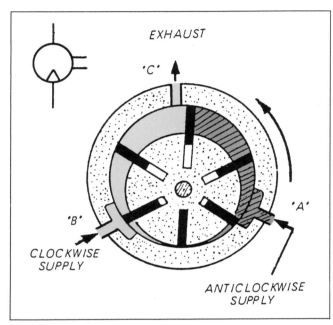

Fig. 5–5 Concept of a reversible vane type air motor.

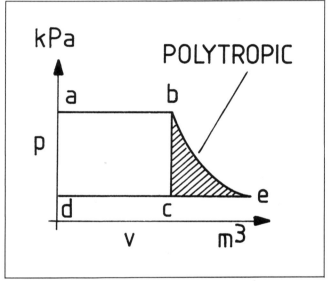

Fig. 5–8 *Above:* Technical work of an air motor with or without utilization of the expansion work. With expansion work it is a, b, e, d; without it is a, b, c, d.

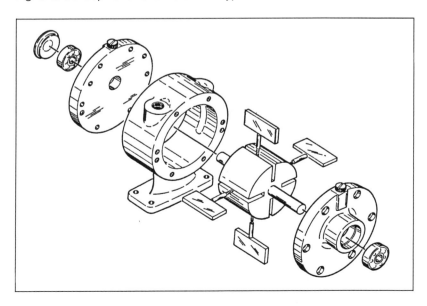

Fig. 5–6 Exploded view of a vane air motor.

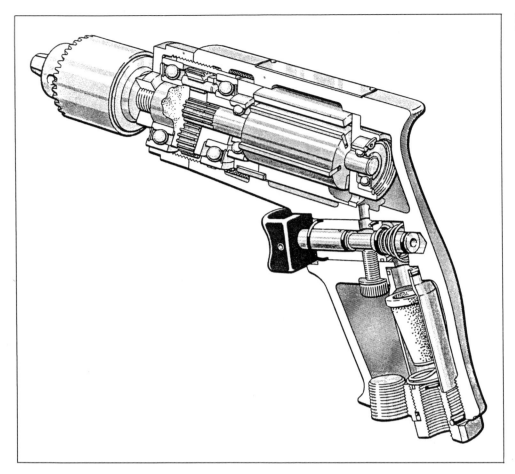

Fig. 5-9 Vane motor driven "ÅTLAS COPCO" hand drill.

Figure 5-7 shows power and speed curves for a typical vane motor and figure 5-9 illustrates a section through a hand drill using a vane motor.

Theoretically maximum torque should be available from the start, but due to internal leakage and depending on the position of the vanes relative to the inlet port, this maximum torque may not be available until the rotor has turned a number of revolutions. Maximum practical torque output, however, is produced at the stall point—used to advantage in many air tools. The highest power output is obtained at approximately half the free running speed of the motor, and this is generally the maximum performance characteristic specified for the power given data of air tools.

The power to weight ratio is good for vane motors, but because of their torque and power characteristics, they are not ideally suited to low speed applications. Where used in air tools, the drive torque is usually increased through reduction gearing to provide the required output torque. Nearly all air tools used on production lines, such as drills, screw drivers and nut runners, are vane motors. Vane motors are also used in sump pumps

and concrete vibrators. By means of various design methods such as the addition of extra rotor vanes, the deficiencies in the starting and slow speed characteristics of vane motors can be overcome sufficiently to allow their use in many heavy duty applications such as hoists, winches and transmission drives for mining equipment.

Gear motors

Gear type air motors consist of two meshing spur gears in a casing which is provided with inlet and outlet ports. (Fig. 5-10.)

The pressurised air entering through the motor housing air inlet where the gears intermesh, impinges on a rectangular gear area and forces the gears to rotate. The air then exhausts through the outlet port after having completed part of a revolution. Gear motors are relatively inefficient due to their high leakage losses between the gear teeth. Free speeds of up to 3000 r.p.m. are possible with a power range of 1 to 15 kW. Their efficiency can be increased by the use of helical instead of spur gears, due to the greater area exposed to air pressure. Reduction gear drives can be used with these motors to increase their output if required. These air motors are not used to any great extent in industrial manufacturing, but are used in applications such as the driving of drill strings in deep bore drilling rigs.

No air expansion takes place in the fixed volume teeth pockets when the air is driven through the motor, thus power output of such motors occurs without utilization of the potential energy (expansion work) stored in the compressed air (see also fig. 5–8).

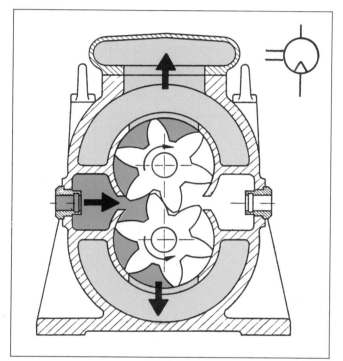

Fig. 5–10 Gear motor principle.

Radial piston motors

For applications where high running torque coupled with excellent starting torque is necessary, the radial piston motor is the ideal choice. These motors are specifically designed for low speed and high power output (with free speed nevertheless up to 4000 r.p.m.). Their power range may reach 25 kW, normally at about 25–40% of their free speed (fig. 5–12).

The cylinders in radial piston motors (up to five or six) are arranged radially around a crankshaft driven by the piston of each cylinder. Compressed air is ported and exhausted to and from each cylinder in turn by a rotary sleeve valve built into the pintle opposite the drive shaft. This causes the crankshaft to rotate. The reciprocating piston

movements can be finely balanced to provide a very smooth torque output, and if required valve timing can be arranged for reverse rotation. This valve timing can also be constructed for modified output characteristics for motors used on special applications.

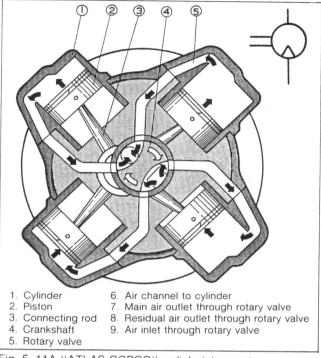

1. Cylinder	6. Air channel to cylinder
2. Piston	7. Main air outlet through rotary valve
3. Connecting rod	8. Residual air outlet through rotary valve
4. Crankshaft	9. Air inlet through rotary valve
5. Rotary valve	

Fig. 5–11A "ATLAS COPCO" radial piston motor.

Description of the ATLAS COPCO piston motor

Depending upon whether the throttle valve is set for clockwise or anti-clockwise rotation, the compressed air passes through one or the other of two channels to the rotary valve ⑤ which is driven by the crankshaft ④. The rotary valve has three channels: air inlet ⑨, main air outlet ⑦ and residual air outlet ⑧. At an alteration of rotation, the channels of air inlet and residual air outlet change functions. When the rotary valve rotates, the compressed air is distributed in turn through channels ⑥ to the cylinders. The air trapped in the cylinder ① expands and presses the piston ② down to its lower position. The air is expelled from the cylinders in two stages, main air outlet ⑦ and residual air outlet ⑧. The main air outlet takes place through a separate outlet channel in the rotary valve ⑤ directly into the motor casing and, via a silencer chamber through the air outlet pipe to the atmosphere. This residual air passes from the rotary valve through the throttle valve and then into the motor casing and out through the air outlet pipe. When the piston has resumed the inital position the working cycle is repeated. The motor is lubricated by oil in the crankcase and by oil that is supplied via the compressed air by a lubricator.

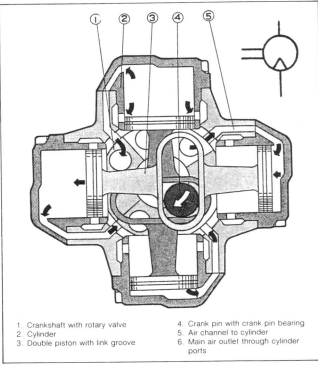

1. Crankshaft with rotary valve
2. Cylinder
3. Double piston with link groove
4. Crank pin with crank pin bearing
5. Air channel to cylinder
6. Main air outlet through cylinder ports

Fig. 5–11B ''ATLAS COPCO'' radial piston motor.

Description of the ATLAS COPCO piston link motor

The piston link type motor, which is reversible, has each opposed piston pair and associated by a link to a so-called double piston. The motor has two identical double pistons ③. The crankshaft ① has one crank pin ④, which runs in needle bearings in the two link grooves. A rotary valve controls the air supply to the cylinders ②, so that the pistons move and drive the crankshaft continuously. Air inlet and residual air outlet function in the same way as for radial piston type motors, while the air outlet passes through cylinder ports. The motor is lubricated by oil that is supplied via the compressed air by a lubricator.

Axial piston motors

These air motors, like the radial motors, are designed with up to six cylinders, arranged not radially with respect to the output shaft, but in line with (or axially to) the drive shaft. The cylinders are connected to the output shaft through an angled ''swash'' plate which in fact tumbles and converts the stroking action of the cylinder pistons into a rotary motion (see schematic diagram of figs. 5–1 and 5–13).

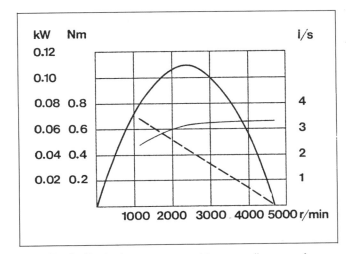

Fig. 5–12 Typical power-speed-torque diagram for an ungeared radial piston motor.

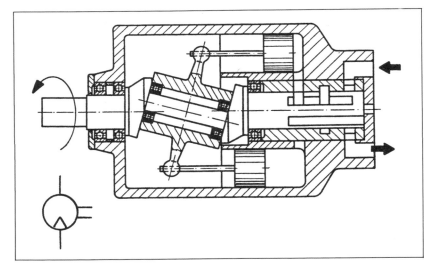

Fig. 5–13 Swashplate or axial-piston motor.

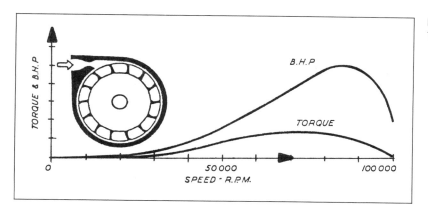

Fig. 5–14 Typical power speed curve for turbine motors.

These air motors are at the bottom end of the torque and power output range of piston motors and provide operation speeds much the same as radial motors. However, their power output is generally not greater than 3kW.

Although there are other design variations in axial piston motors, this brief description refers to the type of motor shown in fig. 5–13. In operation compressed air enters the motor through valve ports incorporated on the output shaft similar to the radial motor. This enables the pistons to be pressurized in turn. The piston rods are fastened to the piston and to the wobble plate with ball bearings so they may be free to assume a non-axial direction. The pistons as they extend and retract in turn, impart a rotary motion to the wobble plate. The wobble plate, connected by bevel gears to the motor drive, transmits this motion to the output shaft.

Turbine motors

The turbine motor consists of a small turbine, where the kinetic energy of an air jet impinging on the turbine blades generates very low torque (in comparison with other air motors), but very high free running speeds of up to 100 000 r.p.m. Their applications are thus distinctly limited to those where very high speeds, but extremely light loads, are encountered (dental drills, small tool grinders, pencil sharpeners). Because there are no sliding parts in the construction of turbine motors, except for the spindle bearings, these air motors can be operated with oil-free air. Reduction gearing to increase the power output is not feasible with turbine motors because the friction losses incurred in the gearing would consume most of the available power output. Fig. 5–2 shows the concept of a turbine motor, and a typical power-speed curve for these motors is given in fig. 5–14.

Pressure and air flow regulation on air motors

Pressure regulation affects the torque output of the air motor, whereas air-flow regulation affects the motor speed (shaft revolutions). The output power, however, is altered if either of these parameters is changed (fig. 5–1). Figure 5–15 shows a motor control circuit which provides high pressure supply for start-up and automatic timer controlled change over to lower running pressure once the load is accelerated. Air motors should be sized on 75% of the available air pressure, leaving 25% available for start up and overload.

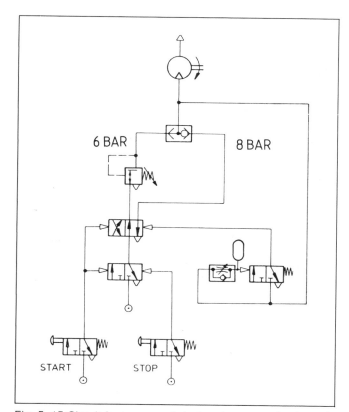

Fig. 5–15 Circuit for increased start-up torque and automatic (time-delayed) changeover to reduced running torque.

Air pressure (e)		Output power (kW)	Speed (r/min)	Torque (Nm)	Air consumption (l/s)
bar	psi				
7	102	1.21	1.03	1.17	1.15
6	87	1.00	1.00	1.00	1.00
5	73	0.77	0.95	0.83	0.82
4	58	0.55	0.87	0.67	0.65
3	44	0.37	0.74	0.50	0.47

Example: The data for a specific air motor at 6 bar (600 kPa) is as follows:

Max. output		Free speed	Speed at max. output	Torque at max. output	Air consumption at max. output
kW	hp	r/min	r/min	Nm	l/s
0.57	0.77	800	400	13.7	14.2

At a pressure (e) of 400 kPa (4 bar) the following approximate data is obtained:

Max. output	Idling speed	Speed at max. output	Torque at max. output	Air consumption at max. output
kW	r/min	r/min	Nm	l/s
0.57×0.55 = 0.31	800×0.87 = 700	400×0.87 = 350	13.7×0.67 = 9.2	14.2×0.65 = 9.2

1 bar = 100 kPa = 1.02 kp/cm² = 14.5 psi	$0.77 \times 0.55 = 0.42$ hp
1 Nm = 0.102 kpm = 0.7376 lbf.ft	1 kW = 1.341 hp
1 l/s = 0.06 m³/min	1 hp = 0.7457 kW

Fig. 5-16 Performance conversion table for different air pressures.

Motor Type	Max Free Speed R.P.M.	Typical Operating Speed Range R.P.M.	Torque Speed Ratio	Initial Torque	Typical Max. Power Outputs kW
Vane	20,000	7,000–15,000	Low	Medium	15
Gear	3,000	1,500–2,000	Low	Low	15
Radial Piston	3,000	600–1,500	Medium to High	High	25
Axial Piston	3,000	600–1,500	Medium to High	High	3
Turbine	100,000	50,000–80,000	Very Low	Negligible	Less than 1

Fig. 5-17 Air motor performance characteristics.

In reversible air motors a suitable valve can accomplish change of rotation. No requirement for slow down or stopping between reversals is needed. If strong enough, an outside force can in fact reverse the motor against its direction of rotation without damaging the motor. In this manner it can be used as a break.

Data at different air pressures

The normal working pressure (e) for air motors is 600 kPa (6 bar). When the air pressure is lower than 600 kPa or somewhat higher, it is then valuable to be able to calculate the performance of the air motor at the air pressure in question.

By means of the table shown in fig. 5-16 it is possible to calculate the approximate speed, output power, torque and air consumption at different air pressures for an ungoverned motor.

Air motor performance comparison

The table (fig. 5-17) gives various characteristics for the types of air motors discussed in the preceding pages.

Air motor sizing and torque calculations

Exercise 1

Rotary power output of pneumatic motors is calculated:

Watt $= 2\pi \times$ revolutions per second $\times$ torque

$$W = 2\pi \times n \times M$$

Torque $=$ newtons $\times$ metres

$$M = Nm$$

Using these formulae, calculate the required torque on the drive shaft of the cable drum in fig. 5–18, if the load shown must be held in suspension.

Torque $=$ newtons $\times$ metres (radius)

Torque $= 100 \times 0.2$

　　　(10 kg $=$ approx. 100 N; precisely 98.1 N)

　　　$= 20$ Nm

Exercise 2

Using the results obtained in exercise 1, select a suitable air motor from the information given in fig. 5–19. The load must be lifted with a speed of 50 m/min. For friction loss and acceleration add approximately 25% to the required torque obtained in exercise 1.

$$\frac{20 \text{ Nm} \times (100 + 25)}{100} = 25 \text{ Nm (including friction loss and acceleration)}$$

$$\text{Required r.p.m.} = \frac{\text{lifting distance}}{\text{circumference of cable drum}}$$

$$\text{Required r.p.m.} = \frac{50 \text{ m}}{0.4 \text{ m} \times 3.1416} = \frac{50}{1.2566}$$

$$= 39.78 \text{ r.p.m.}$$

The selected motor is of the ATLAS COPCO type LMK 33 S002 but the LMK 22 5001 could also be used.

Fig. 5–19 Selection of air motor performance curves.

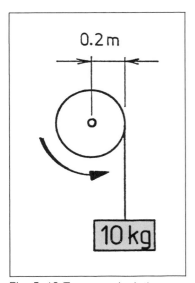

Fig. 5–18 Torque calculation.

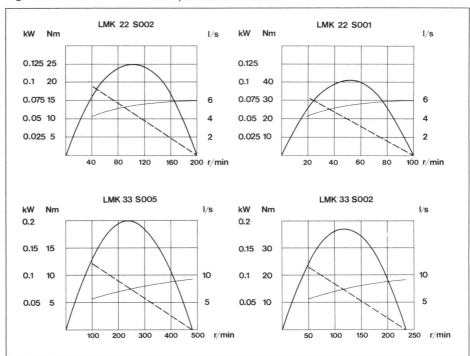

6 Pneumatic Sensors

In order to make pneumatics a better means of machine control, design engineers as well as maintenance engineers should have a thorough knowledge of the vast range and choice of sensing equipment available today. Position sensors are built for two different forms of sensing; contact and non-contact type sensing. A third form of actuator motion or event sensing, is the previously described form of pressure sensing (see Chapter 2).

One-way trip and roller valve for contact sensing

Contact type sensing is the most widely used form of actuator motion or position sensing. It is achieved with mechanically operated valves such as roller valves, plunger valves, one-way trip valves, whisker valves and fluidic back pressure sensors (figs. 2-7, 2-9, 2-31, 6-1, 6-2 and 6-3).

Where the complete linear or swivel movement of a mechanical part operated by an actuator must be detected with accuracy, the one-way trip valve, however, would be completely inadequate. One-way trip valves require an overtravel of the cam of at least 10 to 20 mm to allow the hinged roller lever to swing back.

This means that the signal from such a valve will emerge when the actuator movement tripping it, still has to go on for yet another 10 to 20 mm, thus precise end-position sensing is out of the question (fig. 6-1).

One-way trip valves render a pulse rather than a stationary signal. The pulse disappears as soon as the hinged roller is permitted to swing back (fig. 6-1). The following table compares this (only) advantage with the many disadvantages for the use of one-way trip valves.

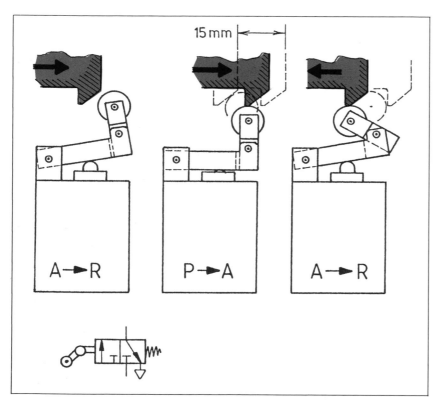

Fig. 6-1 One-way valve: 15 mm overtravel is required to trigger and thereupon cancel the signal.

Roller/plunger valves	One-way trip valves
Advantages: • Signal length not dependent on actuator speed. • Precise end-position control is possible. • Logic linkages of signals is possible. • Time delayed signal processing is possible. Disadvantages: • Opposing signals must be avoided with circuit design method (see: circuit design Chapter 8).	Advantage: • Gives no opposing signals (short pulse only). Disadvantages: • Precise end-position control not possible. • Logic linkages of signals not possible (pulse only). • Time delayed signal processing is not possible. • Fast actuator speed is not possible (signal too short).

In order to gain a fully controlled sequence of all actuators in an interlocking machine program, it is essential to install position sensors (position or limit valves) at all movement end positions (figs. 8–1 and 8–21).

This practice guarantees a completely interlocked and movement end-position depending sequence, where no machine movement or sequence step can occur unless the previous sequence step is completed. Such close machine movement monitoring eliminates costly machine down-time and work damage.

Machine and operator safety in fully automated manufacturing processes always outweighs the minor costs of a few more sensing valves and should never suffer from the "dollar thrifty" attitude on the part of some machine designers and maintenance engineers.

Back pressure sensor for contact sensing

The back pressure sensor is the only pneumatic position sensor which achieves a sensing accuracy better than 0.2 mm. It may be used for extremely short actuator strokes (less than 5 mm) where one or both actuator stroke end positions must be detected, or for applications where the sensing accuracy of roller or plunger valves (1–3 mm) is insufficient (fig. 6–2).

Compressed air of normal system pressure (4–8 bar, 400–800 kPa) is connected to port P. With no object resting against the protruding plunger, the

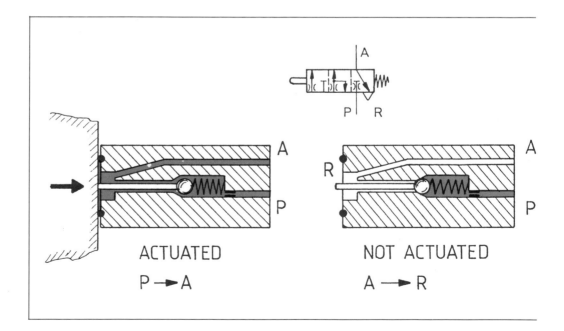

Fig. 6–2 Back pressure sensor.

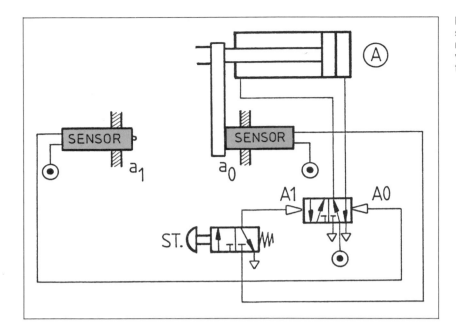

Fig. 6-3 Two back pressure sensors sequence the actuator to extend and to retract. The start valve is series connected with the a_0 back pressure sensor ("AND" function).

check valve will remain closed and hence no compressed air escapes at the sensing nozzle R, but the sensor pilot outlet line A (signal line) is vented to atmosphere through the nozzle opening R (fig. 6-2 and 6-3).

When an object approaches the sensor, the protruding plunger is contacted first. This forces the check valve off its seat and compressed air from port P can now escape through the nozzle R for a short time only, until the approaching object closes the nozzle completely (fig. 6-2). From then on, pressure starts to build up as compressed air flows in from port P and is redirected to port A (for an application circuit see fig. 6-3).

Back pressure sensors must be rigidly mounted (bulk-head mounting with a mounting block) to avoid deflection and thus ensure the extreme sensing accuracy for which they are usually required (fig. 6-3). If properly mounted, back pressure sensors can be used to limit the actuator stroke and can absorb a thrust force of up to 24 000 N. But in order to utilize the full cushioning stroke of the actuator, the back pressure sensor should not diminish the total actuator stroke by more than 1 mm (see Chapter 3, Pneumatic end position cushioning).

Back pressure sensors provide the following sensing advantages:

- Extremely accurate position sensing (better than 0.2 mm)—no other pneumatic sensor can achieve this accuracy!
- Suitable for extremely low actuating force (only 12 newtons at 6 bar supply pressure is required to open the check valve).

- Small dimensions make it suitable for mounting where space is limited.
- The sensor can be used as an actuator stroke limit stop and is easily adjustable (threaded mounting).
- The sensor may be used for extreme ambient temperatures (-10 to $+80°C$) and in extremely dirty environments, since the emitted air cleans the sensor nozzle.

Liquid level sensor

The liquid level sensor operates on a similar principle to the back pressure sensor. The sensor is also supplied with compressed air of either normal system pressure (400–600 kPa, 4–6 bar) or low pressure of 10–30 kPa (0.1 to 0.3 bar). If operated on low pressure, the pilot signal emerging from the sensor must be amplified with a pressure amplifier valve (see figs. 6-5, 6-4 and 6-16).

When the sensor tube is not in contact with the liquid to be sensed, the supply air escapes through the nozzle of the sensor tube (immersion tube). As soon as the rising liquid closes the opening of the immersion tube, pressure starts to build up as compressed air now flows in from port P and is redirected to the signal port A (for application circuit see fig. 6-5). This signal pressure is proportional to the level of the liquid above the tube nozzle and to the specific gravity of the liquid being sensed, and eventually reaches the pressure level of supply at port P (air bubbles are more buoyant in a heavy liquid than in a liquid with lesser specific gravity). The signal pressure remains as long as the outlet R of the immersion tube is closed by the liquid.

Where liquid surfaces subject to wave formation or rapid level fluctuations are to be sensed, a level cushioning jacket should be fitted around the immersion tube. Such a jacket has one or more small holes in the nozzle region which smooth out

the rapid level changes and provide a more stable sensing result.

Proximity Sensor for non-contact sensing

The sensors discussed so far require physical contact with the object being moved. Non-contact type sensors, however, must be chosen wherever physical contact of the sensing mechanism with the object to be detected is either impossible or undesirable. Such applications range from food processing, liquid sensing, extreme lightweight object sensing (such as paper sheets), to the sensing of gauge hands on pressure gauges or weighing machines, and strip material correction when being rolled onto reels or powder storage in tanks and silos (see figs. 6–6 to 6–9 and 10–26, 10–29).

With the proximity sensor (reflex sensor), low pressure compressed air at approximately 30 kPa (0.3 bar) escapes through a ring opening on the front end of the sensor (fig. 6–6). This causes a vacuum pressure cone to form in the centre of the air escape fan. If the escaping air is disturbed by an approaching object, the vacuum cone starts to collapse and turbulent air raises the pressure in the pilot line A. This pilot pressure, however, is extremely low (1–4 kPa, 0.01–0.04 bar) and must be increased with an amplifier valve for use in signal processing (see amplifier valve fig. 6–14 and proximity sensor application circuit fig. 6–7).

Typical air consumption figures for various prox-

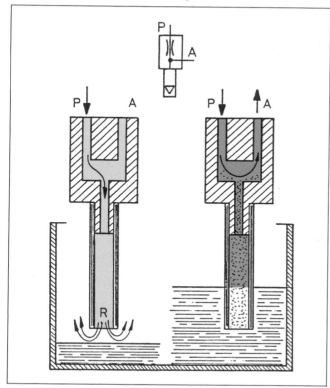

Fig. 6–4 Liquid level sensor.

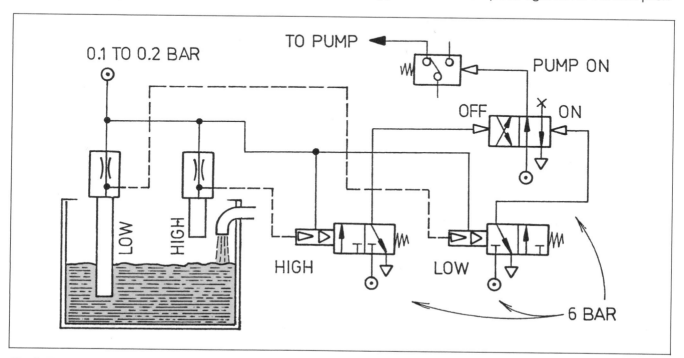

Fig. 6–5 Liquid level sensor application circuit.

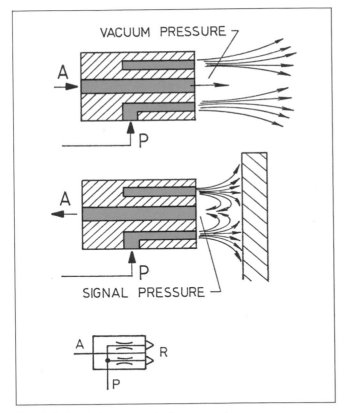

Fig. 6-6 Proximity sensor (reflex sensor).

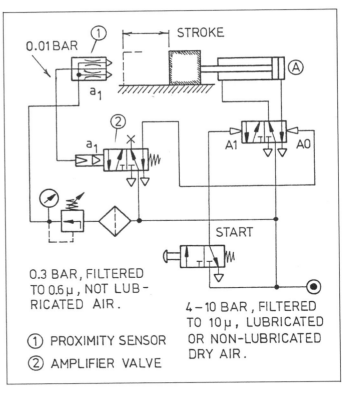

Fig. 6-7 Typical proximity sensor application circuit. The start valve signals extension, the proximity sensor and its amplifier signal retraction of the actuator.

imity sensors as depicted in fig. 6-6 are given in the air consumption graph in fig. 6-8. The air consumption depends on the size of the sensor (sensing distance 3.6 or 12 mm) and the magnitude of the supply pressure being fed into the sensor (see figs. 6-6 and 6-7).

For a 6 mm sensing distance sensor, the air consumption amounts to 30 L/min of free air, if the supply pressure is limited to 0.3 bar, and for a 12 mm sensor with the same supply pressure the air consumption rises to 96 L/min.

To minimize the high air consumption of such

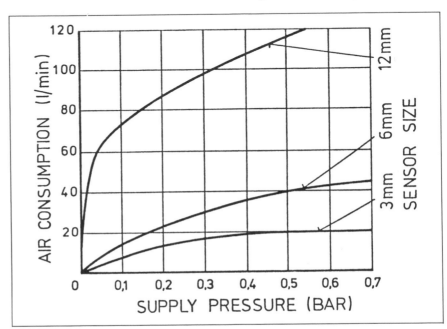

Fig. 6-8 Air consumption figures for proximity sensors 3, 6 and 12 mm sensing distance.

sensors, one can design a circuit interlock to switch the supply air on only when a sensing function is imminent (for a typical circuit of this nature see fig. 6–19).

Figure 6–9 shows how the low pressure pilot signal emerging from the proximity sensor depends on the sensing distance "a" and the supply pressure "P" which is connected to the sensor (see also figs. 6–7 and 6–8). For sensor size 12 mm, with a supply pressure of 20 kPa (0.2 bar), a sensing distance "a" of approximately 10 mm can be achieved if one assumes that the amplifier works on minimal pilot pressure of 1 kPa (0.01 bar). The same sensor may achieve over

14 mm sensing distance if the supply pressure is lifted to 4 kPa (0.04 bar).

Gap sensor for non-contact type sensing

This gap sensor works on the Venturi principle. Compressed air at approximately 30–50 kPa (0.3 to 0.5 bar) flows from port P via jet 1 to outlet 2. At the same time air flow is also diverted via the detour passage to outlet 3. This escaping air from outlet 3 causes a back pressure at outlet 2 which induces a pressure build-up at port A. If the back pressure jet from outlet 3 is interrupted by an approaching object, the back pressure at outlet 2 vanishes and freely escaping air at outlet 2 causes a vacuum to build up at the venturi point 4. This vacuum build-up cancels the signal at port A (fig. 6–10). For circuit see fig. 6–11. Since the approach of an object in the opening of the gap sensor cancels the pilot signal at port A of the sensor, a normally open amplifier is used, which by the principle of signal inversion produces a pilot signal when the sensor detects an object.

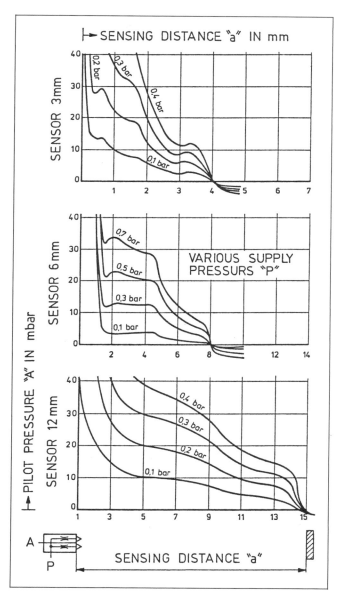

Fig. 6-9 Sensing pressure is dependent on sensing distance.

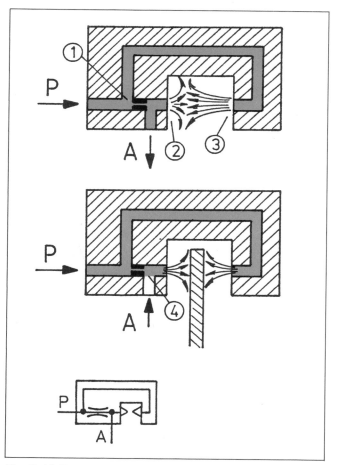

Fig. 6-10 Gap sensor.

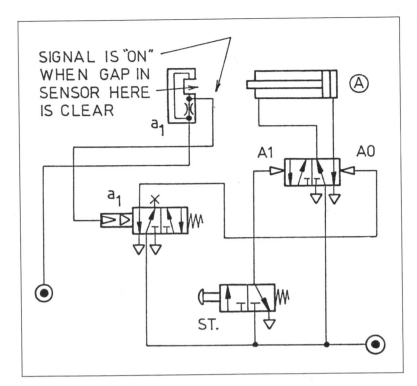

Fig. 6–11 Typical gap sensor circuit.

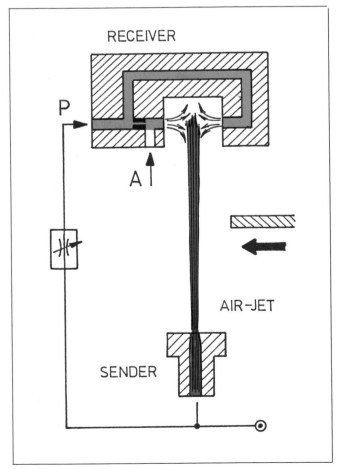

Fig. 6–12 Air barrier.

Air barrier for non-contact type sensing

The air barrier uses the previously explained and depicted gap sensor as a receiver unit on one end of the sensing gap and a jet nozzle as a sender unit on the other end of the sensing gap.

With this arrangement a sensing gap of 200 mm or more may be achieved. By carefully tuning the

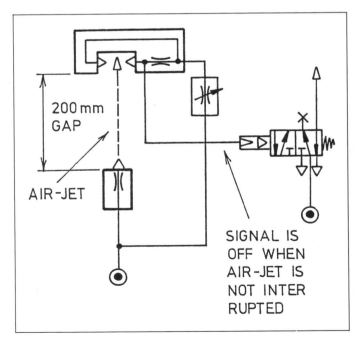

Fig. 6–13 Air barrier with amplifier.

pressures of the gap sensor and jet nozzle using a flow control valve as shown in fig. 6–12, one may achieve much larger sensing gaps (for application circuitry see fig. 6–13). Here too, an amplifier valve is required, but with the air barrier an interruption of the air jet causes a signal at port A to appear (when an object is sensed), and thus a normally closed amplifier valve must be used.

Amplifier valves

Some of the sensors mentioned before, such as the liquid level sensor, the proximity sensor, the gap sensor and the air barrier require an amplifier valve to magnify the extremely low pressure signal from these sensors into a useful pilot signal of at least 200–400 kPa (2–4 bar) or more where required.

Single Stage Amplifier

The single stage amplifier shown in fig. 6–14 amplifies sensor pilot signals from as low as 0.6 kPa into any desirable output pressure between 200 and 1000 kPa (2 and 10 bar). The low pressure pilot signal enters the valve at port X and acts onto the diaphragm. This closes the air escape aperture on the left on the valve spool, so that the vented air flowing from port P through the

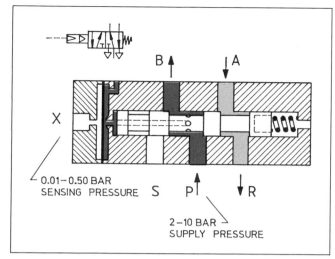

Fig. 6–14 Amplifier valve: the internal air pilot being channelled through the midst of the spool actuates the valve when the low pressure diaphragm closes the vent aperture.

hollowed out spool must build up pressure. This pressure build-up shifts the valve spool and thus compressed air is directed from port P to port A, whereby port B is vented to exhaust port S.

Figure 6–15 shows typical air consumption, pilot pressure and sensitivity figures in relation to the

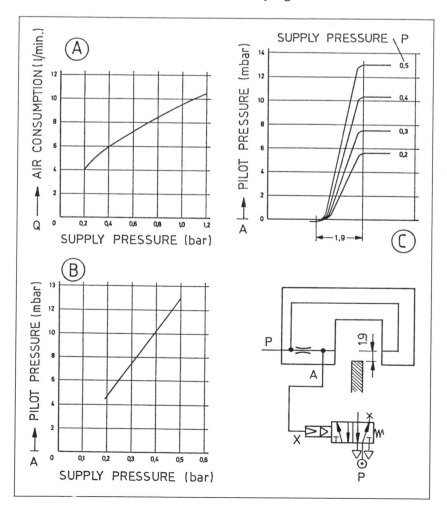

Fig. 6–15 Graph (C) is based on an object sensing distance of 1.9 mm from the centre of the air jet.

supply pressure as measured for the gap sensor depicted in figs. 6–10 and 6–11. Similar values must be used if the gap sensor is turned into an air barrier (fig. 6–12).

Two stage amplifier for high sensitivity

The two stage amplifier shown in fig. 6–16 is more sensitive than the single stage amplifier shown in fig. 6–14. This amplifier must be used for the liquid level sensor depicted in fig. 6–4 or where the pilot pressure from a gap sensor or air barrier is too low for the single stage amplifier.

The two stage amplifier shown in fig. 6–16 amplifies sensor pilot signals from as low as 5 Pa (0.0005 bar) into any desirable output pressure between 200 and 800 kPa (2 and 8 bar). The low pressure pilot signal enters the valve at port X and acts onto the low pressure diaphragm. This closes the medium pressure air escape aperture so that the previously vented medium pressure air is diverted to the underside of the medium pressure diaphragm. This lifts the diaphragm against the opposing spring and opens the poppet thus high pressure air (200–800 kPa) can flow from port P through the valve to port A.

When the low pressure signal at port X disappears, the valve operation is reversed and the high pressure pilot signal can exhaust from port A to port R.

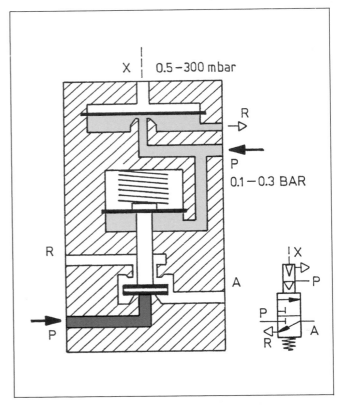

Fig. 6–16 Two stage amplifier for high sensitivity.

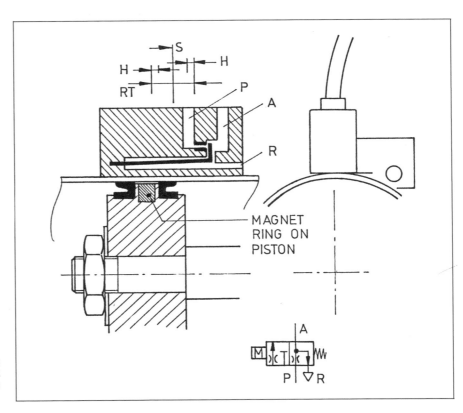

Fig. 6–17 Pneumatic reed sensor. This sensor may also be categorised as a non-contact sensor, and requires an amplifier valve.

Pneumatic reed sensor for non-contact sensing

The pneumatic "reed sensor" is directly mounted onto the actuator barrel. It is actuated by a magnetic ring embedded into the circumference of the piston. When the piston with the permanent magnet approaches the sensor, the reed becomes attracted by the magnet and the sensor is actuated (figs. 6–17 and 6–18).

The sensor operates on a similar principle to the air barrier. When the permanent magnet pulls the reed down, the air jet from passage P blows into air passage A, where the pressure rises. This increased pressure is amplified with an amplifier valve as shown in fig. 6–16. When the piston with its magnet has moved beyond the response travel point RT, the reed springs back again and the tongue on the reed interrupts the air jet again. Thus the pilot pressure in air passage A can exhaust to passage R and out to atmosphere. Any air leakage past and around the tongue can also exhaust to atmosphere.

The switching point S is normally in the centre of the sensor. H denotes the *hysteresis*, a distance in which switching *may* occur but is not guaranteed.

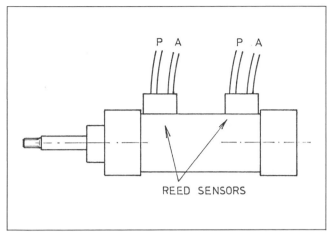

Fig. 6–18 Pneumatic actuator fitted with two pneumatic reed sensors.

Pneumatic reed sensors are either mounted onto the tie rod of the actuator or are fitted onto a specially designed mounting rail which spans from end cap to end cap. Pneumatic reed sensors can only be used with non-ferrous actuator barrels and the piston must be fitted with a permanent magnet.

Fig. 6–19 Air supply interlock.

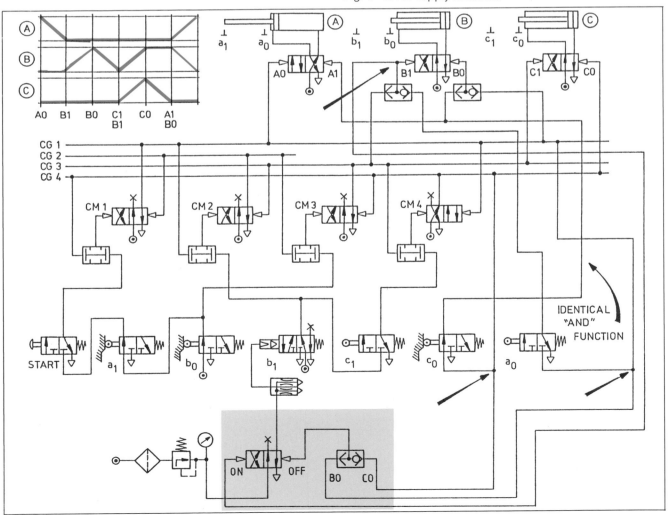

Reduction of air consumption for non-contact sensor

To minimise the high air consumption of non--contact type sensors one can design a circuit interlock to direct supply air to the sensor only, when a sensing function is imminent (figs. 6–19, 6–6 and 6–8).

Valve mounting and actuation methods

Care and accuracy must be applied when valves are mounted onto a machine. If a valve is mounted too close to the actuating cam or actuating machine member, the valve spool or poppet may "bottom out" and the valve mechanism is made to bear the full force of the approaching load driven by the actuator. This often causes return springs to become coilbound, or through constant abuse of the valve mechanism the valve may cease to function altogether.

Cam angles should never have more than a 45° inclination (fig. 6–20). The cam should be made with mounting slots, or the valve mounting bracket should have slots. This permits the valve to be mounted precisely, so that the switching point S is reached approximately 1 mm before the cam stops. The difference between that point and the maximum travel point then acts as a safety margin or safety overtravel so that the valve does not "bottom out" (see fig. 6–21). "Head-on" actuation, as shown in fig. 6–21, gives a more precise actuation (within 1 mm) than overtravelled rollers (fig. 6–20), which may only provide a switching accuracy of about 2-3 mm.

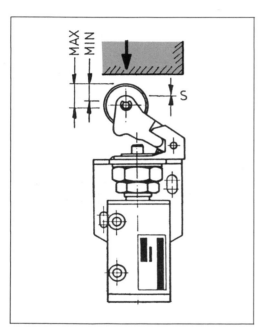

Fig. 6–21 Minimum and maximum travel: the switching point "S" is where the valve opens the desired air passage.

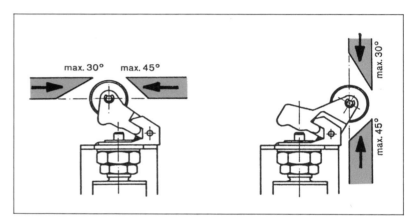

Fig. 6–20 Cam angle.

7 Circuit Presentation and Control Problem Analysis

At present, no international standards exist for uniform circuit array and circuit labelling. However, a well-presented circuit can facilitate circuit construction for the engineer and fault-finding for maintenance personnel, as well as making the total control package a more marketable item.

Circuit layout and presentation

The method of circuit layout shown in fig. 7–1 is still widely used. However, it is very impractical for large and complex circuits, since too many signal lines cross and thus cause confusion. For logically designed circuits, with their large number of "AND" function and "OR" function valves, this type of circuit layout proves totally inadequate. Hence the layout structure shown in fig. 7–2 is recommended. The circuit designer will soon become familiar with this type of layout and appreciate its many real advantages.

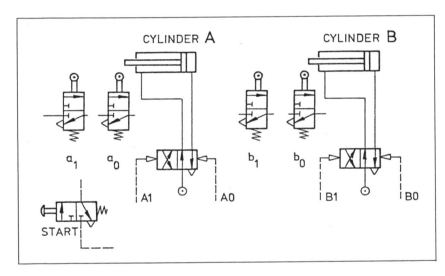

Fig. 7-1 Poor circuit presentation.

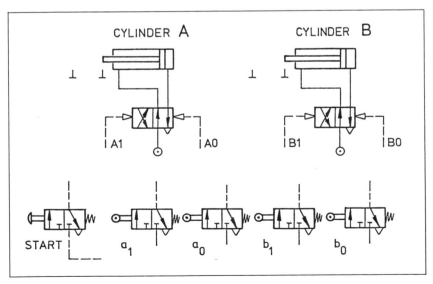

Fig. 7-2 Recommended circuit presentation.

Recommendations for circuit array

There are no international standards for circuit presentation and layout at present, so some well proven principles are listed below. These principles, either in their present form or after input from other experienced circuit designers, may be the forerunner of an international standard.

- Draw all symbols where possible in horizontal position and attach lines to the right hand square for two position valves or to the centre square for three position valves (figs. 2–2 and 7–2).
- Do not mirror-invert symbols if avoidable, since this practice confuses the comprehension of signal direction, fluid-flow and valve function (see fluid power symbols in Appendix).
- Draw all valves in their de-actuated, unpressurised rest position, except for those valves which are actuated either by a lever, cam or machine part, when the machine is at rest, or at the end of a cycle. For such valves one must indicate with a cam or similar device that the valve is actuated and all attachment and pressure lines are to be connected to the left hand square. (See also valve switching positions, Chapter 2 and fig. 7–3.)
- Two-position valves which are not spring biased and can freely assume both switching positions (bi-stable valves, memory valves or detented valves) may be drawn in whatever flow-path configuration is demanded by the actuator position or initial position when the machine is at rest. But here again, the pressure lines should be connected to the right hand square (figs. 7–3 and 2–3). An exception to this rule is frequently made

for memory valves in sequential control circuits which maintain their last selected position to provide the start signal for a new machine cycle (see figs. 7–5 and 8–6).
- Draw all actuators in their machine rest position or initial position. For example, if an actuator is extended at the end of the cycle it should be drawn in the extended position (figs. 7–3 and 2–3).
- Where position sensing valve signals or pressure sensing valve signals have to be distributed to a number of destinations for circuit interlocking (fig. 7–5), it is advisable to adopt the "bus-bar" distribution system (figs. 7–4 and 7–5). The *bus-bar* should be regarded as a manifold line with both extremities closed off.

To ascertain the exact number of T-connections to be used for each valve outlet, one simply counts the number of connection dots on the bus-bar and deducts two. Thus, a bus-bar with four dots requires two T-connectors and a bus-bar with two dots requires none.

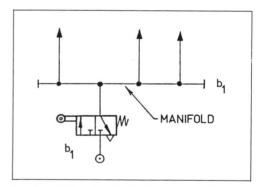

Fig. 7–4 A "bus-bar" with four connection dots requires in reality two T-connectors to distribute the position sensing valve signal to three users.

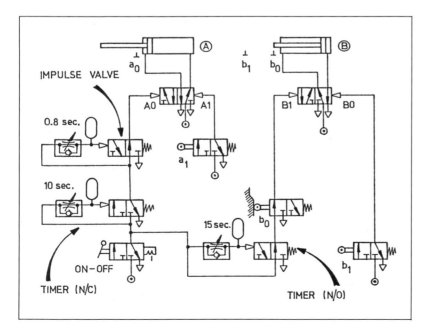

Fig. 7–3 Typical circuit layout showing all the recommendations made for proper circuit array.

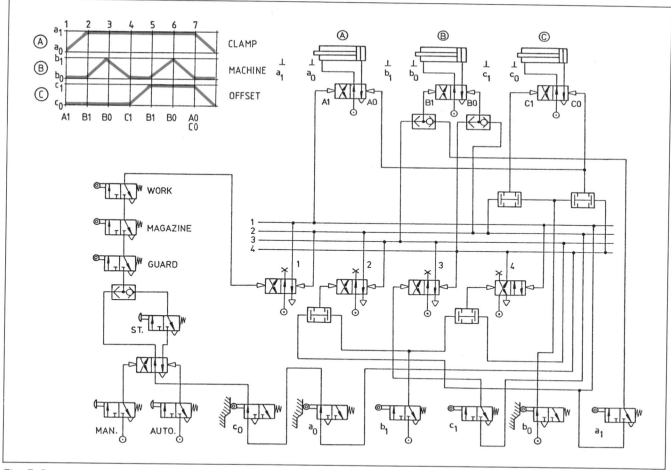

Fig. 7–5

- Clearly separate fringe condition modules such as used for emergency stop, cycle selection and start restriction from the sequential control circuit (figs. 7–3 and 7–6). These modules and the concept for designing sequential circuits are explained in the following chapters.
- Circuit labelling with the binary movement expression and labelling method, as explained in Chapter 2, should be adopted. This eliminates communication difficulties and simplifies circuit installation and fault-finding (figs. 7–3, 7–5, 8–14 and 8–15).
- Sequential control circuits with more than one actuator should include the traverse-time diagram, which depicts the sequence and if possible all specifications and sequence equations derived for the design of the control. Such equations must also include start restriction and combinational equations, where applicable (figs. 7–5, 8–17, 8–18, 8–21, 10–13 and 10–14).

Control Problem analysis

The grouping of the total control circuit into sub-circuits seems to be one of the main difficulties which confront circuit designers. However, these obstacles can be drastically reduced by analysing the total control problem and then splitting the design into four main categories:

- power circuit
- sequential circuit
- fringe condition modules
- combinational circuit

The *power circuit* consists of the actuators and their directional control valves (power valves). It also includes the speed control valves (see Chapter 4) and compressed air lines which connect the actuators to the power valves.

That part of the circuit which controls the stepping or sequencing of the actuators is generally called the *sequential circuit*. It consists of memory valves, "AND" function, "OR" function and time-delay valves (figs. 8–20 to 8–39).

The *fringe condition* circuits, which preferably are arranged in modules, affect the sequential circuit at its fringe and generally affect its start or stop characteristics (start permitted or not, emergency stop, etc.). These fringe condition modules are added in the final part of circuit design and are designed separately (for a detailed description of fringe conditions see Chapters 8 and 10).

Combinational circuits are network-type control circuits and their output signals depend solely on the momentary state of the input signals. Combinational circuits do not consider past actions which would require memory valves. For a detailed description of combinational controls see Chapter 10 and Chapter 11.

Fig. 7–6

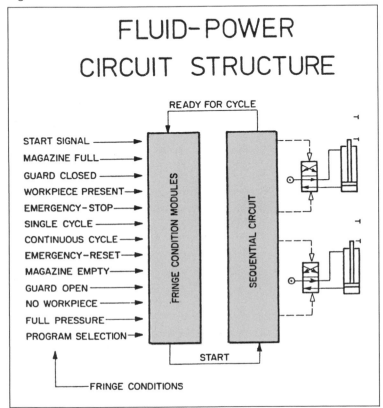

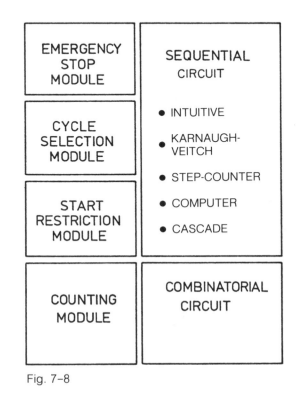

Fig. 7–8

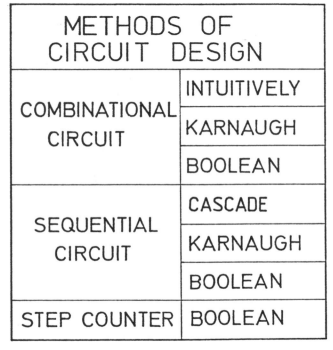

Fig. 7–7

8 Pneumatic Step-Counter Circuit Design

The science or art of designing asynchronous sequential circuits for controlling automation systems has come a long way during the past thirty years. At one time circuits were "put together" more or less intuitively without applying a proper and logical design method. This approach was based on trial and error, although graphical aids such as traverse-time diagrams and step-motion diagrams were often employed to make the search for a viable solution more systematic.

The two methods presented in this book are based on systematic and logic design and are the result of extensive research work undertaken by system design engineers such as D.A. Huffman—The Franklin Institute U.S.A.; Prof. D.W. Pessen—Technion Haifa Israel; Dr. W. Huebl—Martonair Germany; Prof. E.C. Fitch—Oklahoma State University U.S.A.; H. Schwaar—Festo Switzerland and P. Rohner—Royal Melbourne Institute of Technology Australia. Both design methods are well proven and numerous control circuits based on these methods are installed in machines all over the world.

It must also be noted that both methods can be used to program electronic programmable controllers (PLC), and particularly the step-counter method is ideally suited and simple to program electronic controllers as applied for sequential machine control. A separate chapter in this book explains and illustrates how this may be achieved (see Chapter 11).

Traverse-time diagram (step-motion diagram)

The traverse-time diagram is a simple, effective and easily understood graphic representation of the motion sequence (step sequence) at which linear fluid power actuators operate. The machine designer defines the motion sequence and draws the corresponding traverse-time diagram, thus establishing a communication and design basis for all parties involved in the design, implementation and commissioning of the total control system. The following points are to be observed when setting up a traverse-time diagram (see fig. 8–1):

- Step commands are represented by vertical lines ("time lines"). The direction and speed rate of the actuator movements are shown by inclined lines.
- Sequence steps are numbered above the time lines. The beginning of time delay periods are labelled T (fig. 8–19), and stroke length and stroke speed are regarded as equal since these have no bearing on the sequencing of the machine.
- Horizontal lines represent stroke limits and are labelled according to the name given to the position sensing valves (limit valves) located at the stroke end.
- Circuit and signal labelling is based on the binary system whereby actuator extension is denoted with the binary logic 1 and retraction with the binary logic 0. For example, signal command A1 causes actuator A to extend. Limit valve a_1 confirms this extension. Signal command AO causes actuator retraction and limit valve a_0 (feedback signal) confirms the completion of the retraction motion.

Below the traverse-time diagram is the "key". The *key* contains important information which later becomes the basis for circuit design. The key shows which limit valves are actuated at any particular sequence step during the machine cycle. For example in fig. 8–1 the key indicates that at the end of sequence step 2, the limit valves a_1 and b_1 are actuated.

The key may also be used to register an important confirmation signal, or combinations of signals, which are essential to trigger the next sequence step. For example in fig. 8–1 sequence step 3 can only be triggered if the previous step is completed. This completion is signalled by limit valve b_1. Hence limit valve b_1 in the key is circled. This process is called "circling of essential confirmation signals". If these circled essential confirmation signals are now used as the basis for implementing sequence steps, then no action or sequence can ever occur unless the previous sequence step is completed.

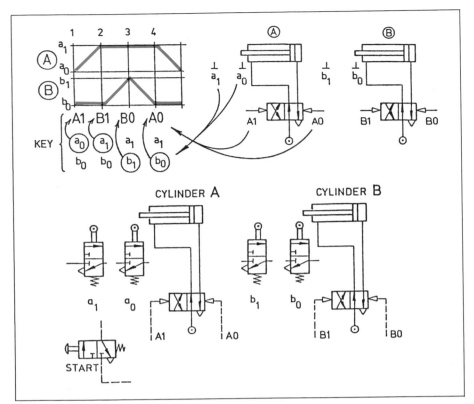

Fig. 8-1 Traverse-time diagram.

Opposing signals

In the design and construction of sequential circuits involving a number of actuators (cylinders) following a given sequence, it is frequently found that circuit design is complicated by the presence of disturbing pilot signals (confirmation signals). Such disturbing pilot signals result from earlier actuator movements and oppose the later arriving pilot signals on the power valve (fig. 8-2). Hence the power valve can't be reversed unless the opposing (disturbing) signal is eliminated. A simple illustration of this problem is given in fig. 8-2. Here

again, all essential confirmation signals are circled in the key and the arrows show how these pilot valves are connected to the power valves (compare also with a circuit given for this sequence in fig. 8-3). A close investigation reveals that the essential confirmation signal a_1 is used to switch sequence step 2 by producing command B1 on the power valve. This means that signal a_1 is used as the signal source to ''SET'' power valve B and thus cause its actuator to extend.

One step later, however, this pilot signal (a_1) is still ''ON'' since its actuator has not moved during

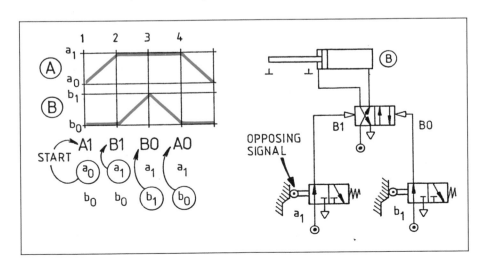

Fig. 8-2 The opposing signal is a_1.

the second time phase of the sequence (see traverse-time diagram in fig. 8–2). Thus when pilot signal b_1 must "RESET" the power valve B to cause actuator retraction, it cannot, because of the opposing signal on the opposite side.

A simple method often applied to rid the circuit of such opposing signals is to use one-way trip valves instead of roller or plunger valves. Such one-way trip valves create only a short pulse while being overtravelled and no pulse when the actuator returns (figs. 8–3 and 6–1). However, the many negative aspects of these valves are discussed in detail in Chapter 6. Another possible solution is to feed the opposing signal through an impulse valve and thus cause the pilot signal to appear for a short time only. This method is shown in Chapter 2, figs. 2–45, 2–47 and 2–52.

The following two circuit design methods automatically eliminate the occurrence of opposing signals, and are therefore highly recommended for use when designing sequential control circuits. These methods are:

1. Step-counter design method (or: step sequencer).
2. Cascade design method (simple version).

Step-counter circuit design method

The step-counter circuit is a modular circuit design concept consisting of a chain of integrated circuit blocks or modules, whereby each sequence step (depicted in the traverse-time diagram) is allocated a module. The simplicity of the step-counter (sometimes called step-sequencer or sequence controller) stems from the construction of its modules and the unique integration of these modules into a sequence chain (fig. 8–4). Each

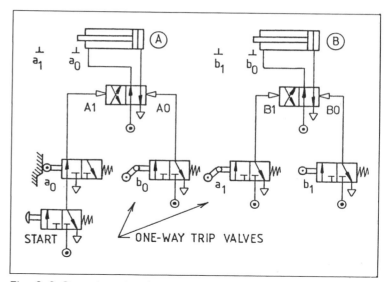

Fig. 8–3 Opposing signals are removed with one-way trip valves.

module is built from a "MEMORY" valve and a pre-switched "AND" function valve. The two signals required to make the "AND" function are the essential feedback signal, confirming the completion of the previous sequence step, and the preparation signal from the memory valve of the previous step-counter module (compare figs. 8–4 and 8–5). The "AND" function sets the memory into the "ON" position. The memory must accomplish three functions: it provides the power valves (D.C.V.) with a signal command to extend or retract an actuator; it re-sets the memory valve in the previous step-counter module; and it provides the next step-counter module in the chain with a preparation signal (on the "AND" valve).

Opposing signals thus cannot occur since the

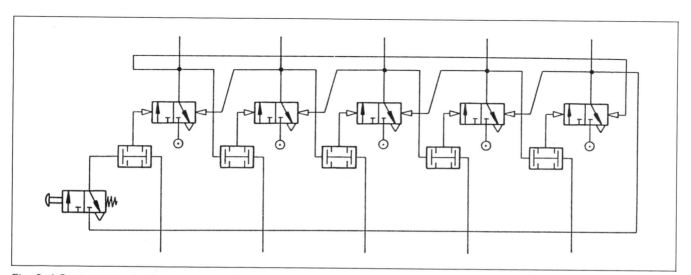

Fig. 8–4 Step-counter for five sequence steps.

previously used confirmation signal (feedback signal) is automatically rendered inactive when the memory, which is "AND" connected to it, is being reset (see fig. 8–4).

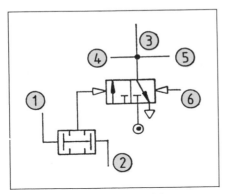

Fig. 8–5 Step-counter module
(1) Preparation signal from previous module
(2) Feedback signal from previous step
(3) Switching signal to power valve
(4) Reset signal to previous module
(5) Preparation signal to next module
(6) Reset signal from next module.

When the sequence program is completed, the last module in the circuit chain remains set until the first module after the start signal has been given, resets it (fig. 8–6). The preparation signal for the "AND" function in module 1 is fed through the start valve (fig. 8–4). Alternatively, one may also place the start valve into the confirmation signal line prior to the "AND" valve in module 1 (see fig. 8–6). Thus

the START valve interrupts continuous cycling and a new cycle cannot commence until the START command is given. A circuit for a clamp and drill application is given in fig. 8–6.

Logic switching equations for step-counter circuit design

Logic switching functions, and this applies to all sorts of control whether this be mechanical, pneumatic, hydraulic, electrical or electronic, can be expressed in equation form. The fundamental principles of logic switching expression use the notations of Boolean algebra, which was discovered and published by George Boole, Professor of Mathematics at Queens College, County Cork, Ireland in 1854. His laws and theorems are still valid today and are the foundation of logic switching from telephones to computers. The five most important logic functions (YES, NOT, AND, OR, INHIBITION) are shown in fig. 8–7. Further logic functions and the use of truth tables and corresponding Karnaugh-Veitch maps are illustrated and explained in Chapter 10. Figure 8–7 shows the name of the logic function in the left hand column, then the pneumatic symbol in the middle and in the right hand column the Boolean switching equation. "A" is signal input 1, "B" is signal input 2 and "S" is the output signal. A dash above the signal means "not on" or signal "not present". The "AND" function is denoted by a dot (•), the "OR" function by a plus sign (+).

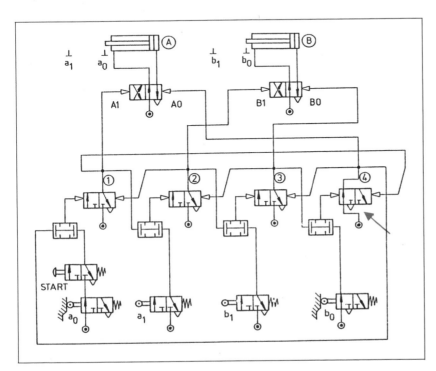

Fig. 8–6 Step-counter circuit for sequence given in fig. 7–1.

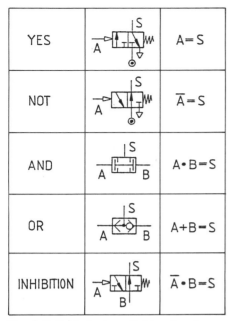

YES		$A = S$
NOT		$\overline{A} = S$
AND		$A \cdot B = S$
OR		$A + B = S$
INHIBITION		$\overline{A} \cdot B = S$

Fig. 8–7 Basic logic switching functions.

Switching equations may be used directly to build a control circuit, thus making a circuit diagram obsolete. But not everybody can read and understand Boolean switching equations, so circuit diagrams are still frequently used in industry. To derive such switching equations one separates the circuit into power valve input signals and step-counter input signals (also called D.C.V. inputs, step-counter inputs). The step-counter is regarded as a chain consisting of individual step-counter modules and each module is given a number according to the step it switches (fig. 8–6).

Switching equations for the circuit shown in fig. 8–6 are:

D.C.V. inputs:

$A1 = $ Step 1 (Module 1)
$A0 = $ Step 4
$B1 = $ Step 2
$B0 = $ Step 3

Step-counter inputs:

Stepping module 1 $= a_0 \cdot$ Start
Stepping module 2 $= a_1$
Stepping module 3 $= b_1$
Stepping module 4 $= b_0$

Switching equations for the circuit shown in fig. 8–11 are:

D.C.V. inputs:

$A1 = $ Step 6
$A0 = $ Step 1

$B1 = $ Step 2 $+$ Step 4
$B0 = $ Step 3 $+$ Step 6
$C1 = $ Step 4
$C0 = $ Step 5

Step-counter inputs:

Stepping module 1 $= a_1 \cdot b_0 \cdot$ Start
Stepping module 2 $= a_0$
Stepping module 3 $= b_1$
Stepping module 4 $= b_0$
Stepping module 5 $= b_1 \cdot c_1$
Stepping module 6 $= c_0$

Repeated actuator movements

Where an actuator must extend and retract more than once during a control cycle, the appropriate output signals from the step-counter chain must be connected with "OR" function valves before being connected to the signal port of the power valve. Thus an actuator may extend and retract several times during a machine cycle but each extension and retraction command must be signalled with a new step-counter module (figs. 8–8 and 8–10).

When pilot signals are to be *gathered*, T-connectors do not substitute for "OR" function valves, since the pilot signal from the step-counter chain would exhaust through the valve on the other end of the T-connection! This is illustrated in figs. 8–8, 8–9, 8–10 and 8–11.

Switching equations for the circuit shown in fig. 8–10 are:

D.C.V. inputs:

$A1 = $ Step 1
$A0 = $ Step 8
$B1 = $ Step 2 $+$ Step 4 $+$ Step 6
$B0 = $ Step 3 $+$ Step 5 $+$ Step 7

Step-counter inputs:

Stepping module 1 $= a_0 \cdot$ Start
Stepping module 2 $= a_1$
Stepping module 3 $= b_1$
Stepping module 4 $= b_0$
Stepping module 5 $= b_1$
Stepping module 6 $= b_0$
Stepping module 7 $= b_1$
Stepping module 8 $= b_0$

Note: the Start is series connected to the essential confirmation signal of sequence step 8 ($a_0 \cdot$ Start)

Fig. 8–8 *Right:* Gathering of signals requires an "OR" function valve.

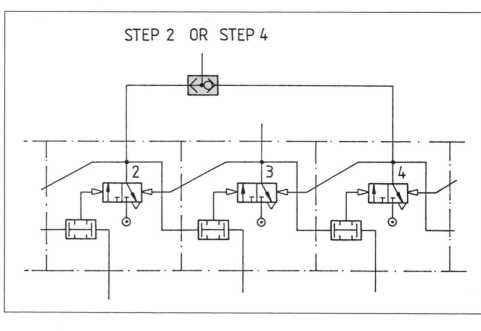

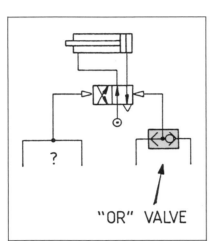

Fig. 8–9 T-connector must not be used to gather signals.

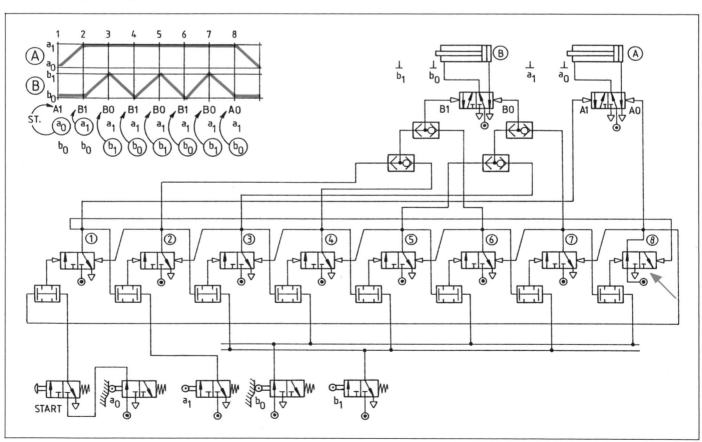

Fig. 8–10 Circuit with repeated actuator movement.

Simultaneous actuator movements

Where a step-counter module actuates more than one actuator (that is, two or more actuators traverse simultaneously), the output signal from the appropriate step-counter module is simply branched out (see arrow in fig. 8–11), and connected to the signal ports of the power valves involved in the simultaneous movements.

The next sequence step which precedes the step with simultaneous actuator movements must not switch, until all these movements are confirmed. Premature switching is prevented by

"AND" function valves or where possible by connecting the appropriate limit valves in series order. Fig. 8–11 illustrates such an application where multiple confirmation signals are series connected. This is indicated by arrows on the lower edge of the circuit. Fig. 8–12 shows how additional "AND" valves can be attached where a series connection of multiple confirmation signals is not possible.

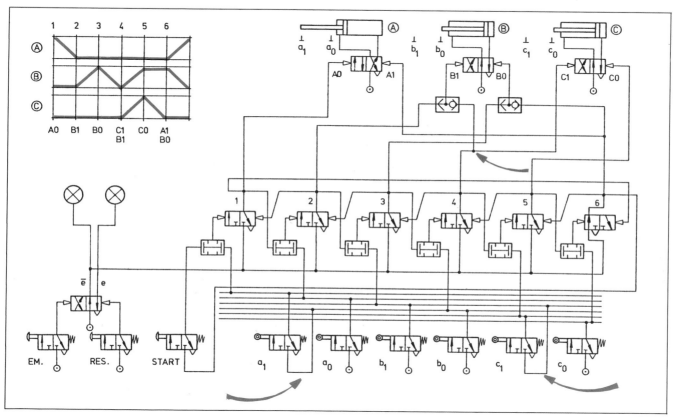

Fig. 8–11 *Above:* Sequential control with simultaneous actuator movements at steps 4 and 5.

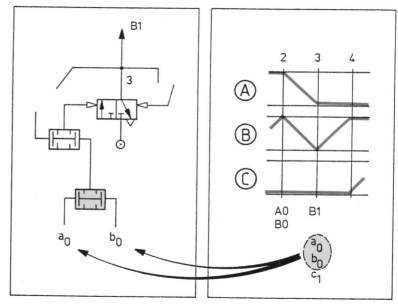

Fig. 8–12 Multiple confirmation of essential limit valve signals after simultaneous actuator movement. The key below the traverse-time diagram shows that the signals a_0 and b_0 are required to confirm the double movement of the actuators A and B.

Series connection of "AND" functions

In cases where a limit valve signal, start valve signal or timer signal is "AND" connected to another such signal, one may use a series connection to combine these signals (see also Chapter 2 fig. 2–34). With a series connection the "AND" valve becomes superfluous (fig. 8–12).

However, a series connection is only permitted if the valve which is downstream in the series connection appears only once in the switching equations for this circuit. The reason for this rule is the fact that any upstream valve "contaminates" the "name" of its downstream valves and thus the "name" of the downstream valves is no longer "pure" and can no longer be used on its own.

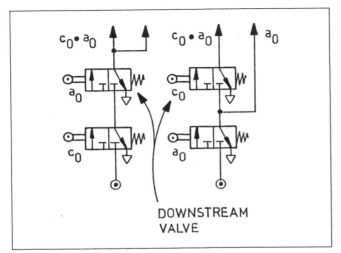

Fig. 8–13 "Contamination" rule for series function. In the left hand circuit the signal a_0 is no longer pure; but if it must stay pure, and still be "AND" connected as shown in fig. 8–14, then the right hand circuit solution must be used.

To demonstrate this rule a traverse-time diagram for three actuators is given in fig. 8–14. All essential confirmation signals are circled in the key and a series function chart is drawn. A series connection may only be envisaged for signals which appear only once in the switching equations (see switching equations and columns 1 and 2 of the series function chart) and are "AND" connected to another signal.

Switching equations for circuit shown in fig. 8–14 are:

Step-counter inputs:

Stepping module 1 = $c_1 \cdot a_1 \cdot$ Start
Stepping module 2 = a_0
Stepping module 3 = b_1
Stepping module 4 = $a_1 \cdot b_0$
Stepping module 5 = $a_0 \cdot c_0$

Connecting the step-counter input circuit from the switching equations and the series function chart is not complicated, if one follows the information established in the key of the traverse-time diagram and the connection column of the series function chart (compare figs. 8–13, 8–14, 8–15 and step-counter input equations). Fig. 8–15 shows only the step-counter input. All other circuit parts such as step-counter and power circuits are not shown.

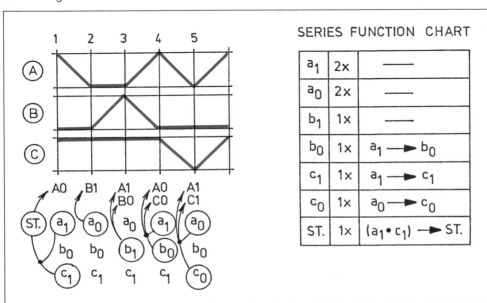

Fig. 8–14 Traverse-time diagram and series function chart.

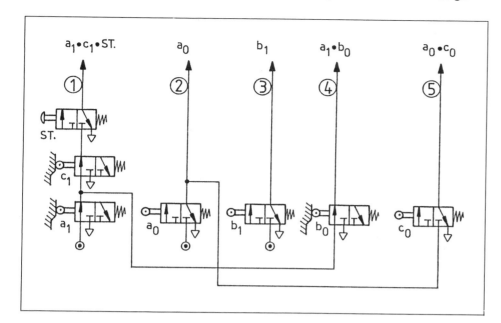

Fig. 8–15 Step-counter input circuit for machine sequence shown in fig. 8–14.

Identical "AND" function

A further reduction of circuit hardware may be achieved when the switching equations contain *identical "AND" functions*. To demonstrate this a circuit segment is given in fig. 8–16. Here the identical "AND" function $c_1 \cdot a_1$ is used by itself in the left hand signal branch and is "AND" connected to signal b_1 for the right hand branch. Even though the signal a_1 seems to appear twice in the series function chart, the fact that it can be grouped into an identical "AND" function (together with c_1) means it need only appear once (see fig. 8–16). Hence it can be further series connected to signal b_1. It is therefore wise to investigate the equations for identical "AND" functions before establishing the series function chart.

You may discover that further series functions may be formed. To demonstrate this, a typical cir-

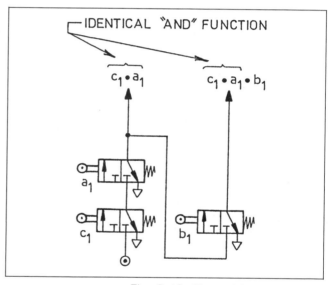

Fig. 8–16 *Above:* Identical "AND" functions need only be connected once and may then be distributed wherever they are required in the circuit.

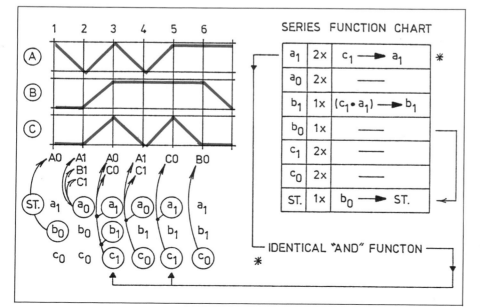

a_1	2x	c_1	→	a_1	✳
a_0	2x	—		
b_1	1x	$(c_1 \cdot a_1)$	→	b_1
b_0	1x	—		
c_1	2x	—		
c_0	2x	—		
ST.	1x	b_0	→	ST.

SERIES FUNCTION CHART

IDENTICAL "AND" FUNCTON

Fig. 8–17 Control with identical "AND" function for $c_1 \cdot a_1$.

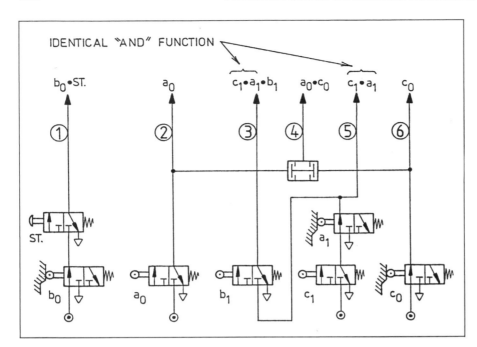

Fig. 8–18 Step-counter input circuit for control shown in fig. 8–17.

cuit sequence is depicted in fig. 8–17 together with the series function chart. The step-counter input circuit for this sequence is also given in fig. 8–18.

Step-counter inputs for fig. 8–17:

Stepping module 1 = b_0 • Start

Stepping module 2 = a_0

Stepping module 3 = c_1 • a_1 • b_1

Stepping module 4 = a_0 • c_0

Stepping module 5 = c_1 • a_1

Stepping module 6 = c_0

Time delayed sequencing

Time delayed sequencing is often encountered in industrial machine control circuits. The pneumatic timer may either be built into the confirmation signal line leading to the step-counter module or into the switching signal line leading to the power valve (see figs. 8–19 and 8–5). Timer placement in the switching signal line delays only the switching of the power valve. Resetting of the previous step-counter module is not delayed. With the timer in the confirmation signal line, both the power valve switching function and the resetting and setting of the step-counter modules is delayed. Circuit designers therefore have to evaluate both timer positions for their merits and select the correct timer position.

A typical time delay circuit where timer placement is important is shown in fig. 8–20. If the pneumatic timer were wrongly placed (in the position indicated by the arrow), then actuator B would overshoot the limit valve b_1 at the completion of sequence step 3.

All circuits with non-memory type power valves (spring centered, three position valves and spring reset four or five port power valves) need therefore careful consideration in the placement of the timer.

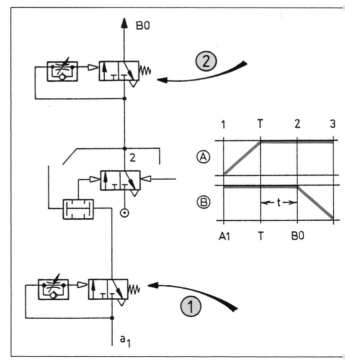

Fig. 8–19 Timer placement options.

Switching equations for circuit shown in fig. 8–20 are:

D.C.V. inputs:

$A1 = $ Step 1

$A0 = $ Step 5

$B1 = $ Step 2 $+$ (Step 3 $\bullet$ Timer)

$B0 = $ Step 4

Step-counter inputs:

Stepping module 1 $= a_0 \bullet$ Start

Stepping module 2 $= a_1$

Stepping module 3 $= b_1$

Stepping module 4 $= b_2$

Stepping module 5 $= b_0$

Woodpecker motion control

Woodpecker motion control may easily be accomplished with the step-counter design concept (see fig. 8–21). Woodpecker motion control is also closely related to the repeated actuator movement control outlined in figs. 8–8 and 8–10. Here again

each extension and retraction command shown in the key of the traverse-time diagram requires a separate step-counter module (fig. 8–21).

Switching equations for circuit shown in fig. 8–21 are:

D.C.V. inputs:

$A1 = $ Step 1

$A0 = $ Step 8

$B1 = $ Step 2 $+$ Step 4 $+$ Step 6

$B0 = $ Step 3 $+$ Step 5 $+$ Step 7

Step-counter inputs:

Stepping module 1 $= a_0 \bullet$ Start

Stepping module 2 $= a_1$

Stepping module 3 $= b_2$

Stepping module 4 $= b_1$

Stepping module 5 $= b_3$

Stepping module 6 $= b_1$

Stepping module 7 $= b_4$

Stepping module 8 $= b_0$

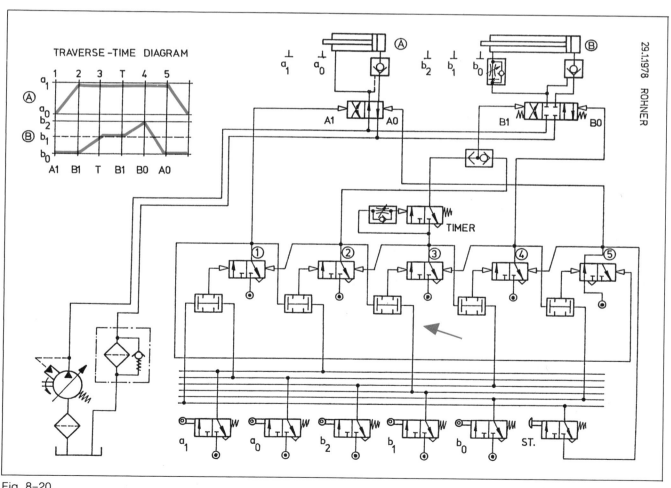

Fig. 8–20

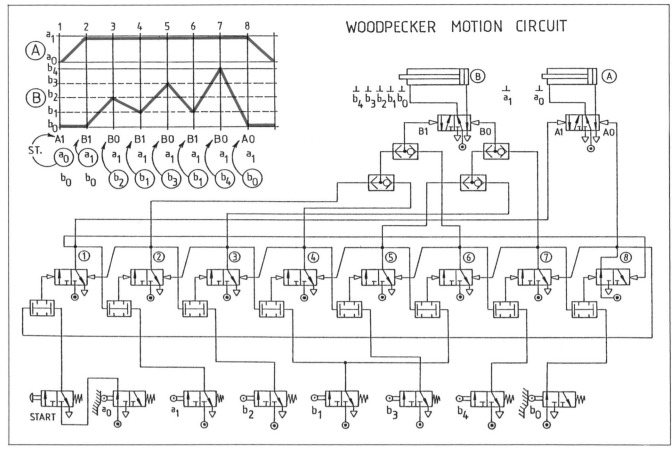

Fig. 8-21

Emergency stop with step-counter

The majority of automated machines operate according to a programmed sequence. Due to mishaps such as a damaged tooling, a misaligned workpiece in an assembly line, an empty workpiece magazine, a sudden drop in supply pressure or an endangered operator, the sequence must be interrupted or the start of a new cycle prevented if automatic cycling was selected.

The nature of such an interruption varies with the anticipated degree of danger and the inherent safety equipment already built into the machine (safety guards, remote control, intermittent operation, various and unskilled operators, etc.). The distinction is basically between the following three emergency stopping modes:

1. Stop instantaneously (use piston rod brakes if necessary).
2. Stop at end of commenced sequence step.
3. Stop at end of sequence cycle (if machine cycles automatically).

Once the machine cycle is interrupted and actuators have stopped, one has several options to choose further sequence or machine action. It is, therefore, the circuit designer's responsibility to decide what this action is to be. His choice is crucial, since it can prevent serious harm to the machine operator or avoid costly damage to workpieces, tooling and machinery. To illustrate: in a machine tool, one has to stop the drilling head and retract it completely before the workpiece may be unclamped and finally the automatic workpiece ejector is permitted to clear the machine (fig. 8-36). It would be disastrous if all these movements would happen simultaneously! However, in another machine, all actuators may quite safely act in unison and move into the position required for them to start a new cycle, once the emergency situation is cleared. In a third application, the machine cycle is brought to a halt with either stop-condition 1 or 2, as enumerated previously (fig. 8-22), and when the emergency situation is cleared the sequence is permitted to continue to the end of cycle (figs. 8-31 and 8-32). Such machine applications are quite common. They

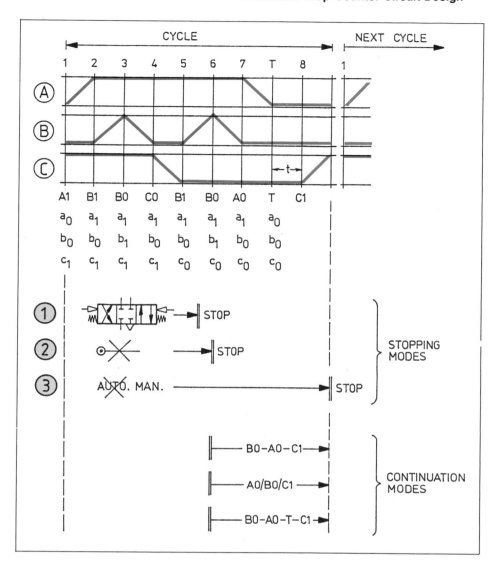

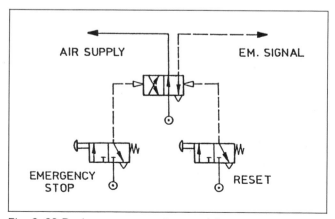

may be regarded as classical "continuation modes" for machine sequence behaviour after emergency stop was signalled (see fig. 8–22). A fourth and not so common continuation mode is to reverse only those actuators which are in motion when the emergency stop is signalled. This emergency stop module is explained later and illustrated in fig. 8–30.

Basic emergency stop module

For maximum operator and machine safety, all emergency signals should be stored in a memory type valve. It is recommended that the emergency reset valve be placed inside the control cabinet, or if possible, that a key operated valve be used, which is only accessible to authorised personnel.

To make the emergency stop functional at all times, its memory valve must never be "AND" con-

nected to any other signal and must therefore have its own air supply (see fig. 8–23)! For a circuit integration of this module see figs. 8–31 and 8–32.

Fig. 8–23 Basic emergency stop module.

Detent-type push button emergency valves are sometimes used but have an inherent danger: the valve can simply be reset by operators unaware of an emergency situation. This could lead to machine damage or, more seriously, cause harm to the operator! The emergency stop memory may be a four port or five port valve and must never be a spring reset type valve.

Emergency stop module with multiple input

Where machinery is operated from several locations, it is imperative to provide an emergency stop valve for all such locations—particularly on large machines, or machines where operators are exposed to dangers, one should use an emergency stop module with multiple emergency inputs (fig. 8–24A).

Emergency stop module with manual and automatic input

Where the emergency signal must be triggered either by an operator or by an external machine operation, such as "PRESSURE INADEQUATE", "MAGAZINE EMPTY", "GUARD OPENED" or a similar stop-demanding signal, the emergency stop module shown in fig. 8–24B is used.

Simple cycle selection module

Although not directly an emergency stop module, the simple cycle selection module may be used to stop a continuous machine cycle (automatic cycling). This module is illustrated in fig. 8–26 (see also fig. 8–22, stopping mode 3).

The simple cycle selection module (see fig. 8–26) provides an inexpensive solution to furnish "low danger" rated machines with automatic

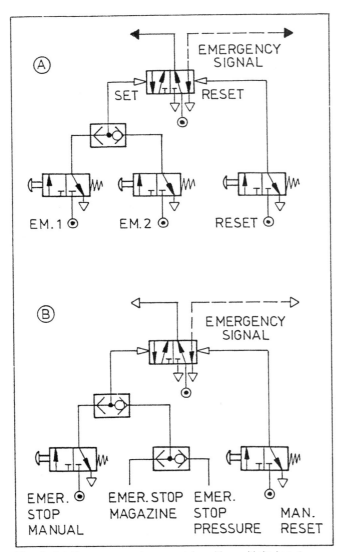

Fig. 8–24 Emergency stop module with multiple input.

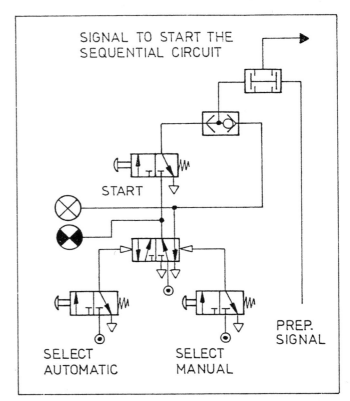

Fig. 8–25 Simple cycle selection module with visual indicators for automatic and manual start selection. Since the preparation or confirmation signal is only momentary (but display is required as permanent) one cannot use series function as in fig. 8–26. An "AND" valve is required to connect the start signal to the preparation signal.

cycling or manual cycling (single cycling). The disadvantage of this module is its combination of auto-selection and auto-start in the same push-button valve, which causes the machine to operate as soon as the "AUTO" selection is made (see fig. 8–26). For the integration of this module see fig. 8–33 where the cycle selection module is enclosed by a frame.

both cycling modes. The additional cost for the extra push button valves and one more memory valve is easily justified for the additional safety feature of this module (see fig. 8–27).

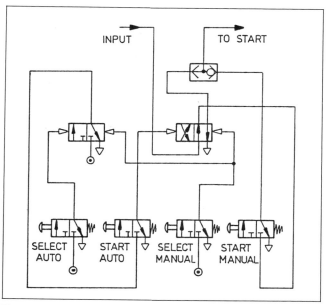

Fig. 8-27 Extended cycle selection module.

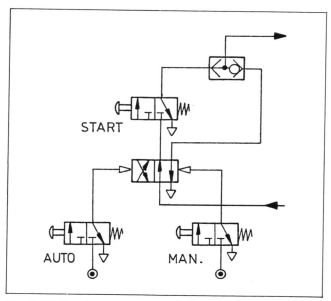

Fig. 8-26 Simple cycle selection module.

Extended cycle selection module

This module is an extension to the simple cycle selection module. It eliminates the disadvantage of the simple cycle selection module by providing separate selection as well as separate start for

Cycle selection module combined with emergency stop module

In order to avoid the cycle starting unexpectedly after the emergency reset valve is actuated, while the cycle selection is in the automatic mode, this combination is furnished with a "LINK" which kicks the cycle selector into manual cycling mode (see fig. 8–28).

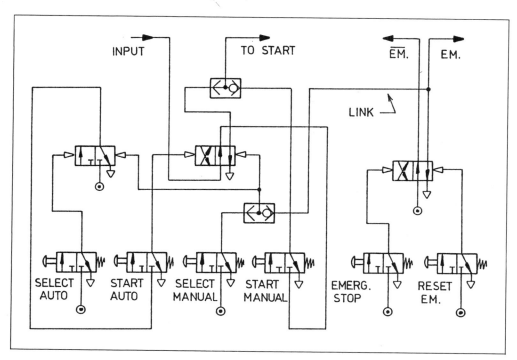

Fig. 8-28 Extended cycle selection module with emergency stop link.

Cycle selection with automatic manual selection

This cycle selection module provides the following switching functions. *Automatic selection* is only possible if the push button selector valve is actuated *and* other fringe conditions are also met. These fringe conditions may entail such inputs as: magazine full, guard closed, system pressure above basic limit etc. *Manual selection* is achieved when the push button selector valve is actuated or when one or more of the fringe conditions are no longer met. This module is similar to the module combination shown in fig. 8–28 where manual selection may be triggered if emergency is signalled.

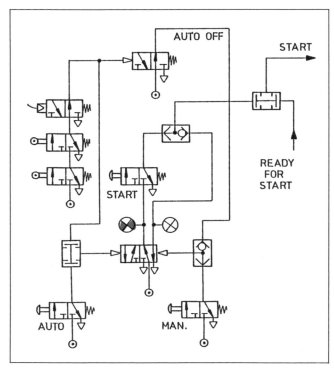

Fig. 8–29 Cycle selection with automatic manual selection.

The "READY FOR START" signal may be series connected through the memory valve. However, permanent auto- or manual indication on the visual indicators would then not be possible. Integration of this module is identical to the two cycle selection modules depicted in figs. 8–26 and 8–27 The fringe conditions may include several inputs.

Reverse actuators while in motion

The "REVERSE ACTUATORS WHILE IN MOTION" module is expensive and complicated and should, therefore, only be used for actuators which abso-

lutely have to be reversed when an emergency is signalled (fig. 8–30). On actuation of the emergency stop, the cylinder which is in motion reverses to its previous position (should it be at rest in an end-position, it remains stationary).

This emergency stop module is very useful, since it is often the actual cylinder in motion which is causing the emergency, and therefore its motion must be reversed.

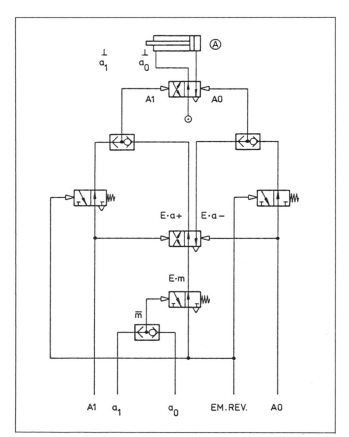

Fig. 8–30 Special module to reverse actuators while in motion.

Stop instantaneously (stopping mode no. 1)

The sequence in this circuit will stop instantaneously when an emergency is signalled. This is achieved by the three-position type power valves which centre automatically when the signals from the step-counter cease. Once the emergency is cancelled (reset), the circuit will continue with the interrupted sequence to the end of the cycle. Since the emergency stop memory valve has its own air supply, one of the two visual indicators will always be "ON". Pneumatic piston rod brakes must be used if rigid position holding is required during the emergency stop (see figs. 3–19 and 8-31).

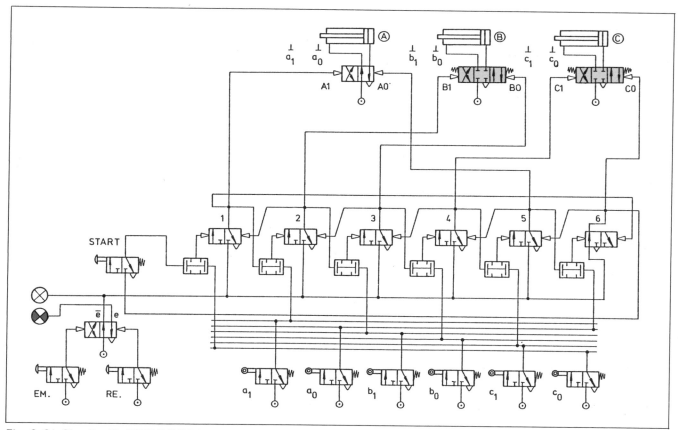

Fig. 8-31 Circuit with "STOP INSTANTANEOUSLY" emergency stop. The sequence for this control is A1-B1-B0-C1-A0-C0.

Fig. 8-32 Circuit with "STOP AT END OF COMMENCED STEP" emergency stop. Once the air supply to the step-counter is denied, the commenced step will be completed but no further steps can occur until the air supply is restored. After the restoration of the air supply the program continues along its normal sequence.

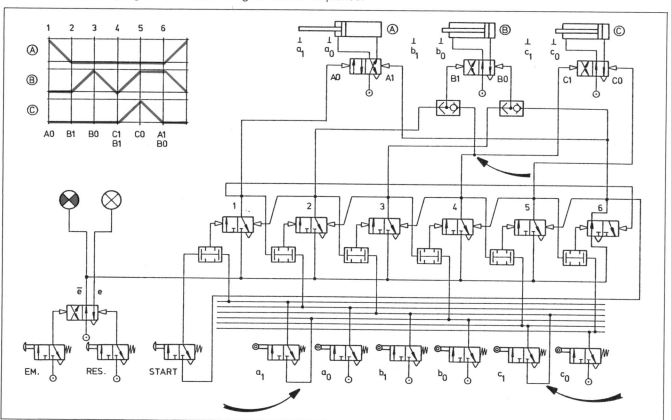

Stop at end of commenced step (stopping mode no. 2)

The normal sequence in this circuit will stop when an emergency is signalled and the current sequence step (emergency) is completed. This is taken care of by memory type power valves which maintain their position even when the pilot signals from the step-counter cease. Thus although the commenced step is completed further sequence steps cannot commence until the air supply to the stepping modules is restored. When the air supply is restored the interrupted sequence will continue to the end of cycle (see fig. 8–32 and compare with figs. 8–22 and 8–31).

Stop at end of cycle (stopping mode no. 3)

Athough not directly an emergency stop circuit, this control does bring about a stop of the ongoing sequencing. Automatic cycling ceases when the manual selection is made. The circuit will always finish a commenced cycle, but no *new* cycle is per-

mitted to commence until the start command is given (see fig. 8–33).

Continuation modes deviating from the normal sequence

The three circuits depicted in figs. 8–31, 8–32 and 8–33 have a continuation mode which follows the normal sequence once the emergency module is reset (see fig. 8–22, continuation mode 3). However where the continuation mode must deviate from the normal sequence (as shown in the traverse-time diagram of fig. 8–36) one can no longer use the step-counter to sequence the actuators, and therefore the step-counter must be reset to "ZERO".

Resetting the step-counter to "ZERO" while the machine-sequence cycle is not yet fully completed, requires an "OR" function valve attached to every step-counter module, except the last one (see fig. 8–34). The last step-counter module logically needs no "OR" function valve since an emergency rarely occurs during the last sequence

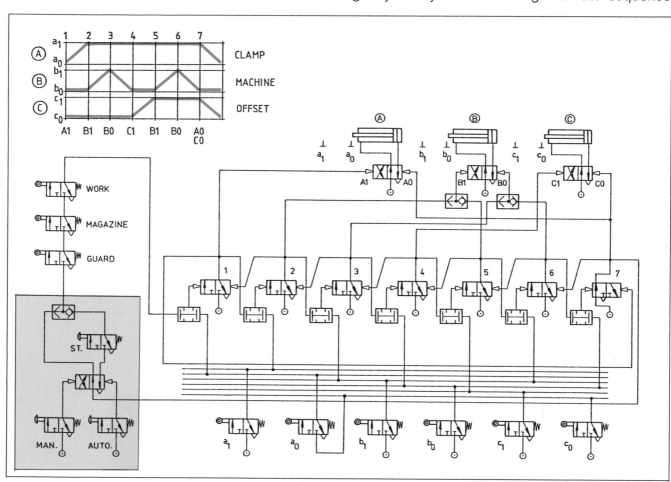

Fig. 8–33 Circuit with "STOP AT END OF CYCLE" feature.

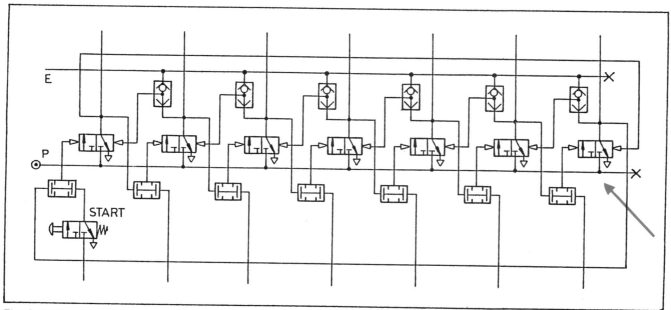

Fig. 8–34 Step-counter with external reset via ''OR'' function valves.

step. However, should there still be a possibility of an emergency, an ''OR'' valve may be simply added to that module.

Under normal cycling conditions the preparation signal for the first step-counter module is provided by the last step-counter module in the chain (fig. 8–34). However, under emergency conditions, the circuit could stop anywhere between the first and the last module in the step-counter chain. The sequence would therefore never reach the last step-counter module in the chain, which means its

memory could not send the essential preparation signal to module 1!! Under these circumstances one could no longer start a new cycle after the emergency has been reset. To overcome this problem, a special circuit modification may be placed between the emergency stop memory and the ''AND'' valve of the first step-counter module in the chain (see fig. 8–35).

Emergency stop modification module

This modification provides the circuit with a

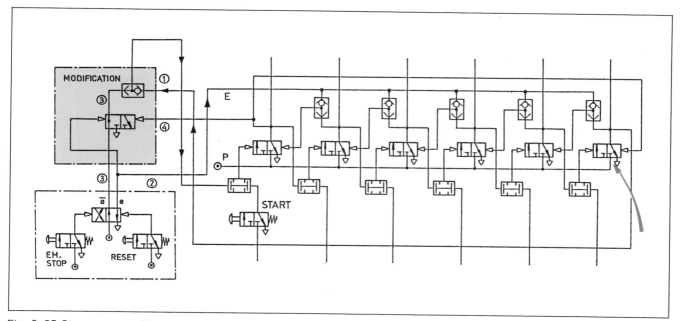

Fig. 8–35 Step-counter with emergency stop modification.

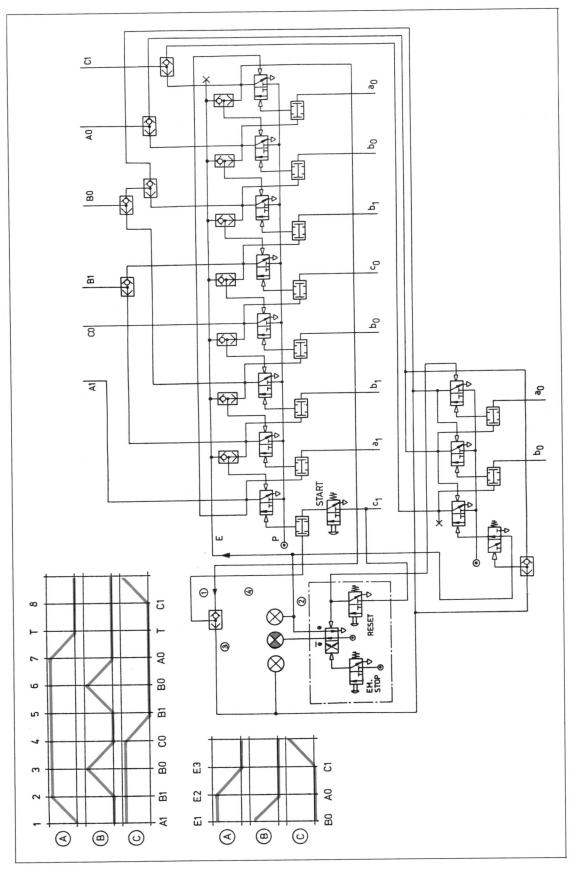

Fig. 8-36 Control with special emergency sequence.

"SUBSTITUTE PREPARATION" signal. The first step-counter module cancels this substitute preparation signal as soon as the first sequence step is being set, thus any interference by accidental actuation of the start valve during cycling is avoided (fig. 8–35).

Once the step-counter is reset to zero by the emergency stop signal "E", the same signal will also reset the actuators to their end-of-cycle position (continuation mode).

Where the continuation mode must follow a special emergency program (fig. 8–22 continuation mode (1)), one may also use step-counter modules to sequence this program (fig. 8–36). However, in this case, the modification as shown in fig. 8–35 is not required, as the preparation signal from the emergency program can be used (in "OR" connection) to provide the necessary preparation signal. The circuit depicted in fig. 8–36 is designed to achieve the sequence shown in fig. 8–22. When an emergency stop is signalled the normal program must stop and the emergency program must take over, with the sequence B0–A0–C1. This emergency program brings the actuators into the condition required to start a normal sequence program (compare figs. 8–22 and 8–36).

Commercial step-counter modules

In recent years complete step-counter modules, consisting of a memory valve, an attached "AND"

function valve and an "OR" function valve for emergency reset, have appeared on the market. These step-counter modules are simply pushed together on a mounting bar and thus interconnections for preparation and reset signals, as well as air supply connections, are made automatically by means of air ducts sealed by O-rings on the connection faces (fig. 8–37). To merge the step-counter with the power valves and the limit valves, one simply has to connect the output ports of the step-counter modules to the appropriate pilots on the power valves, and the step-counter input ports to the appropriate limit valves. End plates connect the reset and preparation signal from the last step-counter module to the first step-counter module via external air tubes (figs. 8–37 and 8–38). These end plates are essential since the modules themselves have no provision to attach fittings and plastic tubing. However, on the outlet ports leading to the power valves, and inlet ports to accept the pilot signal from the limit valves, one finds push-in fittings or barbed sleeves to connect plastic tubing, and also the outer faces of the endplates are provided with fittings.

German design concepts for commercial step-counters

Amongst the many different brands of commercial step-counters one can clearly differentiate two separate design concepts. With the German designs, the last stepping module remains in the

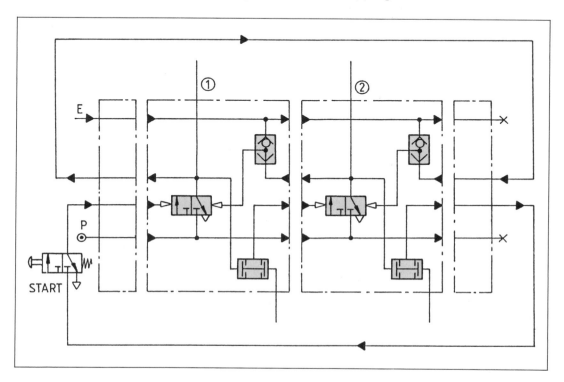

Fig. 8–37 Commercial step-counter modules.

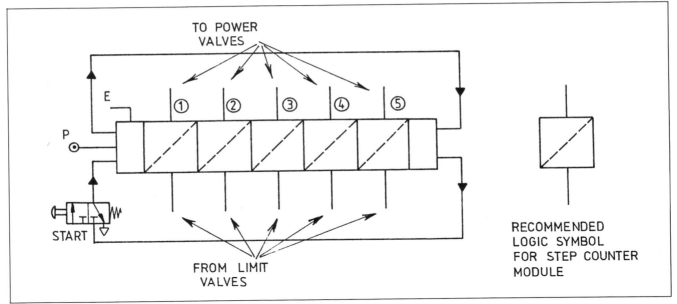

Fig. 8–38 *Above:* Logic symbols for step-counter modules.

Fig. 8–39 *Below:* Step-counter design concepts; A = French design concept B = German design concept C = passive design concept.

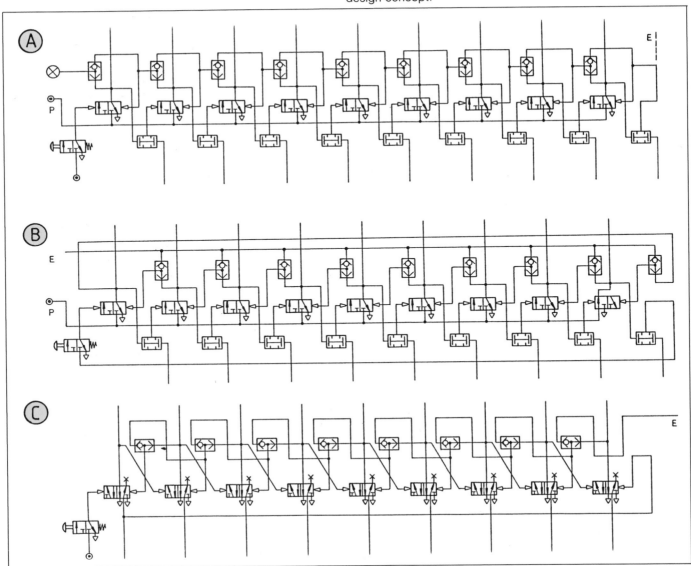

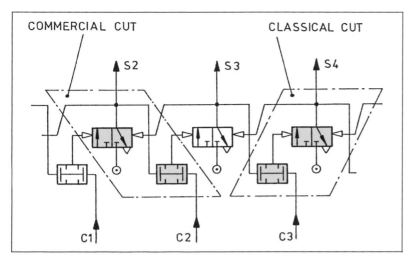

Fig. 8–40 Classical versus commercial step-counter.

"SET" position to provide the first stepping module in the chain with a preparation signal (see figs. 8–31 to 8–36 and 8–39B). However, with the French design concept, the last step-counter module is reset as soon as the last confirmation signal arrives, and a unique interlinkage of the "OR" function valves constantly holds the first memory valve (in module 1) reset, as long as the step-counter chain is in progress of sequencing. Thus any interference from the start valve during cycling is prevented (fig. 8–39A). A third concept is the passive step-counter where the confirmation signal is fed through the step-counter memory valve and so becomes the switching signal for the power valves (fig. 8–39C).

The first two step-counter design concepts have, however, one common design characteristic. They both switch and confirm such switching in the same module block. This means step 2, for example, is switched by block 2 (signal S2) and the confirmation signal from the completion of that actuator movement (signal C2) is also connected to block 2 (compare figs. 8–36, 8–37, 8–39 and 8–40). The philosophy behind this arrangement is to make it easier for the inexperienced circuit designer to remember that each block switches an actuator function and that the confirmation signal of that function is connected to the same block.

French step-counter design concept

With the French step-counter design concept the last module in the chain is reset as soon as the last confirmation signal arrives; this happens when the last sequence step is completed. That is, prior to start no step-counter module is set at all! This concept has several advantages. The most important

Fig. 8–41 Crouzet step-counter modules.

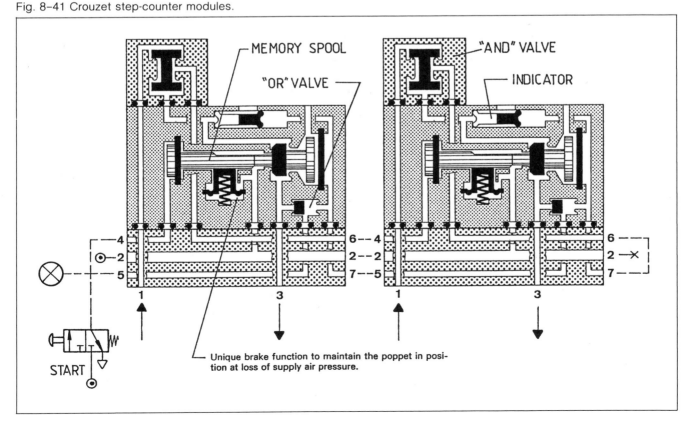

Unique brake function to maintain the poppet in position at loss of supply air pressure.

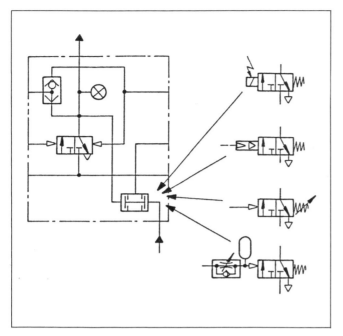

Fig. 8–42 French design step-counter module with detachable "AND" valve. The "AND" valve may be replaced with the valves shown on the side.

advantage is that the emergency stop modification shown in fig. 8–35 is *not required* with the French design. The unique interlinkage of its "OR" function valves holds the first memory valve in module 1 reset as long as the machine sequence is in progress. One may also attach a visual indicator to the "OR" function valve output of module 1 (fig. 8–39). This indicator is in the "ON" position as long as the sequence is in progress. A cross-section through a "CROUZET" step-counter module is shown in fig. 8–41. Another advantage with the French design is that the "AND" function valve is detachable and thus can be replaced with a timer, a pressure sequence valve, a signal amplifier valve, or even with a solenoid valve to interface with an electrical input. All these valves are thus used to render the "AND" function and are designed in such a way that they fit exactly into the position of the "AND" valve.

Step counter with interim start

Interim start may be used where the normal cycle program is divided into two or more cycle phases. Once the first cycle phase is completed the sequencing stops, and is then only permitted to

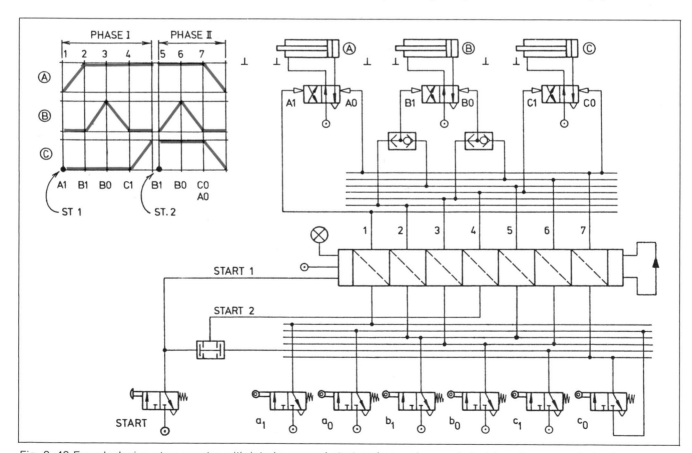

Fig. 8–43 French design step-counter with interim manual start and normal manual start from the same start valve.

continue when an external continuation signal is given from the interim phase being completed (fig. 7–43).

The interim phase can be a foreign machine function such as a forging press, a machining station, or a robot. In a forging machine for example the first few steps of phase 1 load the component into the machine, and the remaining steps of phase 2, which unloads the component out of the forging press, must only occur when the forging process (interim phase) is completed. Thus the forging machine provides the continuation signal.

Step-counter with undefined skipping

Skipping of certain sequence steps may be required if a machine malfunctions. In such a case the beginning of the skipping motion may or may not be previously defined or predictable. The skipping signal, if present, causes the step-counter to omit (skip-over) certain sequence steps, and after the skipping is completed, the program follows its normal sequence. A typical application for a drilling machine, where skipping would be required, is given in fig. 8–44. The machine follows the sequence depicted in fig. 8–36. Here a timer is used to make the skipping signal. Should the drilling signal B1, due to a broken or blunt drill, exceed the predetermined time of the timer, the skipping motion will automatically start and advance the program to the end of step 5. Thus the sequence steps 3, 4 and 5 in case 1 will be rapidly advanced (skipped) to the beginning of step 6. In case 2 the malfunction could occur during step 5. The skipping signal would then also advance the step-counter to the beginning of step 6. The remaining sequence steps 6, 7 and 8 will then follow the normal sequence controlled by the step-counter.

In order to skip sequence steps, one simply "OR" connects the skipping signal to the confirmation signal of these step-counter modules which must be skipped-over. The skipped modules are

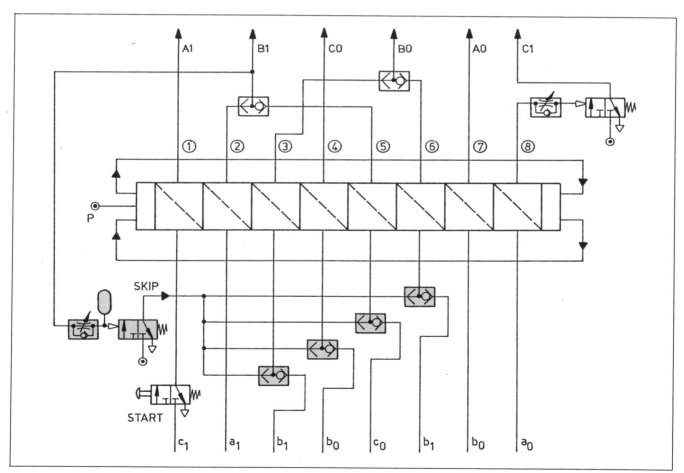

Fig. 8–44 Step-counter with undefined skipping.

then switched in extremely rapid succession, but the output signals (switching signals) of the skipped modules are too short to cause its connected power valves to switch.

Step-counter with defined skipping

Skipping of certain sequence steps may provide a simple solution if two machine programs have some sequence steps in common and are almost identical apart from the sequence steps which can be skipped. To illustrate, the traverse-time diagram of fig. 8–36 may again be used. Program I follows the sequence depicted in fig. 8–36. Program II, however, must only drill one hole into the workpiece and therefore sequence steps 4, 5 and 6 must be skipped (see also fig. 8–45). Since Actuator C never moves in program II, and therefore signal c_1 is constantly given, the sequence step 8 for C to extend is also skipped and therefore needs no extra "OR" valve. Thus the start of a new cycle can begin as soon as step 7 is completed and the timer phase is also skipped.

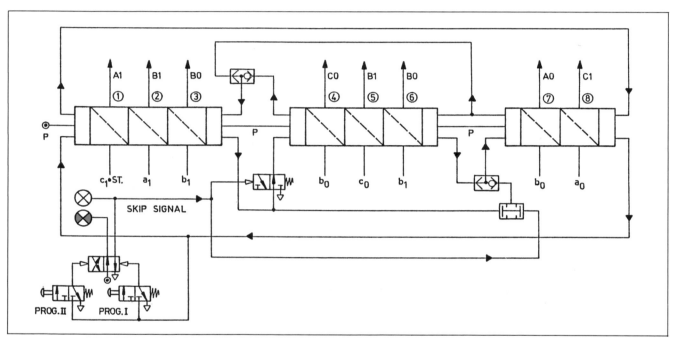

Fig. 8–45 Step-counter with defined skipping.

Fig. 8–46 Step-counter with repeated steps.

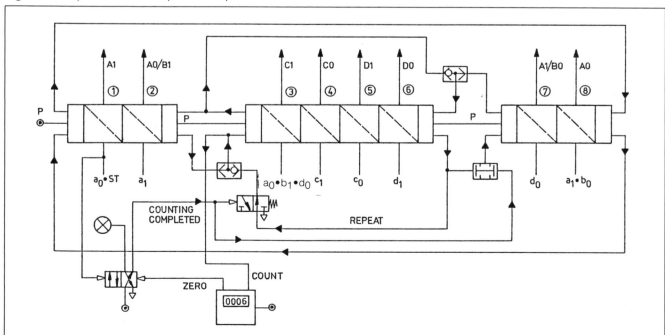

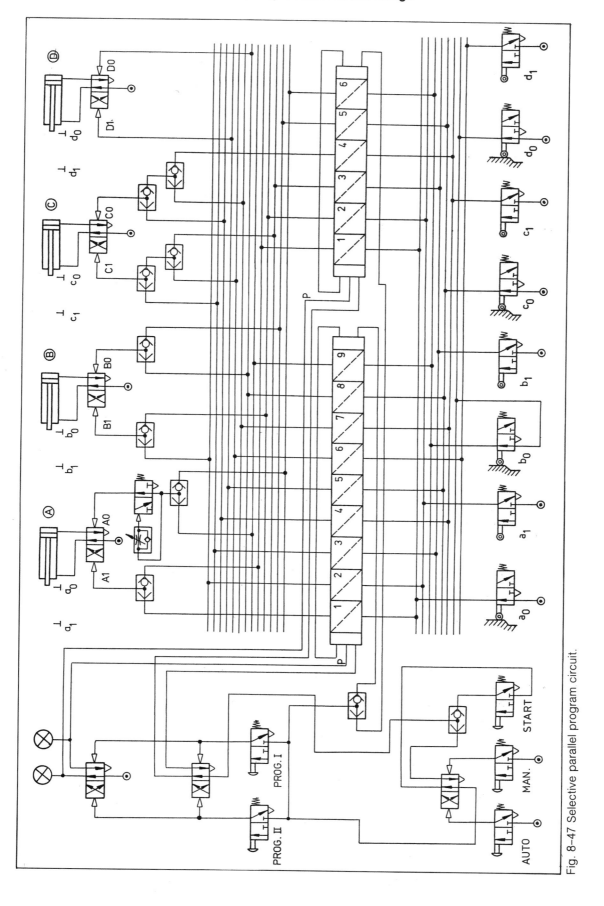

Fig. 8-47 Selective parallel program circuit.

Step-counter with repeated steps

Step repetition is an extremely versatile and frequently used control method, which may simplify the control circuit considerably and thus save step-counter modules. To illustrate this control design concept, a sequence program is given for a disc to be machined. The disc is loaded automatically onto an indexing table by actuator A. The disc is then clamped by actuator B and the first drilling operation is accomplished with the drill actuator C. Indexing (with actuator D) and repeated drilling then must occur five times, thus drilling a hole every 60°. When the last indexing motion is completed, the workpiece must be unclamped with actuator B and automatically removed with actuator A. After the start command is given, a new sequence program may begin (see fig. 8–46).

Step-counter for selective parallel programs

Machine controls with several programs are frequently encountered on machine tools and multipurpose machines. Program selection and program start is made with a "program selector circuit" which may be designed either with the binary or with the cascade method (figs. 8–51 to 8–53).

With parallel selective control, each individual program consists of an independent circuit which may or may not have anything in common with its sister circuits. The sequence used to illustrate this control method in fig. 8–47 is as follows:

Prog. I → A1–B1–C1–C0–B0–A0

Prog. II → A1–B1–C1–C0–D1–C1–C0–B0/D0–A0

Since the last step of either program is maintained until reset by step 1, and therefore the signal A0 issued by this step is also maintained, an impulse valve must be used in line A0 to avoid an opposing signal when signal A1 of the alternative program appears (see fig. 8–47 and figs. 8–2, 8–3, 2–45 and 2–47).

The selective parallel program circuit depicted in fig. 8–47 is also furnished with a simple cycle selection module which replaces the start valve, and two visual indicators display the program in selection.

An other way of achieving selective parallel programs is depicted in fig. 8–48. Here the common steps A1–B1–C1–C0 are grouped together from both programs. Then a program selector shunts the preparation signal through the uncommon group consisting of the signals D1–C1–C0. Leaving this group, the preparation signal merges via an "OR" valve with the other preparation signal branch and actuates the final group consisting of the signals B0/D0–A0 which may again be regarded as common for both parallel programs. Although the signal D0 is not required for program I, its presence at step 8 causes no harm, as the actuator D is already retracted and was anyhow not used in program I.

Although somewhat more complex than the control method used in fig. 8–47, this kind of control is definitely less expensive as it requires only 9 step-counter modules.

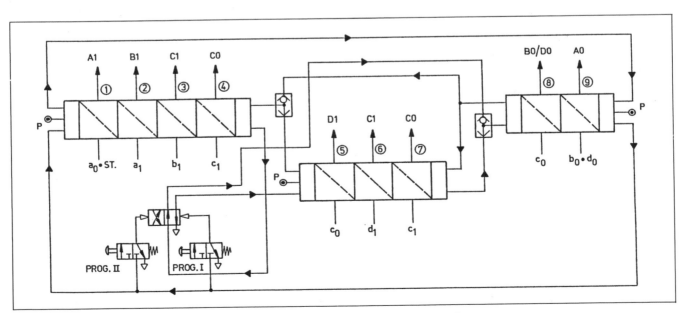

Fig. 8–48 Selective parallel program circuit (grouped approach with common steps).

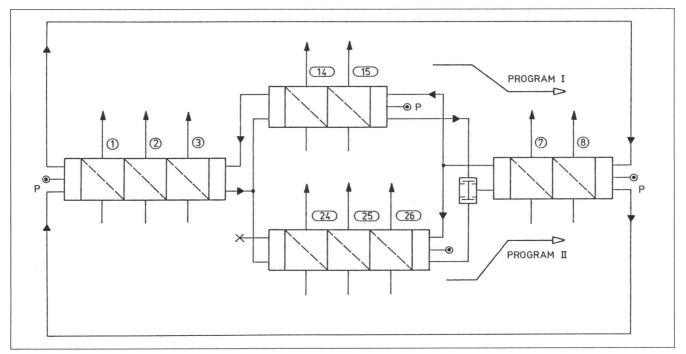

Fig. 8-49 Step-counter for simultaneous parallel programs.

Step-counter for simultaneous parallel programs

Simultaneous parallel controls are often found on transfer and indexing table machines, where each machining station has its own individual program, but the feed-in, transporting, indexing and clamping as well as loading of the work-pieces (components) is a common program. For such application one may select a control as shown by the circuit in fig. 8-49 and a real application for such a control is given in figs. 8-55 to 8-58.

Simultaneous parallel programs may consist of several branches, not just two as shown in fig. 8-49, and may or may not have a common program at the head-end or tail-end of the total cycle. But in each case the preparation signal of all the branches must be collected with "AND" functions, and only one reset signal is required to reset the head-end common line. The reset signals from the other branches must be blocked (fig. 8-49).

Step-counter for parallel programs but different actuator start positions

Control circuits of this nature are almost identical to the controls shown in figs. 8-47 and 8-48. The

Fig. 8-50 Step-counter for parallel programs but different actuator start positions.

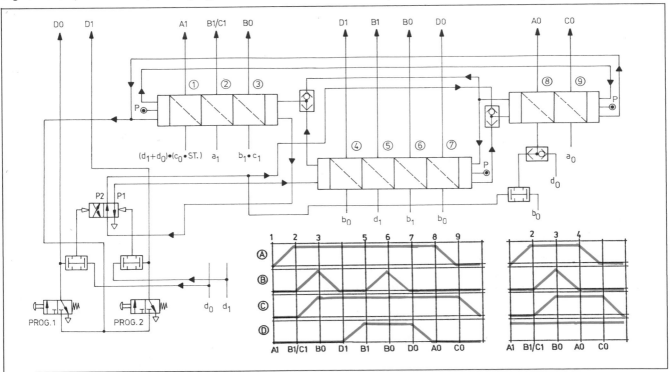

start position for actuator D differs, however, and actuator D in program 2 does not move at all.

It remains extended throughout the entire cycle (fig. 8–50). To accomplish the rectification of actuator D into its new start position, the program selector circuit must be modified as shown in fig. 8–50. When program change is signalled, actuator D will receive a rectification command. This rectification is then confirmed with the ''AND'' valve and, together with the program change signal, causes the program selector memory to shift. The program selector push button pilot valves are series connected to the preparation signal for step-counter module 1. Thus a program change can only be signalled if the cycle is completed.

Program selector circuits

Program selector circuits are closely related to the cycle selection module shown in fig. 8–27. The sole purpose of the program selector circuit is to shunt the start signal to the selected line of step-counter modules in the parallel program (fig. 8–47) and maintain that program selection until another selection is made.

If a step-counter with defined skipping is used to provide alternative programs, then the program selector circuit is used to make or cancel the skipping signal (fig. 8–45).

Program selector circuits may range from a simple hand lever operated memory valve to multiple input and remotely controlled selector subcircuits which may be designed either with the

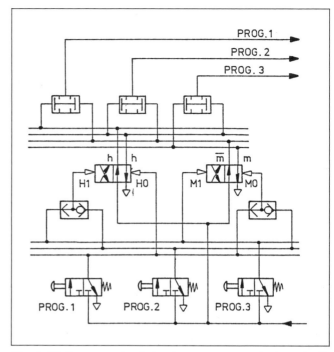

Fig. 8–52 Program circuit for three programs designed with binary methods.

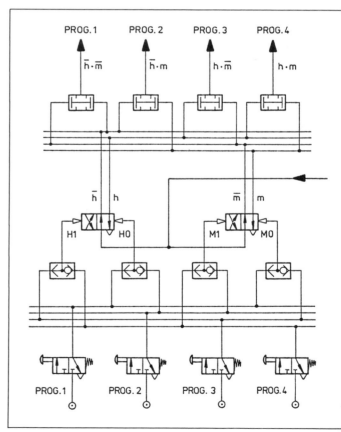

Fig. 8–53 Program selector circuit for four individual control programs.

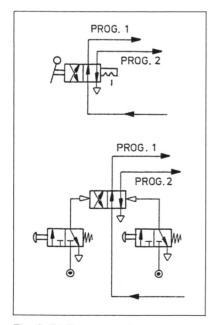

Fig. 8–51 Program selector circuits for two programs.

binary logic or the cascade method (fig. 8–51 to 8–54).

The program selector circuit in fig. 8–51A uses a hand lever valve with detent to shunt the program selection signal to the selected line of the step-counter program. The program selector circuit in fig. 8–51B is remotely controlled, with two pneumatic pilot signals acting alternatively onto the memory shunt valve.

The program selector circuit in fig. 8–52 is designed with the binary method (see Chapter 10, fig. 10–30) and can shunt the selector signal to either of three step-counter lines. A binary program selector circuit for four program choices is shown in fig. 8–53.

Step-counter with stepping module

For non-stepping operation, the stepping selector module provides a permanent signal to all stepping "And" function valves. For stepping operation the select step valve provides the operator-dependent stepping signal to be given for each step. This module and circuit arrangement is mandatory when toolsetters are preparing a machine after changing tools or after necessary maintenance work.

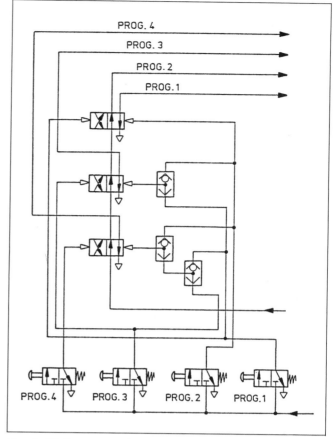

Fig. 8–54 Program selector circuit designed with cascade concept.

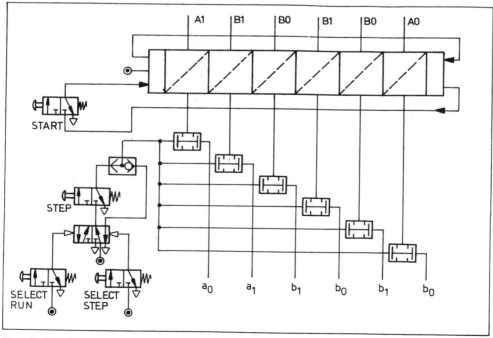

Fig. 8–55 Step-counter with choice of manual stepping or normal cycling (running).

9

The Cascade System of Pneumatic Sequential Control

In Chapter 8 the problem of opposing signals has been discussed, and how the step-counter circuit design method overcomes such opposing signals is outlined on page 96. The cascade method is also designed to remove opposing signals and its structure is closely related to the step-counter method.

The first cascade method was invented and introduced to the field of pneumatic sequential control by Arthur M. Salek, a mechanical engineering graduate of Canterbury University, New Zealand. His paper "Pneumatic Mechanisation for Industry", was published in *New Zealand Engineering* October 15, 1954. Mr. Salek's circuit design method became widely used in New Zealand, U.S.A. and Great Britain and was updated in a later paper in 1961. The cascade method has since been modified many times, and four popular versions (including Mr. Salek's version) are given in this chapter.

The step-counter circuit design method as outlined in the previous chapter allocates a memory valve to *each* sequence step. Thus opposing signals can never occur, since each step-counter module resets the memory valve in the previous step-counter module. The essential confirmation signal relayed through that memory valve is thus automatically cancelled and can no longer cause an opposing signal.

The cascade circuit design method uses fewer memory valves than the step-counter method, and opposing signals are eliminated when the cascade in operation resets the previous cascade. A detailed design and operation procedure for the cascade is given through the following pages.

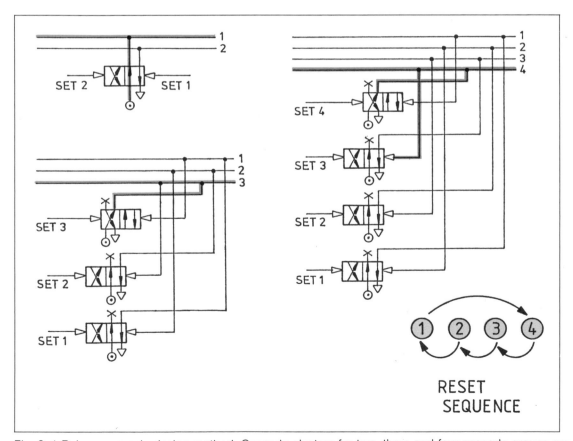

Fig. 9–1 Rohner cascade design method. Cascade clusters for two, three and four cascade groups are shown.

Rohner cascade design method

The "Rohner" cascade design method (design by Peter Rohner) uses one cascade memory valve (Group memory, or GM) for a two cascade group sequence program and thereafter one cascade memory for every cascade group (fig. 9–1). The "Rohner" design method is extremely simple but requires above two cascade groups one more cascade memory than the "MARTONAIR" method (fig. 9–5) or "FESTO" method (fig. 9–4). This, however, is well compensated for by the increased safety aspects of the "Rohner" method (no series and no series-parallel connection to the other cascade memories). The "Rohner" method is also simpler for cluster design and connection to other circuit parts, and because it is not series connected it is easier to faultfind and include emergency stop control.

Figure 9–2 shows how the cascade cluster can also be arranged horizontally, which then proves its remarkable similarity to the step-counter method. With the cascade design method, each cascade memory valve (group memory) performs three main functions (fig. 9–3).

- Function 1 is to switch the power valves either directly from the cascade group output signal or to switch them in "AND" connection with an essential confirmation signal (limit valve signal, see fig. 9–10).

- Function 2 is to reset the previous cascade memory valve. This then automatically cancels all opposing signals arising from that cascade group.

- Function 3 is to prepare the next cascade group set-signal which is "AND" connected to the last essential confirmation signal of the present cascade group. Thus one may conclude that these cascade valves function similarly to the step-counter modules explained and illustrated in Chapter 8, fig. 8–5. A detailed construction algorithm is given through the following pages.

Alternative cascade design methods

Both the "Salek" and "MARTONAIR" cascade design methods are based on the use of five port memory valves and utilize the exhaust flow ports R and S as input ports. Therefore these two methods require spool type valves rather than flat slide valves (figs. 2-4, 9–5 and 9–6).

The "Rohner" and "FESTO" methods can be built from four or five port valves, and both flat slide

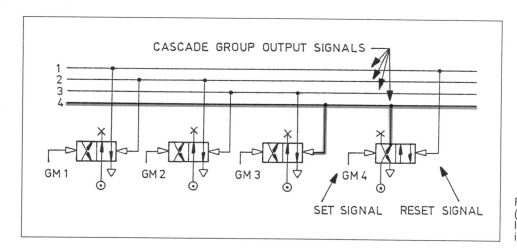

Fig. 9–2 The simple cascade (Rohner method) has only parallel connection and is shown here in horizontal arrangement.

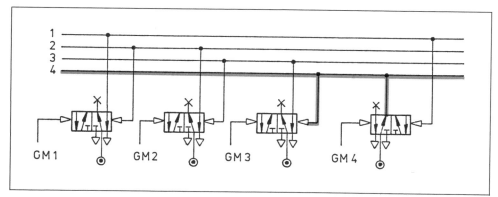

Fig. 9–3 Cascade cluster (Rohner method) built with five-port group memory valves.

or spool type valves may be used (figs. 9–2, 9–3 and 9–4).

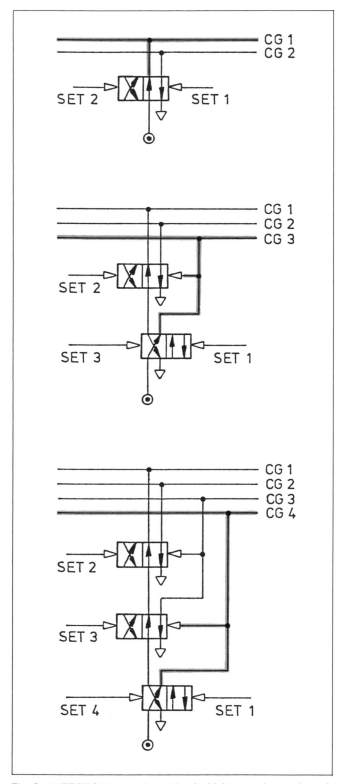

Fig. 9–4 "FESTO" cascade method with long series and parallel connection (four- or five-port valves may be used).

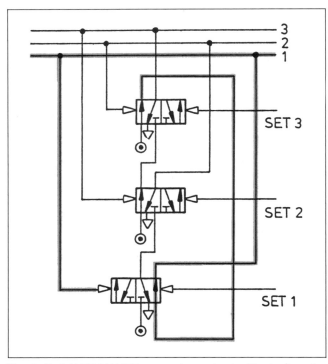

Fig. 9–5 "MARTONAIR" cascade method with short series and parallel connection similar to the "Salek" and "FESTO" methods (see figs. 9–4 and 9–6).

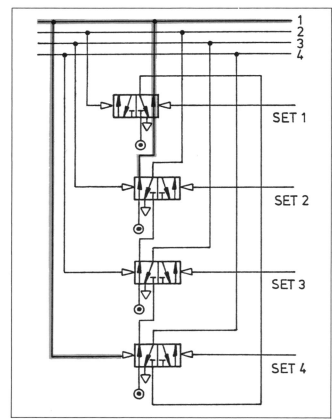

Fig. 9–6 "Salek" cascade method with short series and parallel connection.

Grouping procedure for cascade groups

With the cascade circuit design method the sequence program is divided into cascade groups. These groups are formed in such a way that no group contains an opposite switching signal (for example: A0, A1). When dividing the sequence program into cascade groups, the dividing should always be attempted from both sides (left to right, and thereafter right to left). Sometimes one approach produces less groups than the other, and thus less cascade valves have to be used. One must also remember that a cycle borders onto the beginning of the next cycle. When considering continuous cycling, cycle interruption is only made possible because of the start valve being placed into the switching command line for sequence step 1. Hence, a cycle should be regarded as a part of a continuous production program and therefore a cascade group running to the end of the sequence cycle may continue into the next cycle (over and beyond the start of the next cycle). To illustrate this, the following sequence is investigated:

A1–B1–C1–B0–D1–D0–A0–E1–E0–C0

Dividing the cycle from the front, one obtains:

![Division from the front: group 1 = A1–B1–C1, group 2 = B0–D1, group 3 = D0–A0–E1, group 4 = E0–C0]

Note: The final group out of these four cascade groups cannot be joined to the first group in the program, since there would be a C0 command and its opposite C1 command in the same group! By splitting the cycle from the rear, one obtains:

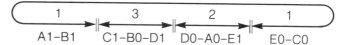

With the grouping pattern above one can join the last two signal commands in the cycle to the first two commands without having opposite commands within one group. Thus dividing from the rear, for this program, produced a simpler result—only three cascade groups instead of four with the first dividing approach!

Cascade circuit construction

Once the grouping procedure is completed and the minimum number of required cascade groups is determined, one can draw the cascade valve cluster as shown in figs. 9–1, 9–2 or 9–3. To illustrate we assume a program sequence as shown in the traverse-time diagram given in fig. 9–7, for which four cascade groups have been formed. The first signal command in each cascade group is now directly attached to the cascade group output

Fig. 9–7 Drilling machine.

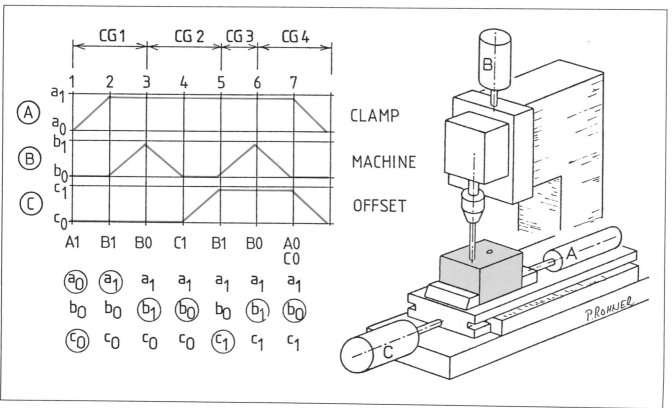

signal. Consecutive signal commands of the *same* group are constructed as an "AND" function. The source signals of this "AND" function are:

(a) The *essential* confirmation signal which confirms the completion of the previous sequence step (see transverse-time diagram).

(b) The *cascade group signal* which ensures that the essential confirmation signal is only active at the correct sequence step (fig. 9–9). Signal command C1 for example is switched by b_0 and so are the signals A0/C0. Signal C1 can only be switched by b_0 if cascade group 2 in "ON" and signals A0/C0 can only be switched by b_0, if cascade group 4 is "ON" (see switching equations and fig. 9–10).

To illustrate: Signal command A1 at step 1 is *directly* attached to cascade group signal CG1, but the next signal command in the sequence (B1) consists of the cascade group signal CG1 *and* the essential confirmation signal a_1 (see figs. 9–7 to 9–9 and equations). Similarly the first signal command in cascade group two (CG2), which is the command B0, is again *directly* attached to its cascade group output signal (CG2). The sources of the next signal command in group two are again "AND" connected and consist of the signals CG2 and b_0 (see also switching equations for this program sequence).

Cascade Memory Set Commands:

$GM1 = CG4 \cdot a_0 \cdot c_0 \cdot Start$
$GM2 = CG1 \cdot b_1$
$GM3 = CG2 \cdot c_1$
$GM4 = CG3 \cdot b_1$

D.C.V. Pilot Signals:

$A1 = CG1$
$A0 = CG4 \cdot b_0$
$B1 = CG3 + CG1 \cdot a_1$
$B0 = CG2 + CG4$ Identical "AND"
$C1 = CG2 \cdot b_0$ function*
$C0 = CG4 \cdot b_0$

a_1	1x	$CG1 \rightarrow a_1$
a_0	1x	$CG4 \rightarrow a_0$
b_1	2x	—
b_0	3x	$CG4 \rightarrow b_0$ *
c_1	1x	$CG2 \rightarrow c_1$
c_0	1x	$CG4 \cdot a_0 \rightarrow c_0$
ST.	1x	$CG4 \cdot a_0 \cdot c_0 \rightarrow Start$

Fig. 9–8 Series function chart.

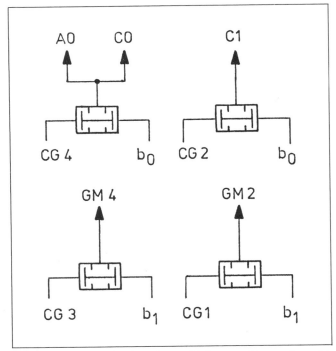

Fig. 9–9 Four "AND" functions which cannot be series connected (see equations).

The cascade memory valves in the cascade cluster are "SET" at the beginning of each cascade group and their "SET" commands may be compared to step-counter module "SET" commands (see fig. 8–6).

Hence, these set commands consist of the previous cascade group signal which may be regarded as the preparation signal. This preparation signal is then "AND" connected to the last essential confirmation signal of the previous cascade group (that sequence step which ends when the new cascade group begins). To illustrate: the cascade set command for the cascade group memory 2 requires signal CG1 *and* the essential confirmation signal b_1. To set cascade group memory 1, the preparation signal CG4 of the previous cascade group and the double confirmation signal of the successful completion of the previous sequence step is required. These signals must also be "AND" connected to the start valve (see switching equations and fig. 9–10).

"Series" functions and identical "AND" functions

The same principles and procedural steps outlined in Chapter 7 for the formation of identical "AND" functions as well as "Series" functions may also be used for the cascade circuit design method. In figures 9–7, 9–8 and 9–9 and their corresponding switching equations, provision is made for a typical

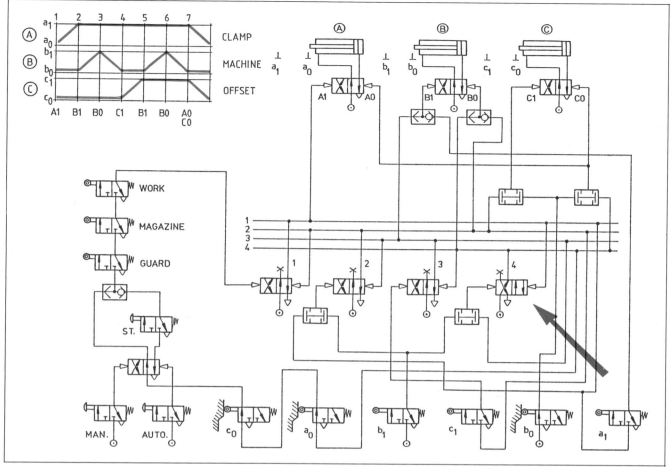

Fig. 9–10 Cascade control circuit for drilling machine shown in figs. 9–7, 9–9, 9–8 and its switching equations shown on page 128. The circuit is also equipped with a cycle selection module and some fringe starting conditions. Note: Group memory 4 remains set until being reset by group memory 1.

application, where the circuit hardware may drastically be reduced if "Series" and identical "AND" functions can be formed. Thus a circuit such as fig. 9–10 can be built with four instead of ten "AND" valves (as shown in the switching equations for this sequence).

Emergency stop for cascade circuits

Emergency stop integration for control circuits designed with the cascade circuit design method is often far more complex and difficult than for the step-counter design method discussed and illustrated in Chapter 8.

Limit valves frequently must be "AND" or "Series" connected to output signals from the cascade memory valves (fig. 9–10, in particular limit valves a_0, c_1 and a_1). Furthermore, cascade memory valves require extra "OR" function valves if the cascade cluster must be reset to "ZERO" (see figs. 8–22, 8–34, 8–35, 8–36 and 8–39 of the previous chapter). These "OR" function valves are already integrated into the step-counter modules if one decides to use commercial modules, but for cascade design such modules are not available.

In control problems where an emergency program is required to sequence the actuators to the end-of-cycle position, one requires limit valve signals to confirm the emergency steps (see fig. 8–36). If these limit valves are series connected to the cascade valves (see fig. 9–10), they are useless, since resetting the cascade valves would render them without air supply. Thus one cannot connect limit valves in series; "AND" function valves must be used (see fig. 9–9).

With all these extra "OR" and "AND" function valves, the hardware cost difference for step-counter versus cascade designed control problems becomes negligible. Step-counter control is preferable overall, due to its simplicity of circuit design, ease of faultfinding and ready application for the programming of PLC electronic controllers (see Chapter 11).

Should the decision still be made to use the cascade circuit design method in conjunction with an emergency stop provision, then the principles discussed and illustrated in the previous chapter would also apply here (see figs. 8–22 to 8–36). These limitations of cascade circuits of course only hold true for control problems with emergency stop integration! For other problems, cascade circuits, although more complex than step-counter circuits, often prove to be less expensive. Three totally different control problems are presented below, to demonstrate the different advantages of step-counter and cascade methods.

Control problem 1

The control problem given here is based on a machine sequence already discussed in the previous chapter (fig. 8–32). The sequence for this control problem is: A0–B1–B0–C1/B1–C0–A1/B0. Dividing this sequence into cascade groups is not complicated and produces the following pattern:

$$\overset{1}{\underset{A0-B1}{\longleftrightarrow}} \Vert \overset{2}{\underset{B0}{\longleftrightarrow}} \Vert \overset{3}{\underset{C1/B1}{\longrightarrow}} \Vert \overset{4}{\underset{C0-A1/B0}{\longrightarrow}}$$

If one chooses to use either the "Rohner" or the "MARTONAIR" cascade design method, then four cascade memories will be required for these four cascade groups, and the following equations may be derived from the traverse-time diagram (fig. 8–32):

Cascade Memory Set Commands

$GM2 = CG4 \cdot a_1 \cdot b_0 \cdot Start$
$GM2 = CG1 \cdot b_1$
$GM3 = CG2 \cdot b_0$
$GM4 = CG3 \cdot c_1 \cdot b_1$

D.C.V. Pilot Signals

$A1 = CG4 \cdot c_0$
$A0 = CG1$
$B1 = CG1 \cdot a_0 + CG3$ Identical "AND"
$B0 = CG2 + CG4 \cdot c_0$ function*
$C1 = CG3$
$C0 = CG4$

a_1	1x	$b_0 \rightarrow a_1$
a_0	1x	$CG1 \rightarrow a_0$
b_1	2x	—
b_0	2x	—
c_1	1x	$b_1 \rightarrow c_1$
c_0	2x	$CG4 \rightarrow c_0$*
ST.	1x	$(b_0 \cdot a_1) \rightarrow Start$

Fig. 9–11 Series function chart. Although limit valve c_0 appears twice it can be series connected, since its signal is used in an identical "AND" function (see fig. 9–12).

The step-counter control circuit designed for the same sequence requires six step-counter modules

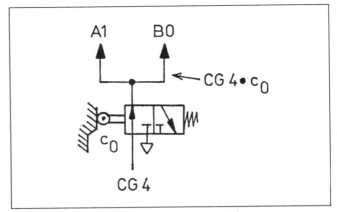

Fig. 9–12 Identical "AND" function for $CG4 \cdot c_0$.

and a "Series" function can be formed for all "AND" functions found in the switching equations. Hardware cost calculations based on commercial step-counter modules, when compared to the four cascade memories and four "AND" function valves for the cascade design, show equal total costs for both designs. The circuit construction costs, however, would be slightly less for the step-counter circuit (see figs. 9–13 and 9–14).

Control problem 2

A circuit is to be designed for a machine which punches holes into aluminium window frames. Two different types of frames must be manufactured with the same machine. The window frames are manually loaded and unloaded to and from the machine.

Type 2 frame requires six holes to be punched, but type 1 frame requires every second hole to be punched—that is, 3 holes. Actuator A clamps the window frame and holes are punched by actuators B1 to B6. A program selector circuit is used to select the punching type (all six, or every second hole, see figs. 9–15 to 9–17).

All actuator end positions must be *sensed* for complete motion confirmation prior to consecutive actuator movements. Control circuits must be designed and drawn using the cascade and the step-counter design methods to permit evaluation of the most suitable method.

Cascade Memory Set Commands:

$GM1 = GM2 \cdot a_0 \cdot Start$
$GM2 = GM1 \cdot b_{1(ODD)} + GM1 \cdot b_{1(ALL)}$

D.C.V. Pilot Signals:

$A1 = GM1$
$A0 = GM2 \cdot b_0$
$B1 = CG1 \cdot a_1 \cdot PROG. 1 (ACT. 1,3,5)$
$B1 = CG1 \cdot a_1 \cdot PROG. 2 (ACT. 2,4,6)$
$B0 = CG2$

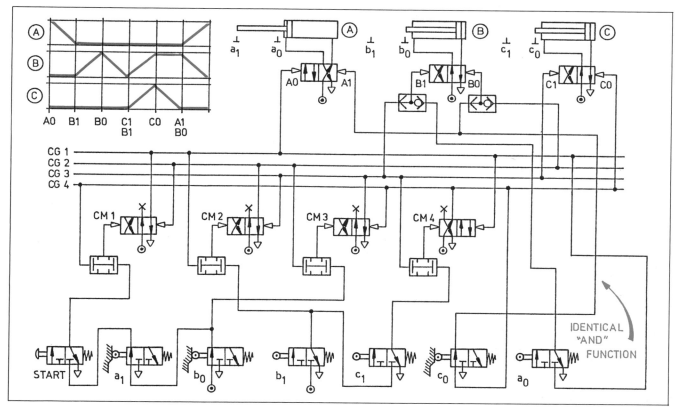

Fig. 9–13 Cascade control circuit to control problem 1.

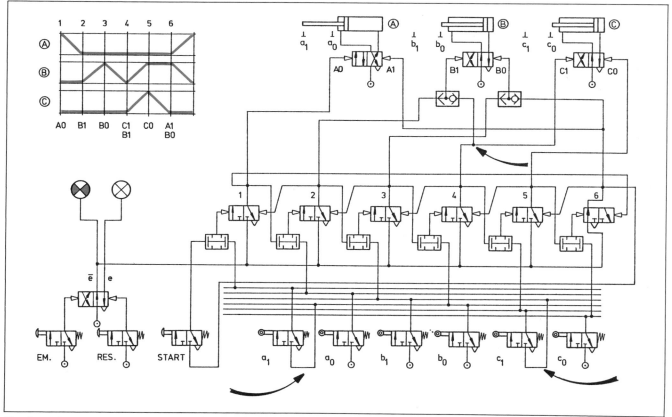

Fig. 9–14 Step-counter control circuit for control problem 1.

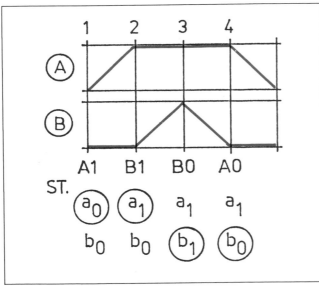

Fig. 9–15 Traverse-time diagram for control problem.

Design Concepts

The use of a restrictor memory ⓡ, when program 1 is selected, inhibits the B1 signal command from reaching the power memories for the punching actuators B2, B4 and B6 to extend, and thus number 2, 4 and 6 holes are not punched with program 1.

The use of a shunting memory Ⓢ, when program 1 is selected, directs the combined b_1 signals of the actuators B1, B3 and B5 to GM2 without being "AND" connected to the other b_1 confirmation signal from the actuators B2, B4 and B6.

When program 2 is selected, the shunting memory directs the combined b_1 signals of the actuators B1, B3 and B5 to the "AND" function valve to be "AND" connected to the combined b_1 signals from the other B actuators, so that *all* six B actuators must be fully extended before they are permitted to retract.

Fig. 9–16 Cascade circuit for control problem 2.

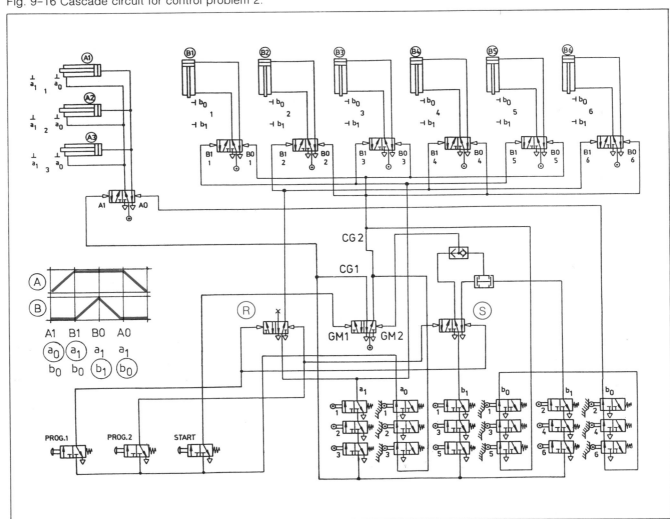

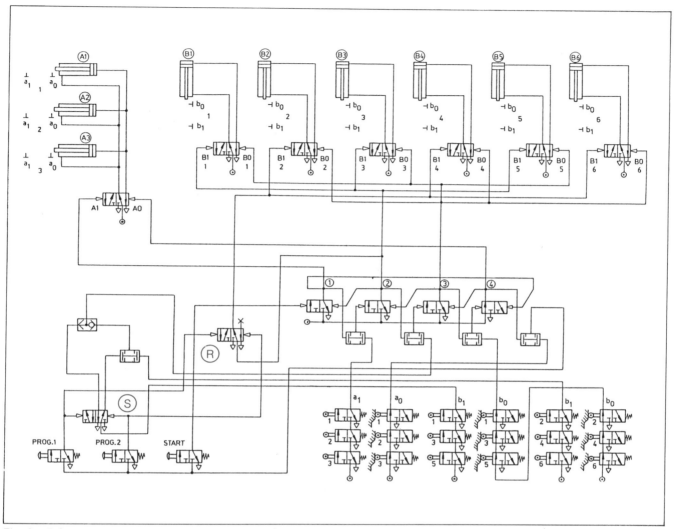

Fig. 9–17 Step-counter circuit for control problem 2.

Review and Comparison of design concepts for control problem 2

Both design concepts employ a shunting memory to direct the essential confirmation signals $b_{1 (ODD)}$ directly to the "OR" function valve. However, if program 2 is selected, these $b_{1 (ODD)}$ signals are shunted to the "AND" function valve where they must be "AND" connected to the $b_{1 (EVEN)}$ signals. Thereafter these b_1 signals are also gathered by the "OR" function valve and are also directed to their destination to command the B actuators to retract. The restrictor memory is also used for both design concepts for the same purpose as explained previously.

The step-counter method uses 4 step-counter modules, whereas the cascade method gets by with only one cascade memory to cater for the two cascade groups. Hence, the cascade method is

less expensive than the step-counter method for this particular control.

Control problem 3

A circuit is to be designed for the control of a silk-screen printing machine which prints and cuts paper labels from paper reels. The paper transport actuator Ⓒ pulls the paper off the reel and feeds it through the printing machine. The paper is gripped by vacuum suction cups. The venturi vacuum generator Ⓥ produces the vacuum for the suction cups. Actuator Ⓐ moves the ink-spreader and actuator Ⓑ operates the guillotine.

There are two machine cycles within the same control program. That is, two labels are produced during the 12 sequence steps of this program. Actuator Ⓐ (the ink-spreader) makes only one stroke per label (cycle), but a full extension and

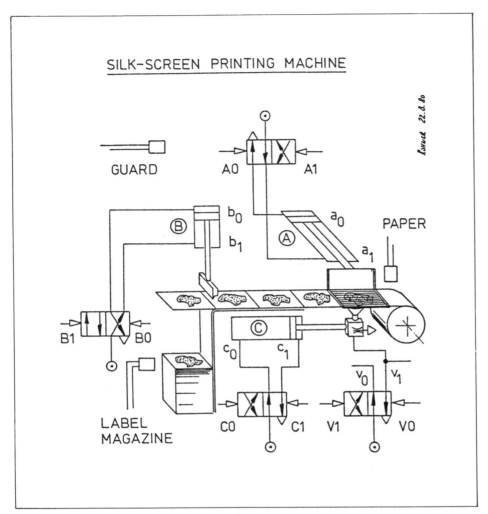

Fig. 9–18 Machine layout for silk-screen printing machine (see also traverse-time diagram Fig. 9–19).

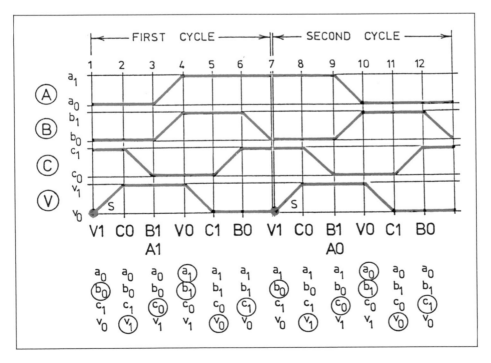

Fig. 9–19 Traverse-time diagram for silk-screen printing machine.

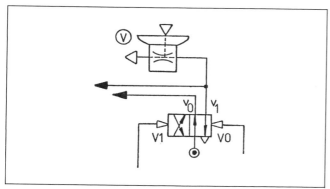

Fig. 9–20 Venturi vacuum generator with suction cup and control valve.

retraction movement for two labels (see traverse-time diagram fig. 9–19 and its associated machine layout fig. 9–18). Designation V in the traverse-time diagram represents the vacuum which must be switched "ON" for the retraction of actuator © and again be switched "OFF" prior to its extension. Figure 9–20 shows how confirmation signals can be achieved from the vacuum control memory. Signal v_0 indicates vacuum "OFF" and signal v_1 vacuum "ON". One may use these signals in the same manner as limit valve signals actuated by the pneumatic cylinders.

The sequence control circuit must include a start restriction control circuit to provide selection and start of the following sequencing modes:

- Push-button selection for sequencing through a complete program: consisting of all twelve sequence steps, and producing two labels (see traverse-time diagram fig. 9–19). The program selector valve selects the program mode, and then the common start valve must be actuated to initiate the program (see figs. 9–21 and 9–22).
- Push-button selection for sequencing through one cycle only (one label only). The cycle selector valve selects the single cycle mode (first or second cycle), then the common start valve must be actuated for each cycle initiation.
- Push-button selection for automatic and continuous sequencing through all twelve sequence steps, with automatic initiation of further programs. To start the first program after automatic cycling has been actuated, the common start valve must be simultaneously actuated.
- Upon actuation of the cycle selector push-button or the program selector push-button, the memorised automatic sequencing selection must be cancelled (fig. 9–21).

Design Concepts

The "MARTONAIR" cascade method has been chosen to design this control circuit. Dividing the sequence into cascade groups produces the following pattern:

$$\overset{1}{\underset{\text{V1-C0-B1/A1}}{\longmapsto}} \;\; \overset{2}{\underset{\text{V0-C1-B0}}{\longmapsto}} \;\; \overset{3}{\underset{\text{V1-C0-B1/A0}}{\longmapsto}} \;\; \overset{4}{\underset{\text{V0-C1-B0}}{\longmapsto}}$$

With the "MARTONAIR" cascade method four cascade memories must be used and the following equation may be derived from the traverse-time diagram given in fig. 9–19.

Cascade Memory Set Commands:

$GM1 = CG4 \cdot b_0 \cdot \text{Start}$
$GM2 = CG1 \cdot b_1 \cdot a_1$ Identical "AND"
$GM3 = CG2 \cdot b_0 \cdot \text{Start}$ function*
$GM4 = CG3 \cdot b_1 \cdot a_0$

Identical "AND" functions may be found for the following functions:

$b_0 \cdot \text{Start}$

$CG1 \cdot c_0$

$CG3 \cdot c_0$

D.C.V. Pilot Signals:

$A1 = CG1 \cdot c_0$ *
$A0 = CG3 \cdot c_0$
$B1 = CG1 \cdot c_0 + CG3 \cdot c_0$ *
$B0 = CG2 \cdot c_1 + CG4 \cdot c_1$ Identical "AND"
$C1 = CG2 \cdot v_0 + CG4 \cdot v_0$ function*
$C0 = CG1 \cdot v_1 + CG3 \cdot v_1$
$V1 = CG1 + CG3$
$V0 = CG2 + CG4$

Series functions may be formed for:

$b_1 \rightarrow a_1 ; \; b_1 \rightarrow a_0 ; \; b_0 \rightarrow \text{Start}$

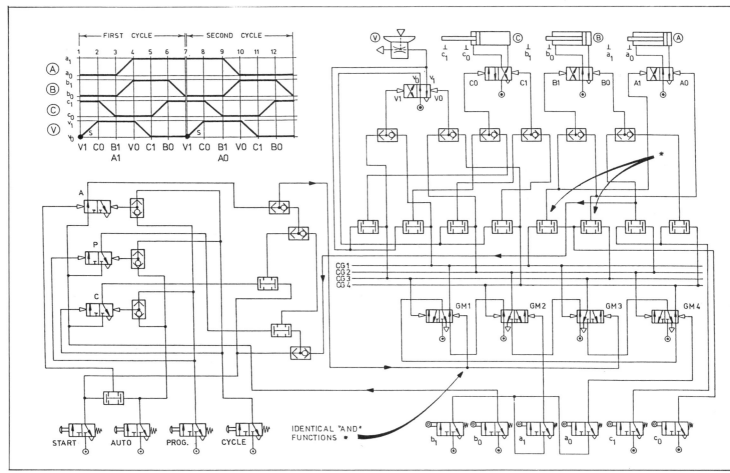

Fig. 9–21 Circuit for control problem 3 designed with cascade method.

Fig. 9–22 Circuit for control problem 3 designed with step-counter method.

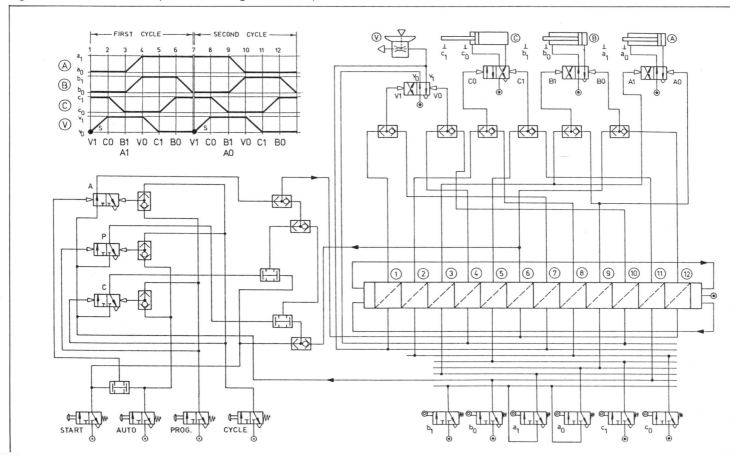

10 Combinational Circuit Design

Combinational circuit output signals depend solely on the momentary state of their input signals or input variables. Time-delay and signal memorisation are not taken into consideration in combinational circuits. This fact makes combinational circuits relatively simple to design. The tools used to design them are:

- Boolean algebra (see page 96 and fig. 8–7)
- Truth tables (see page 139 and fig. 10–4)
- Karnaugh-Veitch maps (see page 142 and fig. 10–9)

Each of these design tools is explained and applied separately and in detail, and application exercises are presented. Combinational circuits are best designed using the following eight logic construction steps:

1. Analyse *all* input signal sources (valves) and determine all output signal destinations (signal commands) and allocate appropriate letters to these signals.
2. Construct and draw a truth table with columns for all input signals and a column for the required output signal (S).
3. Analyse *all* input combinations for those which must provide an output signal. All the remaining combinations must inhibit the output signal.
4. Construct and draw a Karnaugh-Veitch map, then plot all logic 1 outputs from the truth table into the fields of the Karnaugh-Veitch map and label them.
5. Loop all labelled entries in the Karnaugh-Veitch map and closely adhere to the looping rules listed on page 143.
6. Extract the minimised switching equations from the loops in the Karnaugh-Veitch map, and combine them (by "OR" connection) into a collective equation.
7. Knowledge and application of Boolean algebra, particularly the Distributive Law, may bring a further minimisation of the extracted equations obtained in logic step 6.
8. Find where possible series function formations (see Chapter 8, figs. 8–13 and 8–15) which may render a further hardware reduction and draw the graphic symbol circuit.

Boolean logic concepts

To understand the peculiarities of logic switching, whether this be applied to pneumatic or electric or electronic control makes no difference, one must understand some basic rules of Boolean algebra.

These rules are now presented in simple and abbreviated form and no attempts have been made to prove their origin or consecutive logic development.

Boolean postulates

$\bar{1} = 0$	$\bar{0} = 1$
$0 \cdot 0 = 0$	$0 + 0 = 0$
$0 \cdot 1 = 0$	$0 + 1 = 1$
$1 \cdot 0 = 0$	$1 + 0 = 1$
$1 \cdot 1 = 1$	$0 + 0 = 0$

Boolean theorems (for one input signal)

$A \cdot 0 = 0$	$A + 1 = 1$
$A \cdot 1 = A$	$A + 0 = A$
$A \cdot A = A$	$A + A = A$
$A \cdot \bar{A} = 0$	$A + \bar{A} = 1$

Theorems for more than one input signal

Commutative laws

$$A \cdot B = B \cdot A \qquad A + B = B + A$$

Associative laws

$$(A \cdot B) \cdot C = A \cdot (B \cdot C) = A \cdot B \cdot C$$
$$(A + B) + C = A + (B + C) = A + B + C$$

Distributive law

$$(A \cdot B) + (C \cdot B) = B \cdot (A + C) \text{ or } (A + C) \cdot B$$
$$(A + B) \cdot (C + B) = B + (A \cdot C) \text{ or } (A \cdot C) + B$$

Absorption law

$$A + A \cdot B = A$$

De Morgan's theorem

$$\overline{A \cdot B \cdot C} = \bar{A} + \bar{B} + \bar{C} \qquad \overline{A + B + C} = \bar{A} \cdot \bar{B} \cdot \bar{C}$$

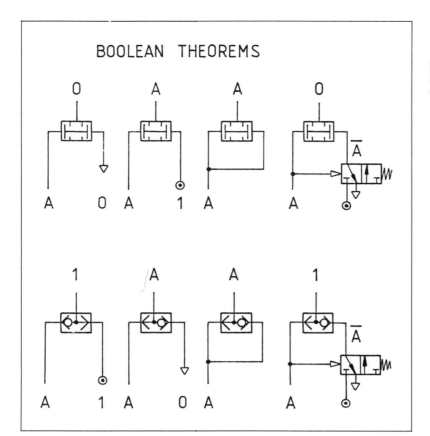

BOOLEAN THEOREMS

Fig. 10–1 Boolean theorems as applied to Boolean algebra for logic combinational and sequential circuit design.

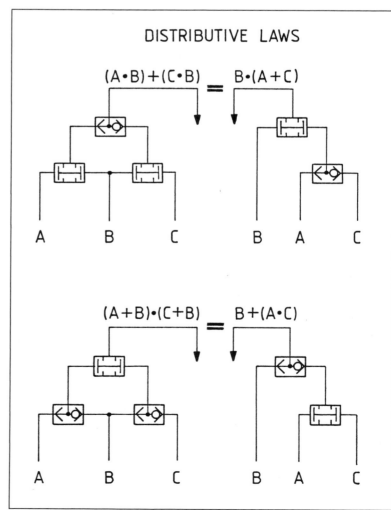

DISTRIBUTIVE LAWS

$$(A \cdot B) + (C \cdot B) = B \cdot (A + C)$$

$$(A + B) \cdot (C + B) = B + (A \cdot C)$$

Fig. 10–2 The distributive law is one of the most important laws in Boolean algebra since it lends itself ideally to minimise pneumatic "AND" and "OR" valves (see also Chapters 8 and 9). This law is an absolute "MUST" for the pneumatic circuit designer and control engineer.

To illustrate some of these laws and theorems it may be best to apply them to simple pneumatic valves and subcircuits, such as figs. 10–1, 10–2 and 10–3.

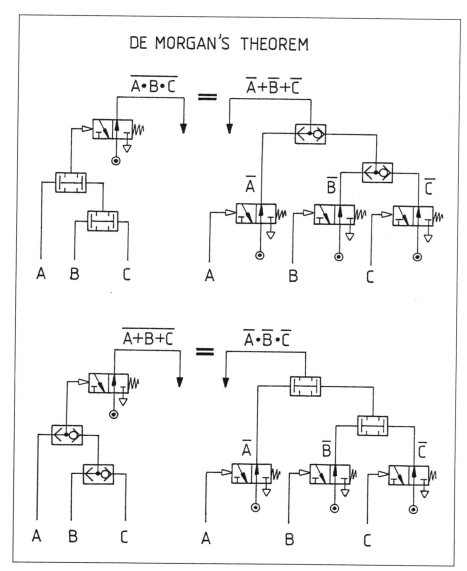

Fig. 10–3 De Morgan's theorem is also an ideal tool for circuit simplification. It may also be used on some programmable logic electronic controllers with limited "AND" input functions but multiple "OR" input instead (use inversion function and De Morgan's theorem).

Truth table construction and usage

To achieve a better understanding of the problem of combinational input relationships, links of the input signals should be tabulated. The table used for such tabulation or grouping is called a value table or, more commonly, a truth table (fig. 10–4).

A truth table consists of columns (vertical) and rows (horizontal). Columns hold input signals which may be labelled x_1, x_2, x_3 and the extreme right hand column shows the resulting output signal for these "AND" connections. Input signals may of course also be labelled with whatever letter is best suited for these signals. Therefore, if a logic function (which is in fact a small combinational circuit, see figs. 10–8 and 10–14) has two input signals, then its truth table will show three columns and four rows. The number of rows is determined by the number of "binary" combination possibilities achieved for its input signals. This number is given the letter K and is calculated with the following formula:

$$K = 2^n \quad (n = \text{input signals, } K = \text{rows})$$

Each input signal in the "binary" system can only have two distinct values. It can either be in the "ON" state (which means a signal is produced by the valve or switch), or in the "OFF" state (which means that no signal is produced by the valve or switch—see figs. 10–5 to 10–6).

TRUTH TABLE

a	b	c	A1
0	0	0	0
0	0	1	0
0	1	0	0
0	1	1	1
1	0	0	0
1	0	1	1
1	1	0	1
1	1	1	1

Fig. 10–4 Truth table for three input signals.

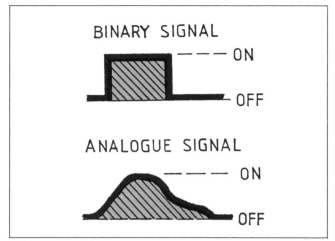

Fig. 10–5 Comparison between binary and analogue switching behaviour.

Pneumatic and electric switchgear have predominantly binary switching characteristics. A valve is said to have "analogue" switching characteristics when it can be selected into any position ranging from fully closed to fully open and/or any in between position. An adjustable orifice or flow control valve is a typical analogue valve and so is a water tap which is slowly opened or closed. A push-button or roller valve, however, which is rapidly actuated into its "ON" or "OFF" position, is effectively a binary valve.

Valves which are used for extremely fast acting combinational logic functions must possess positive overlap for their crossover movement (see Chapter 2 and fig. 10–6). Should for example a valve with negative overlap be used in an inhibition function then the input signal A could still reach the output (S) despite the presence of the inhibition signal B (see fig. 10–8).

Dual coding of truth tables

To bring order and method into the truth table we list all input signals with the binary "DUAL" code. The dual code is also used for tabulation of decimal indications in binary counting systems (see fig. 10–7). Output signals (values) in the far right column of the truth table may be called true (logic 1) or false (logic 0), or optional, as shown in fig. 10–22. Sometimes the degree of trueness is grouped into true 1, true 2, etc., as shown in fig. 10–30. Truth tables can be constructed for any combination of letters, numbers or symbols that pertain to a given control problem.

Only designers familiar with the peculiarities of a particular control problem should fill in the output column of a truth table, since for every row the

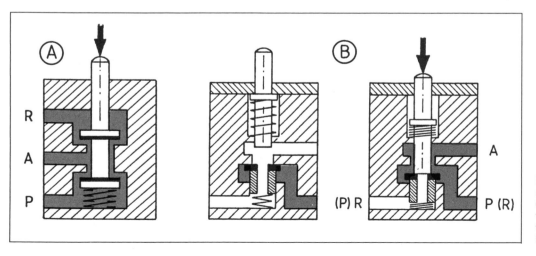

Fig. 10–6A Is a valve with negative overlap. Fig. 10–6B Shows a spool-poppet valve with positive overlap where A -- R closes before P -- A opens.

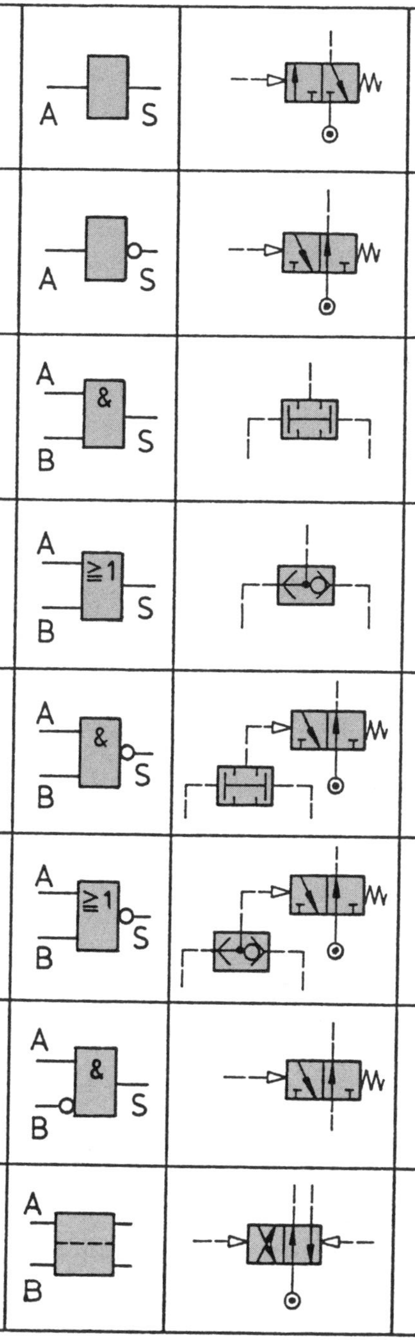

Function	Truth table	Boolean expression
YES	A | S 0 | 0 1 | 1	A=S
NOT	A | S 0 | 1 1 | 0	$\overline{A}$=S
AND	A | B | S 0 | 0 | 0 0 | 1 | 0 1 | 0 | 0 1 | 1 | 1	A•B=S
OR	A | B | S 0 | 0 | 0 0 | 1 | 1 1 | 0 | 1 1 | 1 | 1	A+B=S
NAND	A | B | S 0 | 0 | 1 0 | 1 | 1 1 | 0 | 1 1 | 1 | 0	$\overline{A \cdot B}$=S
NOR	A | B | S 0 | 0 | 1 0 | 1 | 0 1 | 0 | 0 1 | 1 | 0	$\overline{A+B}$=S
INHIBITION	A | B | S 0 | 0 | 0 0 | 1 | 0 1 | 0 | 1 1 | 1 | 0	A•$\overline{B}$=S
MEMORY		A=S_0 B=S_1

Fig. 10-8 Logic functions matched with their truth table, logic symbol, pneumatic symbol and Boolean expression.

0	0	0	0
0	0	0	1
0	0	1	0
0	0	1	1
0	1	0	0
0	1	0	1
0	1	1	0
0	1	1	1
1	0	0	0
1	0	0	1
1	0	1	0
1	0	1	1
1	1	0	0
1	1	0	1
1	1	1	0
1	1	1	1
2^3	2^2	2^1	2^0
8	4	2	1

BINARY EXPRESSION FOR VALUE 13

8 • 4 • 0 • 1 = 13

Fig. 10-7 "DUAL" code applied to a truth table with four input signals.

question must be asked: "Must this combination produce an output signal, and would such an output signal make the machine work correctly and also be safe for the person operating the machine?" Control problem 2 demonstrates this (see figs. 10–16 to 10–19).

Truth tables for basic logic functions

Figure 10–8 illustrates the use of truth tables for all basic logic functions. This figure shows the association of logic functions with their respective truth tables, logic symbols, pneumatic symbols and Boolean algebra switching expressions.

For example, the basic logic "AND" function (see truth table, fig. 10–8) shows the input signals A and B in the two left hand columns. These two input signals render, according to the previously given formula, four distinct combinations ($K = 2^2 = 4$). Therefore, the truth table must show four rows. The output column now indicates an output signal (with the binary value 1) only for the input combination $1 \cdot 1 = 1$, which means that the "AND" function will only render an output signal S if both the input signals A and B show the binary value 1. Mixtures of 1 and 0 values, or both input signals 0, render no output and therefore are given a logic 0 in the output column (see truth tables in figs. 10–4 and 10–8).

Karnaugh-Veitch map

The Karnaugh-Veitch map is by far the fastest and simplest tool to minimise Boolean switching equations. Actually, the Karnaugh-Veitch map is but another graphical form of the truth table, as well as a graphical representation of a Boolean equation. Each field in the Karnaugh-Veitch map represents a row in the truth table and thus is also equivalent to a binary combination ("AND" combination or "AND" function). Furthermore, for each row found in the truth table, the Karnaugh-Veitch map will also show a field, and therefore the number of fields may again be determined by the formula:

$K = 2^n$ (n = input signals, K = fields)

Since the Karnaugh-Veitch diagram is used to minimise switching equations, it has been decided to designate its fields with the "GRAY" code. When moving from one field to the next, the "GRAY" coded fields change only one of their compounded designations (from logic 1 to logic 0 or the reverse). For example the compounded designation for the bottom right-hand field in fig. 10–9 reads:

$a \cdot \overline{b} \cdot c$; which would be equivalent to a = logic 1, and b = logic 0, and c = logic 1.

This unique property of the Gray code enables the user, when minimising with the Karnaugh-Veitch map, to eliminate unnecessary "AND" functions with ease.

The Gray code is sometimes also called "MIRROR CODE". The mirror lines can be regarded as adjacent lines, indicating that the signals divided by them are adjacent (mirror 2 divides signal b but not signal a).

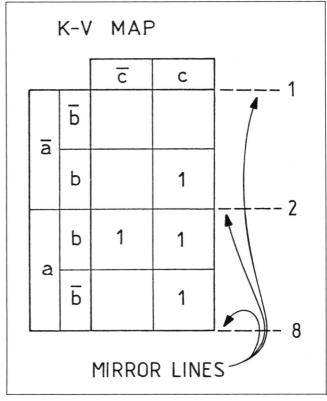

Fig. 10–9 Karnaugh-Veitch map for the truth table shown in Fig. 10–4. The logic 1 output signals from the truth table have been transferred as "ENTRIES" into the map.

This adjacency rule makes the Karnaugh-Veitch map, when rolled together so that mirror line 1 touches mirror line 3, three dimensional (fig. 10–10). Larger Karnaugh-Veitch maps with more than one signal listed on top of the map can also be rolled together in the horizontal direction so that its vertical mirrors 1 and 3 can be joined (figs. 10–9 and 10–10).

Although the map may be regarded as three-dimensional, it must always be drawn as a flat diagram or map with only two dimensions. This assists the process of looping (equation minimisation) and renders the Karnaugh-Veitch map an ideal and highly esteemed design tool for the system design engineer.

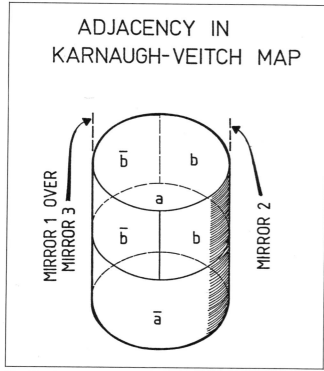

ADJACENCY IN KARNAUGH-VEITCH MAP

Fig. 10–10 By rolling the map so that mirror 1 overlaps mirror 3, input signal 5 becomes also adjacent (see also fig. 10–9).

Rules for looping the Karnaugh-Veitch map

Once all these "AND" combinations rendering a logic 1 output signal in the truth table have been transferred (plotted) into the Karnaugh-Veitch map, the looping process can be started. Seven basic looping rules however must be carefully followed to ensure a complete and correct minimisation of the plotted equation. These fundamental rules are:

1. Loops should encompass as many plotted entries as possible to produce maximum minimisation of the final equation (less "AND" functions and "OR" functions). The map in fig. 10–9 shows these entries.
2. Fields not containing a plotted entry must not be included into the loops, since these fields show clearly that their combination will not give an output signal (see truth table fig. 10–4 and compare with fig. 10–11). (Exceptions to this rule are the "don't care" combinations explained in control problems 3, 4 and 5.)
3. The looped entries in a single loop must always be a power of two (for example 1, 2, 4, 8, 16 etc. entries), thus a loop with six entries would be wrong.

4. Entries which are already looped may be looped again to form other loops (fig. 10–13). The number of times an entry may be used for looping is unlimited.
5. Loops which extend beyond a mirror line must always be symmetrical about this mirror line (for mirror lines and loop symmetry see fig. 10–13).
6. Part-loops which extend beyond an edge of the map (edge mirror line) join the part loop on the opposite end of the map and form one loop (figs. 10–17 and 10–31).
7. Part or full loops, which precisely and fully overlap when the map is folded about one or more of its mirror lines, may be regarded as one single loop. A typical example for this rule is presented in fig. 10–31 in the Karnaugh-Veitch maps on the left (1–6).

Extraction of minimised equations after looping

Input signals which are represented with both binary values (logic 1 opposed by logic 0) in the same loop cancel each other out and thus disappear from the name of this loop. Examples: the horizontal loop in the Karnaugh-Veitch map of fig. 10–11 encompasses two fields and therefore the unminimised equation for this loop reads:

$$\bigcirc = a \cdot b \cdot \bar{c} + a \cdot b \cdot c$$

With the signal cancellation principle applied, one can without difficulty see that signal c in this loop changes its binary logic value from c to $\bar{c}$ (logic 1 to logic 0), and thus is represented with both binary states in the same loop! Therefore, the signal c is cancelled and the remaining equation for this loop reads:

$$\bigcirc = a \cdot b$$

Individual loops are now linked with "OR" functions and are then grouped into the collective equation. Thus each individual loop represents only a part of the collective equation (this part is an "AND" expression), and in the final circuit these loops must be "OR" connected with "OR" function valves to form the collective combinational circuit (fig. 10–14).

To let the reader make personal observations and gain confidence and practice five typical control problems for combinational circuits or sub--circuits are presented and solved. Some of these control problems are exclusively combinational, others which also include memories and timers are

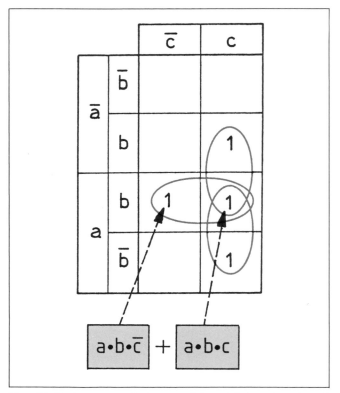

Fig. 10–11 Loops are being used to minimise the plotted entries. Individual loops must be "OR" connected to form the collective equation.

mixed controls which means they become an integral part of a sequential circuit.

Control problem 1

A sorting machine attached to a packaging machine requires an automatic start-stop control.

It was found that under part-load conditions the sorting machine runs very uneconomically. "START" must therefore only occur, when at least two of the three feeding conveyors are running, and automatic "STOP" must always be signalled when less than two conveyors are running. This implies that the "START" command must then automatically disappear.

- First logic step: The signal designations which signal the running of the three conveyors are: conveyor one = a, conveyor two = b, conveyor three = c. The signal which must start the sorting machine is designated A1.
- Second logic step: The truth table, according to the construction rules given before, will show eight rows, three input signal columns and one output signal column (2^3 = 8 rows, for truth table see fig. 10–12).
- Third logic step: Row 1, according to the given logic switching specifications, must not start the machine since none of the three conveyors are running. For rows 2, 3 and 5 where only one conveyor is running, the output signal column must also be filled with a zero (logic 0). In rows 4, 6 and 7 two conveyors are running and in row 8 all three conveyors are in operation. Hence, the output columns for the rows 4, 6, 7 and 8 must be given a logic 1 which renders an A1 signal for the sorting machine to start.
- Fourth logic step: The bottom right hand field in the Karnaugh-Veitch map previously prepared, carries the name of its coordinate signals and therefore is called field a • $\overline{b}$ • c. This field represents row 6 of the truth table and must therefore be plotted with a logic 1. In that sense, only

Fig. 10–12 *Left:* Truth table for control problem 1.

Fig. 10–13 *Right:* Karnaugh-Veitch map for control problem 1.

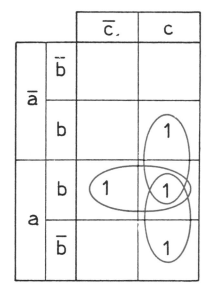

a	b	c	A1
0	0	0	0
0	0	1	0
0	1	0	0
0	1	1	1
1	0	0	0
1	0	1	1
1	1	0	1
1	1	1	1

these combinations with a logic 1 output must be transferred from the truth table into their corresponding Karnaugh-Veitch map fields (fig. 10–13).

- Fifth logic step: The Karnaugh-Veitch map illustrated in fig. 10–13 shows the best possible looping of the four plotted logic 1 entries. The entry in field a • b • c (Row 8 of the truth table) serves all three loops (see looping rule no. 4).

- Sixth logic step: The looping method has now produced a minimized equation which consists of loop 1 "OR" connected to loop 2 "OR" connected to loop 3.

$$A1 = \Big] + \Big] + \bigcirc$$

With logic values inserted this equation reads:

$$A1 = b{\cdot}c + a{\cdot}c + a{\cdot}b$$

- Seventh logic step: Knowledge of Boolean algebra will bring a further minimisation and reduction of switching hardware (valves). Fully minimized with the so called distributive law of Boolean algebra, the equation now reads:

$$A1 = a{\cdot}(b+c) + b{\cdot}c$$

- Eighth logic step: Input signal a appears only once in the afore presented equation; it can

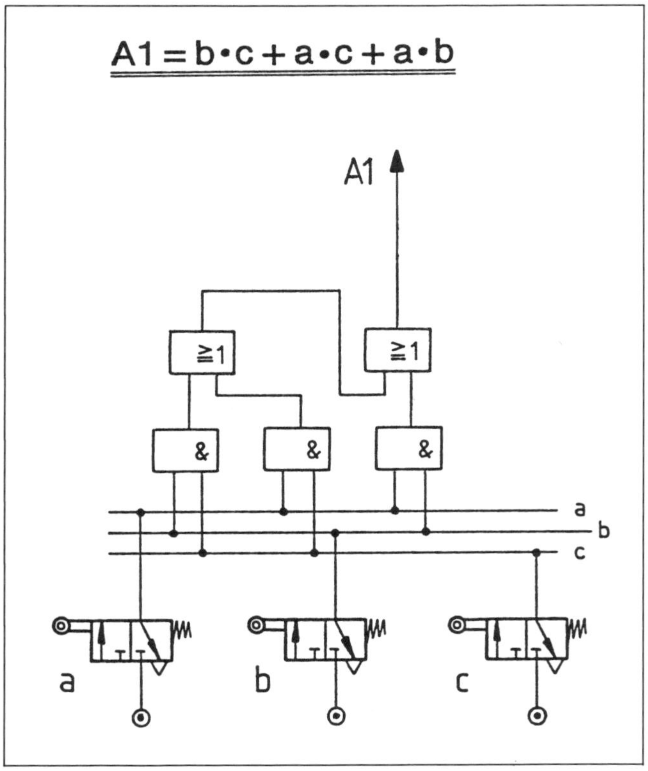

therefore be series connected to the "OR" expression (b + c). This brings a further hardware reduction in the circuit (see fig. 8–13).

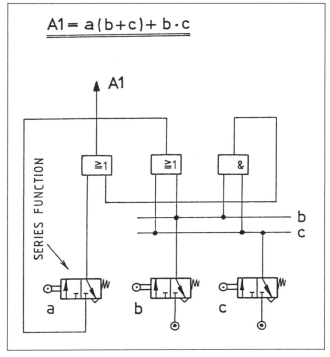

Fig. 10–15 Circuit with algebraic reduction and series connection for signal a.

Control problem 2

A special purpose milling machine (machine tool) must only be permitted to start if certain safety and machine loading conditions are met (fringe conditions).

Two permanent visual indicators must be built into the signal output of the combinational circuit to indicate either "START PERMITTED" or "START NOT PERMITTED". Should the machine fail to start, such visual indicators would then narrow faultfinding procedures to either the combinational circuit or to the sequential circuit or to the external start prerequisites giving the "start permitted" output signal.

- First logic step: The signal designations which render a start signal are: G for the guard on the machine, T for the tool setter and W for the work piece correctly placed into the machining station.

- Second logic step: The truth table according to the three required input signals must show four columns and eight rows (see fig. 10–16).

- Third logic step: The tool setter signal input, which is a key operated push-button valve, must

Fig. 10–14 Circuit for equation obtained in sixth logic step without algebraic reduction.

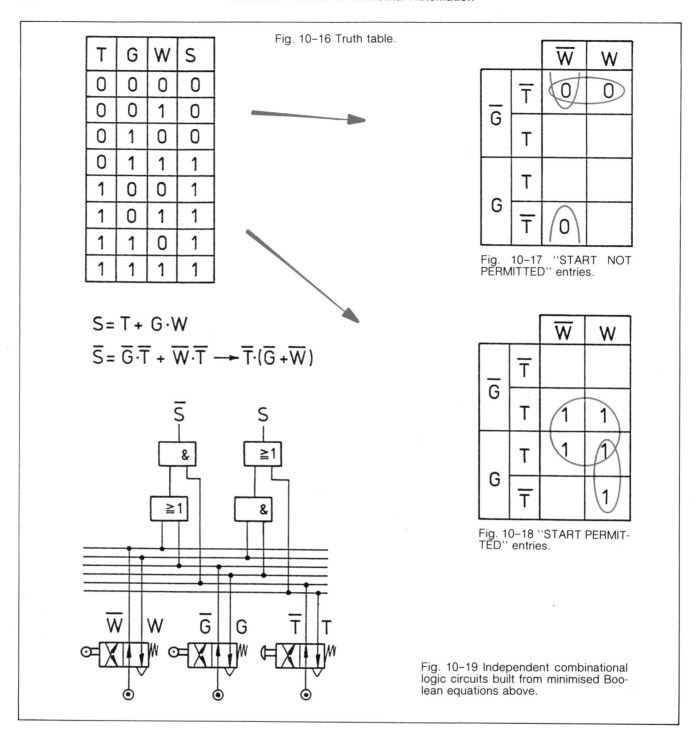

Fig. 10–16 Truth table.

T	G	W	S
0	0	0	0
0	0	1	0
0	1	0	0
0	1	1	1
1	0	0	1
1	0	1	1
1	1	0	1
1	1	1	1

$$S = T + G \cdot W$$
$$\overline{S} = \overline{G} \cdot \overline{T} + \overline{W} \cdot \overline{T} \longrightarrow \overline{T} \cdot (\overline{G} + \overline{W})$$

Fig. 10–17 "START NOT PERMITTED" entries.

Fig. 10–18 "START PERMITTED" entries.

Fig. 10–19 Independent combinational logic circuits built from minimised Boolean equations above.

always start the machine regardless of presence or absence of the other two input signals. With the absence of signal T, a workpiece must be properly located in the machine (signal W to be logic 1) and the guard must be fully closed (signal G to be logic 1) to cause "start".

- Fourth logic step: All logic 1 output signals from the truth table must now be transferred and plotted into the prepared Karnaugh-Veitch map, and these entries are given the logic 1 designation (fig. 10–18).
- Fifth logic step: All logic 1 entries are now being looped to obtain minimised equations for each loop. These equations may then be used for further minimisation if possible (see Karnaugh-Veitch map fig. 10–18).

- Sixth logic step: Extracting the minimised logic equation renders the following result:

$$S = T + (G \cdot W) \longrightarrow \text{START PERMITTED}$$

This equation can be no further minimised with Boolean algebra and thus remains in this state.

- Special logic steps: The specification for this control states that the "START NOT PERMITTED" condition must also be displayed. Looping all logic 0 entries will now render the minimised equation for "START NOT PERMITTED", since all logic 0 entries correspond to the logic 0 output rows in the truth table (fig. 10–17). Extracting the minimised logic equation for the looped logic 0 entries in the Karnaugh-Veitch map renders the following result:

$$\overline{S} = (\overline{G} \cdot \overline{T}) + (\overline{W} \cdot \overline{T}) \longrightarrow$$
$$\text{START NOT PERMITTED}$$

Applying Boolean algebra rules to this equation renders the following result:

$$\overline{S} = \overline{T} \cdot (\overline{G} + \overline{W}) \longrightarrow$$
$$\text{START NOT PERMITTED}$$

This equation can now be no further minimised with Boolean algebra rules or series function (fig. 10–17).

- This "START NOT PERMITTED" signal is the so called "FALSE OUTPUT FORM" of a truth table or Karnaugh-Veitch map. The false output form may also be obtained through the application of "DE MORGAN'S" theorem which is explained later in this chapter.

- Final logic steps: Each circuit obtained from the equations extracted in figs. 10–17 and 10–18 represents an independent combinational circuit. These two circuits are depicted in fig. 10–19.

Since this control problem requires input signals to be in their binary 1 state as well as in their binary 0 state, one may use valves which render a signal for both of these logic states (a signal for "ON" and a signal for "OFF", see fig. 10–19). If such valves are not available one may obtain the same result using two valves (fig. 10–24). This solution would obviously be much more expensive.

Valves which render a signal for their logic "ON" state as well as their logic "OFF" state are said to be *complementary valves* (see fig. 10–20. Such complementary valves may have four or five flow ports and may be manually, pneumatically, electrically or mechanically actuated.

Fig. 10–20 shows the complete combinational

start-stop control circuit as described in the control specification to control problem 2. In this format the circuit is called a "START-STOP MODULE".

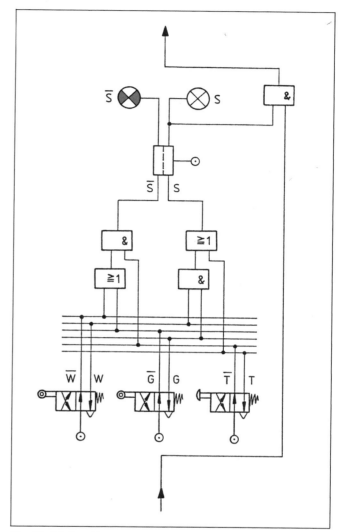

Fig. 10–20 Combinational "START-STOP MODULE" with permanent display (visual indicators) of the "START PERMITTED" and the "START NOT PERMITTED" output signals.

Control problem 3

A stamping press with a pneumatically operated clutch must only operate when safe input combinations exist and *must not start* if any other combination exists. The press and its input signal locations are depicted in fig. 10–21 and the truth table is shown in fig. 10–22.

$$\left.\begin{array}{l} \overline{G} \cdot \overline{R} \cdot \overline{W} \cdot T \\ \overline{G} \cdot \overline{R} \cdot W \cdot T \\ G \cdot \overline{R} \cdot W \cdot T \\ G \cdot R \cdot W \cdot \overline{T} \end{array}\right\} = START$$

Fig. 10–21 Stamping press.

Symbol explanation:

G = Signal from guard when fully closed.
T = Signal from toolsetter or operator valve.
W = Signal from workpiece in the machine.
R = Signal from remote start position.

G	R	W	T	ST.	
0	0	0	0	0	
0	0	0	1	1	→ START
0	0	1	0	0	
0	0	1	1	1	→ START
0	1	0	0	0	
0	1	0	1	0	
0	1	1	0	0	
0	1	1	1	0	
1	0	0	0	0	
1	0	0	1	0	----→ OPTIONAL
1	0	1	0	0	
1	0	1	1	1	→ START
1	1	0	0	0	
1	1	0	1	0	----→ OPTIONAL
1	1	1	0	1	→ START
1	1	1	1	0	----→ OPTIONAL

Fig. 10–22 Truth table for control problem 3.

The combinations rendering a logic 1 output in the truth table are as follows:

$$START = \overline{G} \cdot \overline{R} \cdot \overline{W} \cdot T + \overline{G} \cdot \overline{R} \cdot W \cdot T +$$
$$G \cdot \overline{R} \cdot W \cdot T + G \cdot R \cdot W \cdot \overline{T}$$

The combinations called "OPTIONAL" in the truth table may be regarded as "relatively safe" combinations which could occur but would not be used for the operation of the press (fig. 10–22). Such optional combinations are also called "DON'T CARE" combinations, which means combinations causing no harm. To simplify the problem these combinations are not included in the valid start conditions but their inclusion would result in a much shorter collective equation!

To minimise the combinations with a logic 1 output signal in the truth table, a Karnaugh-Veitch map is constructed (fig. 10–23). This map is shown as a square but it could also be drawn as a single column map with sixteen fields in vertical direction or as a map with two columns and eight fields in vertical direction. The form of the map or its labelling order does not affect the minimisation process and will always render identical results.

For combinational circuit design all the plotting entries must be taken into consideration and this applies also to plotted entries which cannot be looped to other entries. The Karnaugh-Veitch map for control problem 3 (fig. 10–23) shows such a single entry. All individual loops and the single entry being circled are now linked with "OR" functions and grouped into the collective equation which reads:

$$START = \bigcup + \bigcirc + \bigcirc$$

$$START = \overline{G} \cdot \overline{R} \cdot T + W \cdot \overline{R} \cdot T + W \cdot G \cdot R \cdot \overline{T}$$

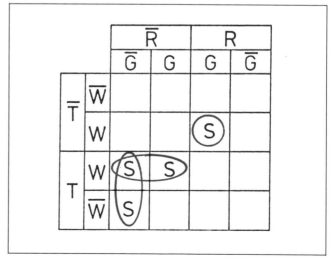

Fig. 10–23 Karnaugh-Veitch map for control problem 3.

This collective equation minimised and extracted from the Karnaugh-Veitch map can now be further minimised with Boolean algebra. As a result the final equation now reads:

$$\text{START} = \overline{R} \cdot T \cdot (\overline{G} + W) + W \cdot G \cdot R \cdot \overline{T}$$

This control problem also requires valves rendering complementary signals for the input signals R, T and G. But to illustrate the logic principle of "signal inversion" it has been decided that valves rendering only the "YES" function must be used (see control circuit fig. 10–23 for control problem 3).

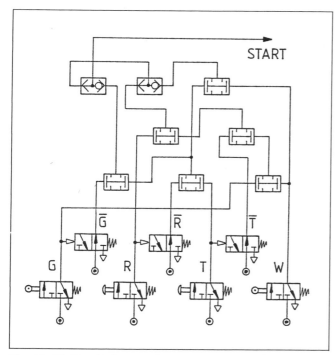

Fig. 10–24 Circuit diagram for control problem 3.

Control problem 4

A bottle sorter must be pneumatically powered and controlled. Glass bottles of four different sizes are to be analysed and then according to size pushed on to separate conveyors for filling and packaging. Four pneumatic air barrier sensors are used to detect the size of the passing bottles. Air barrier sensors were described in Chapter 6. The passing bottles must be counted according to size.

A front view of the bottle size detector is given in fig. 10–25 and a plan view of the bottle sorting machine including attached conveyors is given in fig. 10–26.

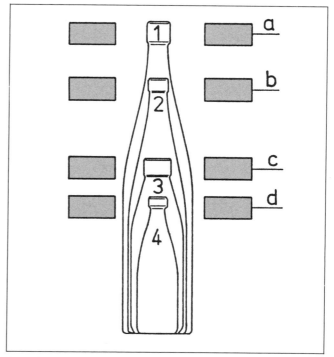

Fig. 10–25 Bottle size detector consisting of four air barrier sensors.

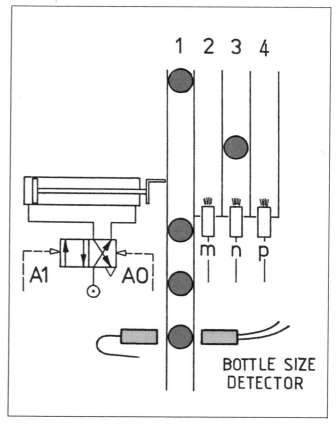

Fig. 10–26 Bottle sorting machine with pneumatic sorting actuator.

Design considerations for bottle sorter

It is advisable to construct a Karnaugh-Veitch map and a truth table to assist in the problem analysis and equation minimisation. The uppermost combination of the truth table stands for "NO BOTTLE IN THE DETECTOR". This combination must not render an output signal. The same may be said for the lowermost combination, which stands for bottle size 1. Size 1 bottles continue on the feed conveyor and therefore require no sorting movement (see fig. 10–26).

All combinations with an A1 output are numbered according to their bottle size. All remaining combinations marked with an X may be used for looping, since they represent "DON'T CARE" combinations (fig. 10–27). So the actuator must only extend when a bottle of size 2, size 3 or size 4 is detected, and thus the minimisation equation consists of the three loops shown in the Karnaugh-Veitch map (fig. 10–28).

If bottles of sizes 2, 3 and 4 were only to give a counting pulse for separate counting, and then had to be shifted on to a common conveyor, the entries in the Karnaugh-Veitch map could be looped into a common loop rendering the equation $A1 = \bar{a} \cdot d$. The collective final equation extracted from the three loops in the Karnaugh-Veitch map reads:

$$A1 = b \cdot \bar{a} + \bar{b} \cdot c + d \cdot \bar{c}$$

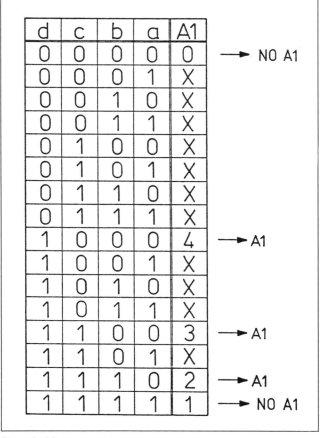

d	c	b	a	A1	
0	0	0	0	0	→ NO A1
0	0	0	1	X	
0	0	1	0	X	
0	0	1	1	X	
0	1	0	0	X	
0	1	0	1	X	
0	1	1	0	X	
0	1	1	1	X	
1	0	0	0	4	→ A1
1	0	0	1	X	
1	0	1	0	X	
1	0	1	1	X	
1	1	0	0	3	→ A1
1	1	0	1	X	
1	1	1	0	2	→ A1
1	1	1	1	1	→ NO A1

Fig. 10–27 Truth table for bottle sorter.

The bottles pass very rapidly through the size detector gate and therefore signal combinations obtained from these detectors disappear just as rapidly. It is therefore imperative that such combinations are stored in a memory valve to make them available for actuator extension limitation.

Such extension limitation is essential for the correct sorting of the various bottle sizes. Bottles of size 4 may be moved over the full actuator extension stroke but bottles of size 3 must only be pushed to limit sensor n and bottles of size 2 only to limit sensor m. Hence, the retraction signal for bottle size 2 consists of the memorised (stored) bottle size combination $\bar{a} \cdot b$ and the limit sensor signal m. The retraction signal for bottle size 3 consists of the bottle size combination $\bar{b} \cdot c$ and the limit sensor signal n.

The memory valves are reset as soon as the power valve (D.C.V.) causes actuator retraction and since the reset signal is a continuous signal which would cause memory blockage an impulse valve is used to transform it into a momentary signal (pulse). For circuit construction and Karnaugh-Veitch map see fig. 10–28.

Control problem 5

A packaging machine for washing powder packets is to be automated. Five filling machines, filling the packets with washing powder, feed onto a common conveyor. This conveyor then transports the packets to the packaging machine where they are wrapped into parcels of 4 kg. Four out of the five filling machines process washing powder packets of 1 kg. Machine A, however, produces 2 kg packets and feeds these always into line A of the conveyor. The other four filling machines (B, C, D and E) feed also into their allocated lines (fig. 10–29). All five filling machines are synchronised and push their washing powder packets simultaneously onto the conveyor, but not all filling machines may operate at any one time.

A weight detector station controlled by a combinational logic signal evaluation system evaluates the weight units of the passing washing powder packets and permits only clusters of a total weight of four kilograms to move on into the wrapping machine.

Powder packet clusters with less or more than four kilograms are ejected into separate chutes

PROBLEM ANALYSIS

$$a \cdot b \cdot c \cdot d = 1$$
$$\bar{a} \cdot b \cdot c \cdot d = 2 \to A1$$
$$\bar{a} \cdot \bar{b} \cdot c \cdot d = 3 \to A1$$
$$\bar{a} \cdot \bar{b} \cdot \bar{c} \cdot d = 4 \to A1$$
$$\bar{a} \cdot \bar{b} \cdot \bar{c} \cdot \bar{d} = 0$$

$$A1 = b \cdot \bar{a} + \bar{b} \cdot c + d \cdot \bar{c}$$

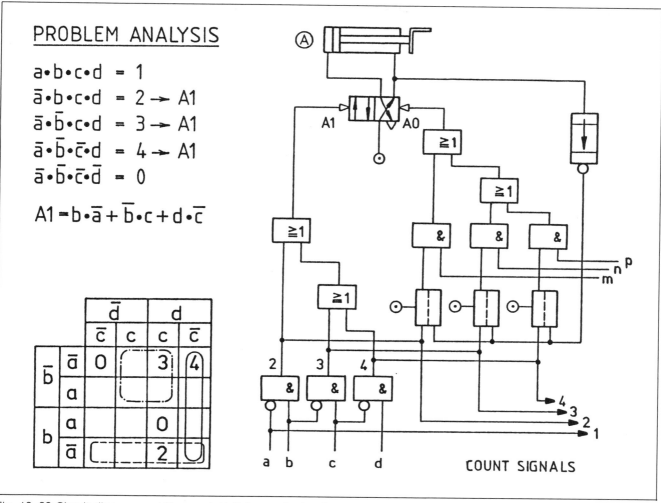

Fig. 10–28 Circuit diagram and Karnaugh-Veitch map for bottle sorter. Inhibition functions may be used for the collection A1 equation (the sensors are not shown).

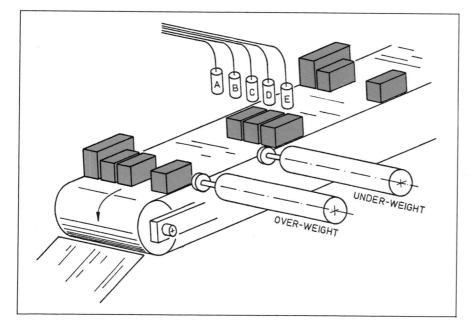

Fig. 10–29 Weight detector station with ejector actuators.

(underweight clusters separately and overweight clusters separately, see fig. 10–29). Ejector actuators must only be extended when a rejected cluster has arrived at its respective eject position. Overweight or underweight commands must therefore be memorised (similarly to control problem 4) and then "AND" connected to their respective eject position signals.

Design considerations

The truth table for this control problem must differentiate between underweight, overweight and correct weight signal combinations; hence, this truth table is to be furnished with three output columns; one for each weight cluster group. Underweight is designated "U", overweight "O", and correct weight "W". Washing powder packets detected in line A of the conveyor weigh 2 kg and are therefore given two columns in the truth table.

A	B	C	D	E	W	U	O
0	0	0	0	0		X	
0	0	0	0	1		1	
0	0	0	1	0		1	
0	0	0	1	1		1	
0	0	0	1	0		1	
0	0	0	1	0	1		
0	0	0	1	1	0		
0	0	0	1	1	1		
0	0	1	0	0	0		1
0	0	1	0	0	1		1
0	0	1	0	1	0		1
0	0	1	0	1	1		1
0	0	1	1	0	0		1
0	0	1	1	0	1		1
0	0	1	1	1	0		1
0	0	1	1	1	1	1	
1	1	0	0	0	0		1
1	1	0	0	0	1		1
1	1	0	0	1	0		1
1	1	0	0	1	1	1	
1	1	0	1	0	0		1
1	1	0	1	0	1	1	
1	1	0	1	1	0	1	
1	1	0	1	1	1			1
1	1	1	0	0	0		1
1	1	1	0	0	1	1	
1	1	1	0	1	0	1	
1	1	1	0	1	1			1
1	1	1	1	0	0	1	
1	1	1	1	0	1			1
1	1	1	1	1	0			1
1	1	1	1	1	1			1

Fig. 10–30 Truth table for control problem 5.

The top row in the truth table represents a combination where there are no clusters under the detector station. Since ejector signals must be "AND" connected to an eject position signal rendered by a passing cluster, this combination cannot cause an ejector to extend. Therefore this combination may be called a "DON'T CARE" combination which can be used for the minimisation of any loop and as many times over as required (see looping rule 4 and figs. 10–11, 10–29, 10–30 and 10–31).

To establish the logic values in the output columns of the truth table, one simply adds all the logic 1 entries in each row. If the sum of these entries is precisely four, then the "W" output gets the logic 1. If the sum is more than four, this would denote an overweight cluster which therefore produces a logic 1 entry in the "O" column of this row. If the sum is less than four, the underweight column gets the logic output 1 (see fig. 10–30).

Because of the large number of underweight combinations in the truth table, it may be wise to draw three Karnaugh-Veitch maps for the underweight entries, and a separate Karnaugh-Veitch map for overweight and correct weight (see fig. 10–31). Looping of large Karnaugh-Veitch maps with more than two or three input signals is of course somewhat more complex and requires skill, but if the looping rules are correctly applied one can always count on a satisfactory and optimal minimisation result. The control circuit is bound to work even with extracted equations based on less than optimal looping!

The minimised equations extracted from the loops in the Karnaugh-Veitch maps for the output signal "U" (underweight) read as follows:

$$U = \overline{A} \cdot \overline{B} + \overline{A} \cdot \overline{C} + \overline{A} \cdot \overline{D} + \overline{C} \cdot \overline{D} \cdot \overline{E} + \overline{A} \cdot \overline{E} + \overline{B} \cdot \overline{C} \cdot \overline{D} + \overline{C} \cdot \overline{B} \cdot \overline{E} + \overline{B} \cdot \overline{D} \cdot \overline{E}$$

Further minimised with the distributive law of Boolean algebra this equation reads as follows:

$$U = \overline{A} \cdot (\overline{B} + \overline{C} + \overline{D} + \overline{E}) + \overline{C} \cdot \overline{D} \cdot (\overline{E} + \overline{B}) + \overline{B} \cdot \overline{E} \cdot (\overline{C} + \overline{D})$$

The minimised equation extracted from the loops in the Karnaugh-Veitch map for the output signal "O" (overweight) reads as follows:

$$O = A \cdot B \cdot C \cdot D + A \cdot B \cdot C \cdot E + A \cdot C \cdot D \cdot E + A \cdot B \cdot D \cdot E$$

Further minimised with the distributive law of Boolean algebra this equation reads as follows:

$$O = A \cdot B \cdot C \cdot (D + E) + A \cdot D \cdot E \cdot (C + B)$$

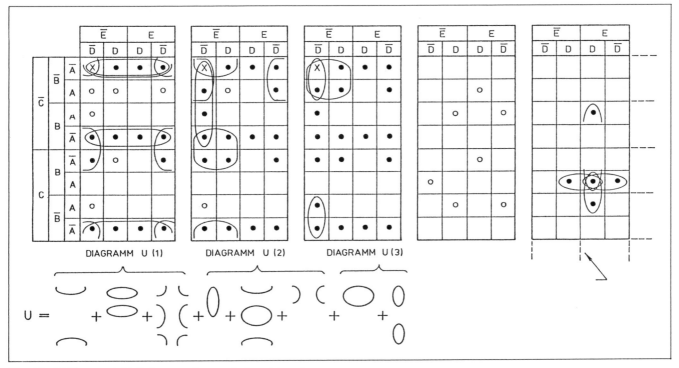

Fig 10-31 Karnaugh-Veitch maps for control problem 5.

Since all output signals in the truth table are filled either with an output 1 or an X (DON'T CARE) and none of the three output signals share their output with a letter in another column of the same row, we can regard these output signals as overlap free and combinationally complete.

This implies that the minimised "O" and "U" output equations when connected with an "OR" function may be regarded as the "true output form" of the truth table (these must render an output or eject signal). The remaining output combinations represent the "false output form" of the truth table (these combinations must not render an eject signal). To achieve an output signal for the purpose of counting the clusters moved through the wrapping station, one now simply has to use the "NOR" function which renders an output when "U" or "O" are not present (see also fig. 10–8).

The weight detector renders an $\overline{A} \cdot \overline{B} \cdot \overline{C} \cdot \overline{D} \cdot \overline{E}$ signal as soon as a cluster has left the detectors. Such a combination would however set the underweight signal! To avoid such pseudo-underweight signals reaching the memory, a logic filter has been built into the signal line leading to the underweight memory (fig. 10–32).

Another precautionary measure has been taken to avoid spurious underweight signals reaching the underweight memory. If, for example, a washing powder packet is slightly in front of the other

packets in a correct weight cluster, its presence signal would then send an underweight signal which would be evaluated as "ONE PACKET ONLY". This misleading signal would simulate line two of the truth table, although it was a correct weight cluster. To avoid such mishaps a timer is used which stops any spurious signals until the complete cluster is under the weight detector (see timer in circuit of fig. 10–32).

Review of control problems

Control problems 4 and 5 were mixed controls, consisting of a combinational control driving a sequential control. Pure combinational controls and pure sequential controls are seldom required in industry. It is therefore important that control designers and service personnel know how to design such controls and therefore understand the use of Boolean algebra (at least the theorems and laws presented in this chapter), truth tables, and Karnaugh-Veitch maps.

Combinational pattern circuits

To round this chapter off, some combinational pattern circuits are given. The circuit in fig. 10–34 is a previously introduced program selector circuit (see Chapter 8), designed with combinational logic principles.

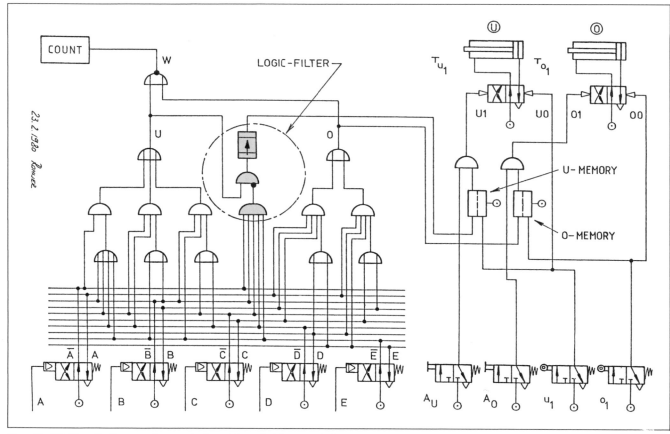

Fig. 10-32 Circuit diagram for control problem 5. The proximity sensors are not shown.

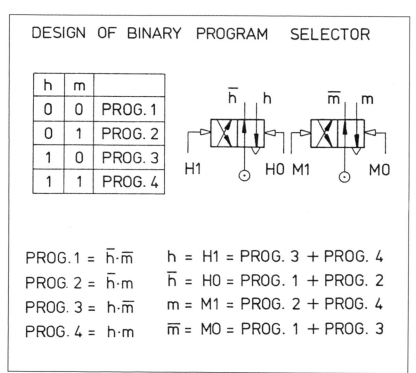

DESIGN OF BINARY PROGRAM SELECTOR

h	m	
0	0	PROG. 1
0	1	PROG. 2
1	0	PROG. 3
1	1	PROG. 4

$PROG. 1 = \overline{h} \cdot \overline{m}$ $h = H1 = PROG. 3 + PROG. 4$

$PROG. 2 = \overline{h} \cdot m$ $\overline{h} = H0 = PROG. 1 + PROG. 2$

$PROG. 3 = h \cdot \overline{m}$ $m = M1 = PROG. 2 + PROG. 4$

$PROG. 4 = h \cdot m$ $\overline{m} = M0 = PROG. 1 + PROG. 3$

Fig. 10-33 Combinational logic design for program selector.

If two memory valves with two output states are used, one can obtain four unique output combinations ($2^2 = 4$). The truth table shows these combination possibilities. The memory input signals are also taken from the truth table by connection with "OR" functions (see figs. 10-33 and 10-34).

T-Flip Flop

This type of memory is used for *binary counting* or binary reduction control. It may also be used to cause an actuator to extend and thereafter retract (or vice versa), but only one command valve is required since every "T input signal" changes the output of the memory on top ($\overline{A}$, A). Such valves are also commercially available in single block form.

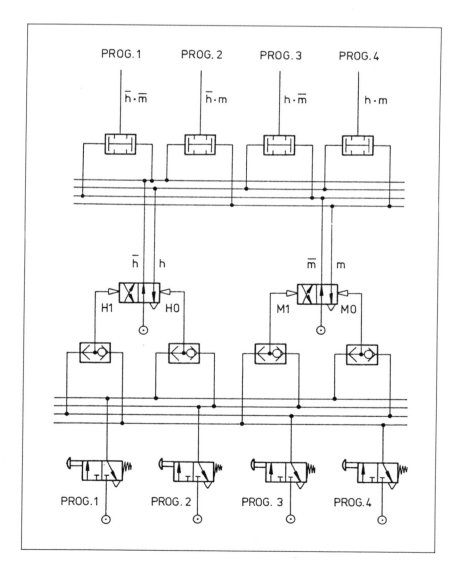

Fig. 10-34 Binary program selector circuit.

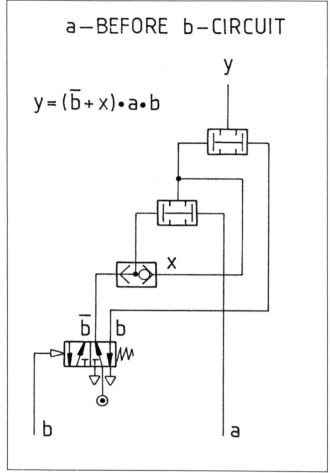

Fig. 10-35 a-before-b circuit.

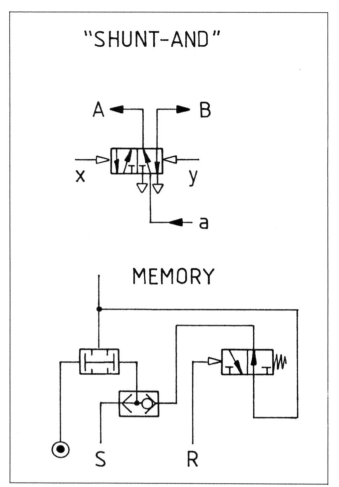

Fig. 10-36 "SHUNT-AND" circuit; memory circuit built with non-memory type valves.

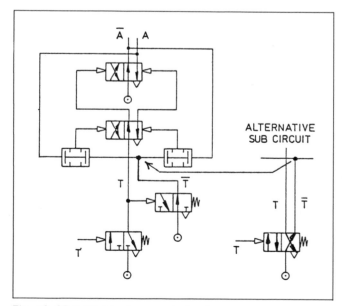

Fig. 10-37 T-flip flop (memory) circuit.

BOOLEAN IDENTITIES

1. (a) 0 is unique
   (b) 1 is unique

2. (a) $A + 0 = A$
   (b) $A \cdot 1 = A$

3. (a) $A + \overline{A} = 1$
   (b) $A \cdot \overline{A} = 0$

4. (a) $A + 1 = 1$
   (b) $A \cdot 0 = 0$

5. (a) $A + B = B + A$
   (b) $A \cdot B = B \cdot A$

6. (a) $A + (B \cdot C) = (A + B) \cdot (A + C)$
   (b) $A \cdot (B + C) = A \cdot B + A \cdot C$

7. $\overline{\overline{A}} = A$

8. (a) $A \cdot A = A$
   (b) $A + A = A$

9. (a) $0 + 0 = 0$
   (b) $1 \cdot 1 = 1$
   (c) $1 + 0 = 1$
   (d) $1 \cdot 0 = 0$

10. (a) $\overline{0} = 1$
    (b) $\overline{1} = 0$

11. (a) $A(A + B) = A$
    (b) $A + AB = A$

12. (a) $A(\overline{A} + B) = AB$
    (b) $A + \overline{A}B = A + B$

13. If $A = B$ then $\overline{A} = \overline{B}$

14. (a) $\overline{A} (A + B) = \overline{A} B$
    (b) $\overline{A} + A B = \overline{A} + B$

15. (a) $(A + B) \cdot (\overline{A} + C) = A \cdot C + \overline{A} \cdot B$
    (b) $(A + B) \cdot (C + D) = A \cdot C + A \cdot D + B \cdot C + B \cdot D$
    (c) $(AB + CD) = (A + C)(A + D)(B + C)(B + D)$

16. (a) $\overline{AB} = \overline{A} + \overline{B}$
    (b) $\overline{A + B} = \overline{A} \cdot \overline{B}$
    (c) $A B = \overline{\overline{A} + \overline{B}}$
    (d) $A + B = \overline{\overline{A} \cdot \overline{B}}$

17. (a) $\overline{AB + CD} = (\overline{A} + \overline{B})(\overline{C} + \overline{D})$
    (b) $\overline{(A + B)(C + D)} = \overline{A} \cdot \overline{B} + \overline{C} \cdot \overline{D}$

18. (a) $(A + B)(A + \overline{B}) = A$
    (b) $AB + A\overline{B} = A$

19. (a) $A + (B + C) = (A + B) + C = A + B + C$
    (b) $A(BC) = (AB)C = ABC$

20. (a) if $A = B$ then $A + C = B + C$
    (b) & $A \cdot C = B \cdot C$

21. (a) if $A + B = 0$ then $A = B = 0$
    (b) if $A \cdot B = 1$ then $A = B = 1$

22. (a) $(A + B)(C + D) = (A + B)C + (A + B)D$
    (b) $AB + CD = (AB + C)(AB + D)$

11 Electronic programmable controllers for fluid power

Over recent years electronic programmable logic controllers have sprung up like mushrooms, and over three hundred different types and brands are presently available. Electronic programmable logic controllers have undoubtedly become a very valuable and indispensable part of the ever advancing industrial automation.

These controllers evolved as industry sought more economical ways to automate their production lines, particularly those involved in the manufacturing of equipment, consumer goods and heavy industry products. Thus the electronic programmable controller has replaced relay based, hard wired electrical systems, and more recently it has also made significant inroads into the traditional domain of pneumatic logic circuitry. These inroads unfortunately have not gone unnoticed and have brought about some problems:

- programming difficulties caused through non-standardised and varying programming techniques;
- programming difficulties caused through the use of varied logic element names and element description for elements within the controller;
- demarcation problems in factories between electrical- and metal-worker unions when maintenance on machines with fluid power control and electronic control is required;
- extreme shortage of skilled personnel who can program and install such electronic controllers and interface them with fluid power circuitry.

What is an electronic programmable controller?

Electronic programmable logic controllers are often abbreviated with the letters P.C. This abbreviation is misleading, since the letters P.C. are also used for "personal computers". It is therefore suggested that the abbreviation P.L.C. (programmable logic controller) should be used instead.

Programmable logic controllers (P.L.C.) operate by monitoring input signals from such sources as push-button switches, proximity, heat or level sensors, limit switches and pressure sensors. When binary logic changes are detected from these input signals, the P.L.C. reacts through a user-programmed internal logic switching network and produces appropriate output signals. These output signals may then be used to operate external loads and switching functions of the attached fluid power control system (see figs. 11–1 and 11–17).

Programmable logic controllers eliminate much of the wiring and rewiring that was necessary with conventional relay-based systems. Instead the programmed "logic network" replaces the previously "hard wired" network. This logic network may be altered as required by simply programming or reprogramming the P.L.C. Thus, the automated processes of a production line or complex manufacturing machine can be controlled and modified at will for highly economical adaptability to a rapidly changing manufacturing environment.

How does an electronic programmable controller work?

A typical programmable electronic controller (P.L.C.) has four separate yet interlinked components; an input/output section, a central processing unit which is microprocessor based, a programming device (console), and a power supply.

The input section is often powered by a 24 volt power supply built into the controller. This supplies the limit switches, sensors and push-button switches with the necessary power. The P.L.C. also reads all the on/off conditions of all input terminals and memorises them in its input image memory before executing the program.

The central processing unit reads the stored information from the image memory and processes that information according to the control plan programmed into the central processing unit (C.P.U.). The C.P.U. is microprocessor based and may be regarded as the "BRAIN" of the controller. The main purpose of the C.P.U. is to continuously scan (monitor) the status of all input signals and thus direct the status of all output signals. Such an internal control plan may include numerous "MEMORY" functions, logic "AND", "OR", as well

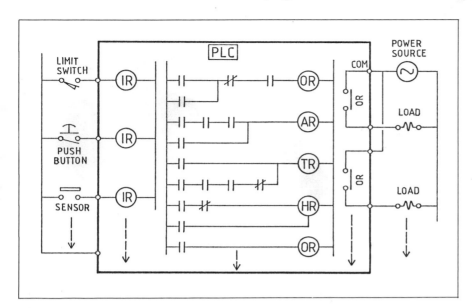

Fig. 11-1 Internal configuration of a P.L.C.

as "INHIBITION" functions, arithmetic computations, timers and counter functions. The program is entered with ladder diagram or other programming concepts and remains in the C.P.U. until deliberately changed by the user with one of the programming devices. Such execution results are then written internally (electronically) into the element image memory. The element image memory then drives the output relays of the P.L.C.

Upon completion of the internal functions the P.L.C. starts again and repeats these processes endlessly. The programming device may be a push-button console or a tape recorded program.

To summarise: A programmable controller is an aggregate control mechanism made up of multiple electronic relays, timers and counters used to execute the internal logical wiring by the programming panel. A conventional hard wired relay panel basically differs from a P.L.C. in the sequence execution method. Its sequences are executed in parallel while the P.L.C. executes its sequences in the order of the program and cyclically as the scanning reveals any input changes.

This book is not intended to teach any particular brand or type of programmable controller and, therefore, care has been taken to make the necessary information as unbiased as possible. To make the applications nevertheless meaningful some practical programming procedures are given, and have been based on the "OMROM C20" programmable controller which is basically identical to the "FESTO FPC 201" and closely related to the "MITSUBISHI Melsec F series". These are all in fact small type controllers, ideally suited for fluid power sequential and combinational control.

Programming the electronic controller (P.L.C.)

A most misleading impression for many control engineers and electricians when confronted with a programmable controller is that these machines (P.L.C.) can easily be programmed with the "ladder logic" method. Ladder logic *is not a method of circuit design*; it is a method of circuit presentation, like a pneumatic or hydraulic circuit. For this reason numerous "amateur" programmers had insurmountable difficulties and frustrations whilst programming their little electronic "wizard"; some to such an extent that they gave up before they ever saw their controller work!

The author of this chapter has therefore made attempts to present a programming and design method which works for most types and brands of P.L.C. This method is of course not new to the experienced circuit designer of pneumatic sequential controls. It is called the "step-counter design method" and is explained and applied in Chapter 8. In fact most principles outlined in Chapter 8 can be directly applied to P.L.C. programming.

Electronic switching peculiarities

Compared with pneumatic valves, the only electronic counterpart to a pneumatic memory valve is the "KEEP RELAY". The keep relay is sometimes also called "holding relay" or "retentive relay", and maintains its logic status (set or reset) even during power failure! For pneumatic memory valves, this memory or retentive behaviour is achieved with the mechanical friction between the spool and the valve body. For electronic memories (relays) the latch-in must be maintained with a battery back-up when power fails. This demands that every logic switching function which in a pneumatic circuit would normally end up as a pneumatic pilot signal to a memory valve, when electronically achieved, also has to end up driving an electronic memory which is in fact a keep relay or holding relay (see figs. 11–8 and 11–9).

Another peculiarity is the terminology used for normally open and normally closed pneumatic valves compared to electronic contacts and switches. A normally closed pneumatic valve does not pass any flow in its non-actuated position. A normally closed contact, however, or a normally closed switch, does pass a signal in its normally closed state (see figs. 2–5 and 11–2)

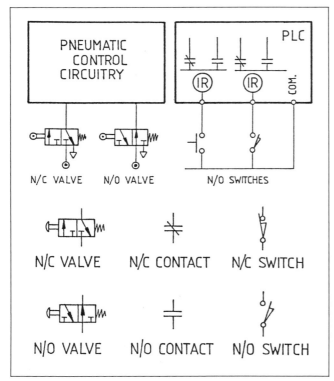

Fig. 11–2A A normally closed valve does not pass a signal but a normally closed contact does. Electronic controllers make all relay signals available in their logic 1 and also logic 0 from (normally closed and normally open contacts; see also fig. 11–9.)

The third peculiarity is related to series connection of pneumatic valves, as explained in Chapter 2, figs. 2–34 and 2–35 and also in Chapter 8, figs. 8–12 to 8–18. The electronic and electric ladder diagram presentation shows only series connection and no "AND" connection by means of "AND" gates. Electronic input relays as well as internal relays drive a vast number of normally closed and normally open contacts. Therefore, one never has

to worry about reducing a complex logic input equation to its fully minimised form, as this would be necessary for a pneumatic control. Thus, when programming an electronic controller, one may safely assume that each relay has an unlimited number of contacts which can always be "AND" connected (see figs. 11–1 and 11–5). These contacts may be called upon at any time, as long as the relay driving them is set and these contacts are always freely available in the logic 1 form or logic 0 form (which means inversion of signals is no longer required — fig. 11–2).

Relay types

A Programmable Logic Controller (P.L.C.) has basically four different types of relays. Figure 11–22 lists these relays in conjunction with their channel allocation number. These relays can be grouped into the following four categories:

1. Input Relays [IR]
2. Output Relays [OR]
3. Internal Auxiliary Relays [AR]
4. Holding Relays [HR]

- Input relays (IR) are used to receive incoming signals and distribute them wherever these input signals are required. These relays may be compared to a limit valve or push-button start or stop valve in a pneumatic circuit. The signal emerging from such relays may be processed in normally open (N/O) or normally closed (N/C) form (see fig. 11–2 and 11–22). The "OMRON C 20" has 16 input relays and may be extended to 80.
- Output relays (OR) are the only relays which can be used to drive a load outside the P.L.C. (for example a solenoid, a siren, start a motor or turn on an oven). The "OMRON C 20" has only 12 output relays, but these may be expanded to 60 (fig. 11–5). Output relays are not memory retentive during power failure (fig. 11–3).
- Internal auxiliary relays (AR) are only used for internal logic signal processing and cannot be used to drive an outside load! The "OMRON C 20" has 136 internal auxiliary relays. Internal auxiliary relays are not memory retentive during power failure (fig. 11–3) and may be compared to an air pilot operated, spring reset pneumatic valve.
- Holding relays (HR) are also used for internal logic signal processing, much the same way as internal auxiliary relays. Holding relays are memory retentive during power failure if they are combined with the Fun. 11 instruction (see fig. 11–5). Holding relays are basically set-reset type relays. The

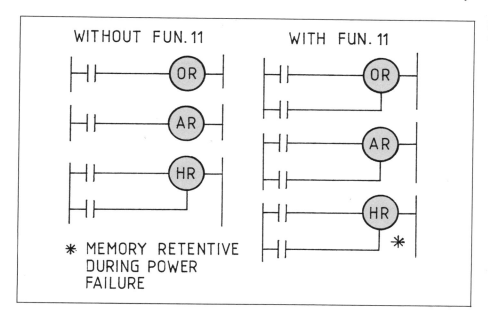

Fig. 11–3 Relay types and integration of "Fun.11."

WITHOUT FUN. 11

WITH FUN. 11

* MEMORY RETENTIVE DURING POWER FAILURE

Latching function (FUN. 11)

"OMRON C 20" has 160 holding relays (fig. 11–22). Only holding relays have true memory capacity which makes them memory retentive during power failure (see figs. 11–3 and 11–5). Such relays are also called "Keep Relays."

Output relays (OR), as well as internal auxiliary relays (AR) may also be combined with the latching function (Fun. 11) to turn them into set-reset relays with a separate set and reset input signal (see fig. 11–3). This, however, does not make them memory retentive during power failure!

Fig. 11-4 Step-counter module comparison for P.L.C. programming, instruction ladder logic diagram and pneumatic circuit diagram. The designation HR stands for holding relay (keep relay) and "OUT" stands for the output relay which is not a memory retentive relay.

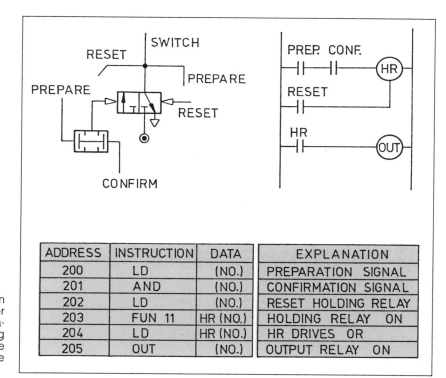

ADDRESS	INSTRUCTION	DATA	EXPLANATION
200	LD	(NO.)	PREPARATION SIGNAL
201	AND	(NO.)	CONFIRMATION SIGNAL
202	LD	(NO.)	RESET HOLDING RELAY
203	FUN 11	HR (NO.)	HOLDING RELAY ON
204	LD	HR (NO.)	HR DRIVES OR
205	OUT	(NO.)	OUTPUT RELAY ON

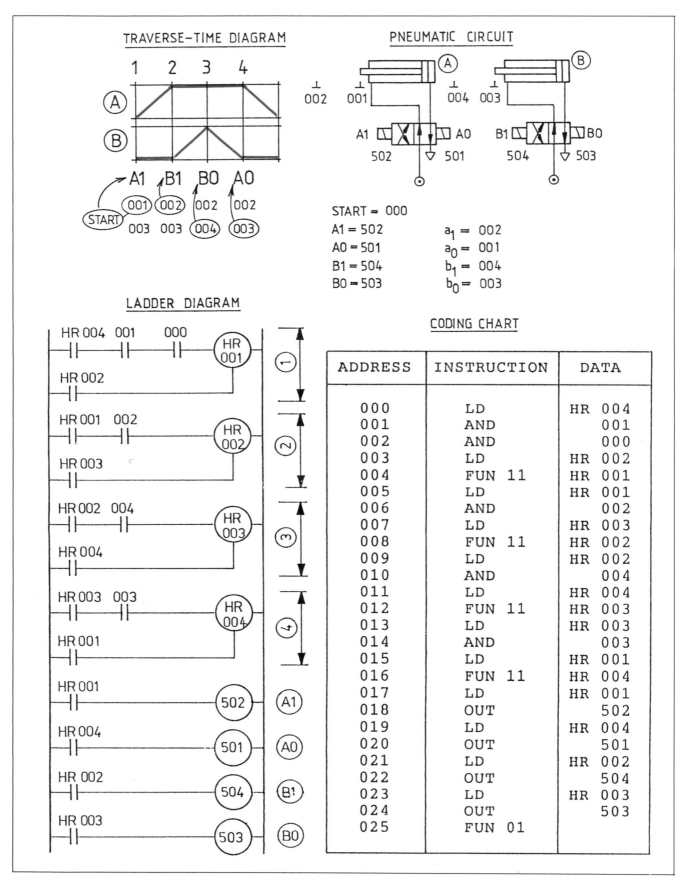

Fig. 11–5 Ladder diagram and coding chart for clamp and drill control sequences as depicted in the traverse-time diagram.

Step-counter programming for a P.L.C.

The step-counter circuit design method may also be used to program a P.L.C. (see Chapter 8, figs. 8–4 and 8–5). Each step-counter module is "built" from a keep relay (HR) and a preswitched "AND" function. The two signals required to make the "AND" function are the essential confirmation signal, confirming the completion of the previous sequence motion, and the preparation signal rendered by the keep relay of the previous step-counter module (or sequence step).

The essential confirmation signal is of course only a contact which is driven by the input relay (IR). The input relay is actuated by the limit switch on the machine (see figs. 11–1 to 11–4). The preparation signal is also only a contact which is driven by the keep relay (HR) of the previous step-counter module (fig. 11–4).

The P.L.C. integration of the individual step-counter modules into the step-counter chain follows the same pattern as established in pneumatic circuit design. The first rung in the ladder diagram includes the start switch or a contact from a cycle selection module (figs. 11–5 and 11–12).

To demonstrate the correlation and integration of the individual step-counter modules, a ladder diagram is now established for a clamp and drill machine with the sequence: A1–B1–B0–A0. The pneumatic power circuit is given and all necessary limit switches and solenoid signal commands from the output relays leading to the solenoids are allocated with their appropriate relay numbers.

It must be noted that the keep relay (HR) in a P.L.C. step-counter program performs the same functions as the memory valve in a pneumatic step-counter. These functions are:

1. To switch the output relay. The output relay then drives the solenoids on the power valves (figs. 11–4, 11–5 and 11–13).

2. To reset the keep relay (HR) of the previous electronic step-counter module.

3. To prepare the "AND" function which switches the keep relay of the next electronic step-counter module.

At this stage it may be appropriate to mention that immediately after the last sequential address (address 024 in the given exercise) an "END" instruction must also be programmed into the P.L.C. program. For the "OMROM C20" this instruction is function 01 (fig. 11–5). An error message would usually be displayed on the P.L.C. if this end instruction was omitted. The C.P.U. of the programmer scans the program data from address 0000 to the address with the end instruction in the sequence (fig. 11–6).

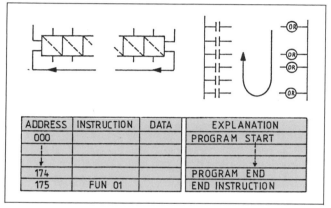

ADDRESS	INSTRUCTION	DATA	EXPLANATION
000			PROGRAM START
174			PROGRAM END
175	FUN 01		END INSTRUCTION

Fig. 11–6 At the end of each sequence program an end instruction must be given to the program. This end instruction brings the program back to step one.

Logic function for P.L.C.

Figures 11–7, 11–8 and 11–9 depict a few logic functions which have not been explained in much detail previously. The structure of the output relay in fig. 11–9 is identical to the structure of all other internal auxiliary relays. These also have a multitude of N/C and N/O contacts to drive them.

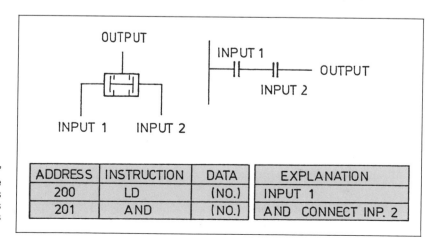

Fig. 11–7 Programming the electronic "AND" function. Although the "AND" function in the electronic ladder diagram is shown as a series function inside the P.L.C., these input signals are "AND" connected by electronic "AND" gates and not in series order!

ADDRESS	INSTRUCTION	DATA	EXPLANATION
200	LD	(NO.)	INPUT 1
201	AND	(NO.)	AND CONNECT INP. 2

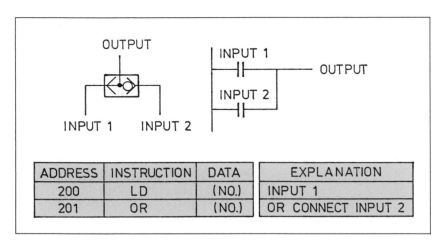

Fig. 11–8 Programming the electronic "OR" function.

ADDRESS	INSTRUCTION	DATA	EXPLANATION
200	LD	(NO.)	INPUT 1
201	OR	(NO.)	OR CONNECT INPUT 2

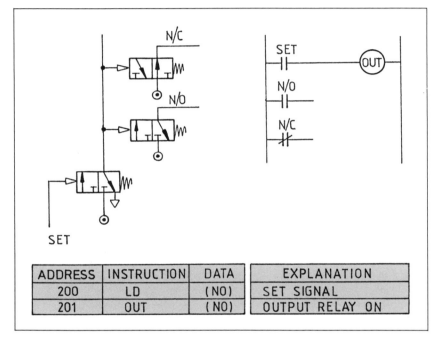

ADDRESS	INSTRUCTION	DATA	EXPLANATION
200	LD	(NO)	SET SIGNAL
201	OUT	(NO)	OUTPUT RELAY ON

Fig. 11–9 Programming the output relay. This is the same for internal auxilary relays.

Timer functions

With timer functions one may again follow the structure given for internal auxiliary relays. A timer may be programmed for a set time ranging between 0 and 999.9 seconds. Timer number allocations for the "OMROM C20" are from 00 to 47. Timers and counters may not be allocated the same number.

The timer works by decrementing and produces an output signal when the time has elapsed to 0000 sec. The timer starts when its timer "ON" signal is logic 1 and resets when its "ON" signal is cancelled. The timer output is transmitted externally through an output relay as shown in fig. 11–10.

The timer appears separately in the ladder diagram and its contact (or contacts if several times

used) may be used in the ladder diagram wherever required (see figs. 11–10 and 11–13). Timer contacts may be normally open or normally closed (see fig. 11–10).

The timer is in reality a counter in which an internal clock pulses the timer from its "set value" to count down to zero. This internal clock works on a pulse rate of 0.1 per second. Thus a 10.6 second delay is programmed as # 106.

Shunt module (selector module)

The shunt module may also be regarded as an independent module similar to the step-counter module discussed and illustrated in fig. 11–4. The shunt module can be used for program selection in selective parallel control applications, as illustrated in figs. 8–47 and 8–48.

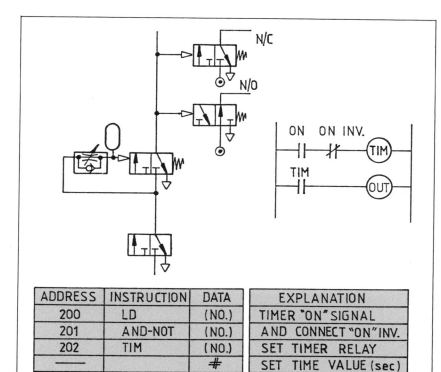

Fig 11-10 On-delay timer. The timer requires an output relay when used externally to drive a load. Timer set value does not require an address. Timer values are set in increments of 0.1 seconds (pulse rate of the internal clock). Thus a 2 second time delay is programmed as value 20 (# 20).

ADDRESS	INSTRUCTION	DATA	EXPLANATION
200	LD	(NO.)	TIMER "ON" SIGNAL
201	AND-NOT	(NO.)	AND CONNECT "ON" INV.
202	TIM	(NO.)	SET TIMER RELAY
———		#	SET TIME VALUE (sec)
203	LD	TIM (NO)	TIM DRIVES OR
204	OUT	(NO.)	OUTPUT RELAY ON

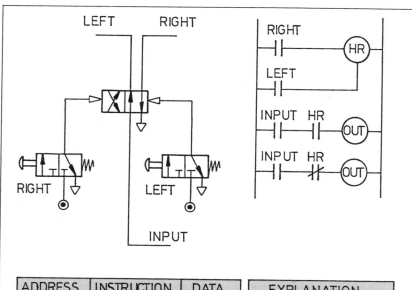

ADDRESS	INSTRUCTION	DATA	EXPLANATION
200	LD	(NO.)	SELECT RIGHT
201	LD	(NO.)	SELECT LEFT
202	FUN 11	HR (NO.)	HOLDING RELAY ON
203	LD	(NO.)	INPUT
204	AND	HR (NO.)	AND CONNECT HR
205	OUT	(NO.)	OUTPUT RIGHT ON
206	LD	(NO.)	INPUT
207	AND-NOT	HR (NO.)	AND CONNECT $\overline{HR}$
208	OUT	(NO.)	OUTPUT LEFT ON

Fig. 11-11 Shunt module as used for program selection in multi-program control applications (see Chapters 8 and 9).

Simple cycle selection module

The simple cycle selection module has already been explained and applied in Chapter 8 (figs. 8–26 and 8–33). This module is used to furnish a sequentially controlled machine with a permanent start signal when "AUTO" cycling is selected. When "MANUAL" cycling start is selected, the operator is required to actuate the start valve each time a cycle is to begin (see figs. 11–12 and 11–13). An application circuit which includes a time-delay function and a cycle selection module is given in fig. 11–13.

Extended cycle selection module

The extended cycle selection module has also been explained and applied in Chapter 8 (fig. 8–27). This module provides separate selection as well as separate start for both cycling modes. It therefore requires two holding relays (keep relays), four push-button or key operated switches, and a number of logic functions (see fig. 11–14).

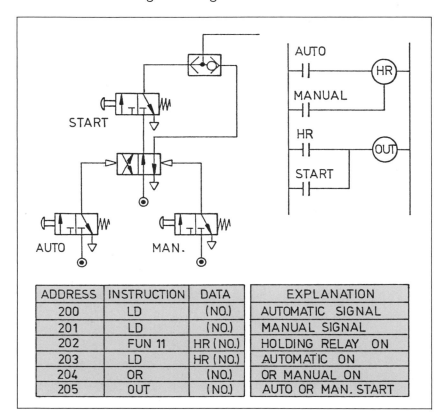

ADDRESS	INSTRUCTION	DATA	EXPLANATION
200	LD	(NO.)	AUTOMATIC SIGNAL
201	LD	(NO.)	MANUAL SIGNAL
202	FUN 11	HR (NO.)	HOLDING RELAY ON
203	LD	HR (NO.)	AUTOMATIC ON
204	OR	(NO.)	OR MANUAL ON
205	OUT	(NO.)	AUTO OR MAN. START

Fig. 11–13 *Right:* Circuit with time delay function and cycle selection module. The sequence for this control is: A1–B1–B0–C1–B1–B0–T–A0/C0. The confirmation from sequence step 7 must be an "AND" function consisting of the signals a_0 and c_0 (001 • 005). Allocation of relay numbers is as follows:

A1	= 502	b_1	= 004
A0	= 501	b_0	= 003
B1	= 504	c_1	= 006
BO	= 503	c_0	= 005
C1	= 506	AUTO	= 013
C0	= 505	MAN.	= 014
a_1	= 002	START	= 015
a_0	= 001	START	= 1000

Fig. 11–12 Simple cycle selection module to provide either automatic cycling or manual cycling. To achieve a true memory function, a holding relay must be used. The output relay (OR) is driven either by the contact of the holding relay (HR) or by the start switch (see also "OR" function depicted in fig. 11–8).

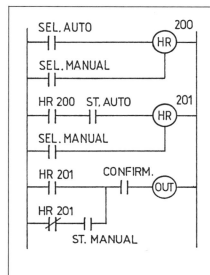

Fig. 11–14 Extended cycle selection module (see also figs. 8–27 and 11–17). This module provides separate selection and start for both cycling modes.

ADDRESS	INSTRUCTION	DATA
200	LD	012
201	LD	014
202	FUN 11	HR 200
203	LD	HR 200
204	AND	013
205	LD	014
206	FUN 11	HR 201
207	LD	HR 201
208	LD-NOT	HR 201
209	AND	015
210	OR-LD	
211	AND	HR 300
212	OUT	500

SELECT AUTO = 012 START AUTO = 013
SELECT MAN = 014 START MAN = 015

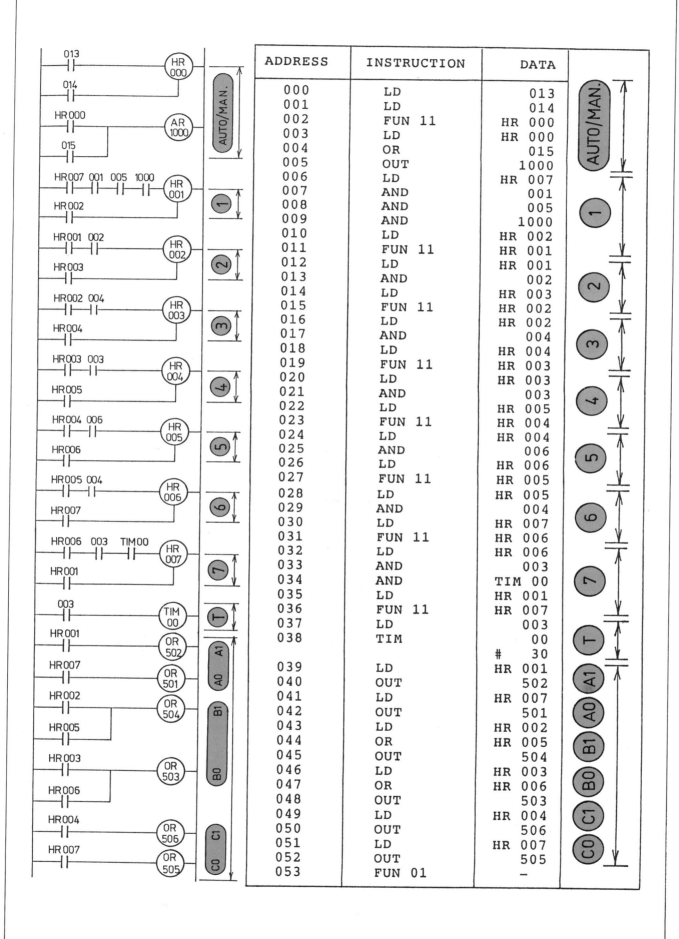

ADDRESS	INSTRUCTION	DATA
000	LD	013
001	LD	014
002	FUN 11	HR 000
003	LD	HR 000
004	OR	015
005	OUT	1000
006	LD	HR 007
007	AND	001
008	AND	005
009	AND	1000
010	LD	HR 002
011	FUN 11	HR 001
012	LD	HR 001
013	AND	002
014	LD	HR 003
015	FUN 11	HR 002
016	LD	HR 002
017	AND	004
018	LD	HR 004
019	FUN 11	HR 003
020	LD	HR 003
021	AND	003
022	LD	HR 005
023	FUN 11	HR 004
024	LD	HR 004
025	AND	006
026	LD	HR 006
027	FUN 11	HR 005
028	LD	HR 005
029	AND	004
030	LD	HR 007
031	FUN 11	HR 006
032	LD	HR 006
033	AND	003
034	AND	TIM 00
035	LD	HR 001
036	FUN 11	HR 007
037	LD	003
038	TIM	00
		# 30
039	LD	HR 001
040	OUT	502
041	LD	HR 007
042	OUT	501
043	LD	HR 002
044	OR	HR 005
045	OUT	504
046	LD	HR 003
047	OR	HR 006
048	OUT	503
049	LD	HR 004
050	OUT	506
051	LD	HR 007
052	OUT	505
053	FUN 01	–

Emergency stop for P.L.C. step-counter

The principles for emergency stop control explained and illustrated in Chapter 8 without exception also apply for P.L.C. programming. Figure 11–15 illustrates the concept and P.L.C. programming instructions used to obtain a basic emergency stop module (see also Chapter 8 fig. 8–23).

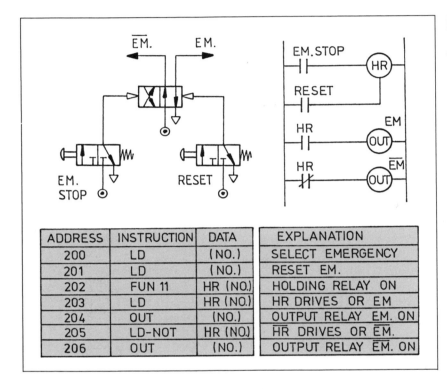

ADDRESS	INSTRUCTION	DATA	EXPLANATION
200	LD	(NO.)	SELECT EMERGENCY
201	LD	(NO.)	RESET EM.
202	FUN 11	HR (NO.)	HOLDING RELAY ON
203	LD	HR (NO.)	HR DRIVES OR EM
204	OUT	(NO.)	OUTPUT RELAY EM. ON
205	LD-NOT	HR (NO.)	HR DRIVES OR EM.
206	OUT	(NO.)	OUTPUT RELAY EM. ON

Fig. 11–15 Basic emergency stop module. This module is useful for all emergency stop modes except emergency stopping mode 3, which requires a cycle selection module (fig. 11–12).

Emergency stop and cycle selection

This module too may be linked to a cycle selection module to shift the cycle selector into manual cycling mode after emergency stopping was selected. The pneumatic version of this arrangement is shown in fig. 8–28 and the P.L.C. version in fig. 11–16.

To obtain stopping modes 1 and 2, which are called "STOP INSTANTANEOUSLY" and "STOP AT END OF COMMENCED SEQUENCE STEP", one must cancel all electrical signals leading to the solenoid valves of the fluid power circuit (see figs. 8–31 and 8–32). With the P.L.C. this may be achieved as is shown in figs. 11–15 and 11–17. Stopping mode 3 is illustrated in fig. 11–13, in conjunction with the basic cycle selection module. A detailed explanation of these stopping modes is given in Chapter 8.

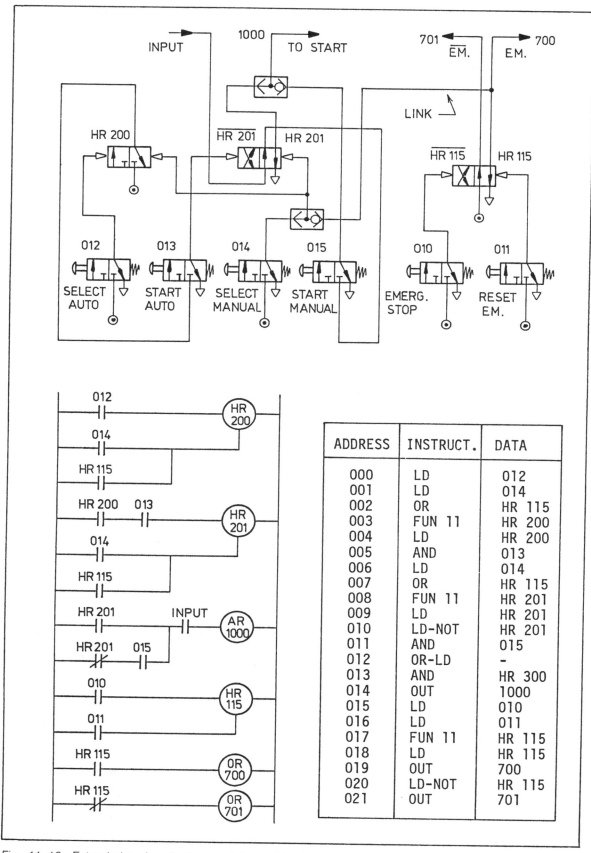

Fig. 11-16 Extended cycle selection module combined with emergency stop module.

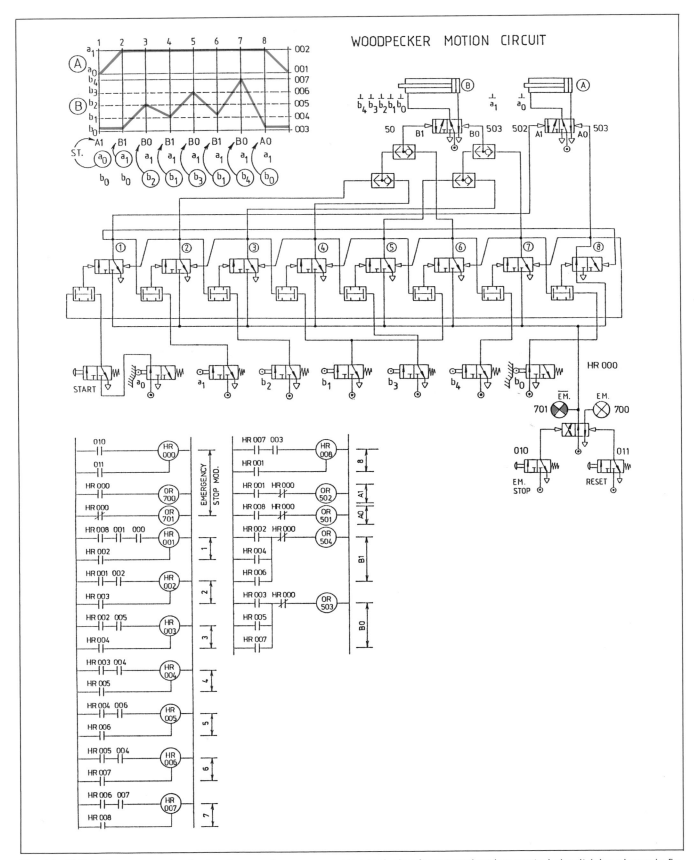

Fig. 11–17 P.L.C. programming instructions and emergency stop inclusion for a woodpecker control circuit (also shown in fig. 8–21). Stopping mode no. 2 is used for this control example.

Continuation modes deviating from the normal sequence

Where the continuation mode (fig. 8–22) must deviate from the normal machine sequence, as shown in the traverse-time diagram, one can no longer use the step-counter to sequence the actuators, and therefore the step-counter must be reset to "ZERO".

To accomplish this, the stepping modules require an "OR" function on every reset rung in the ladder diagram leading to the keep relays (holding relays; see figs. 8-34 and 11–18). The last step-counter module logically needs no "OR" emergency reset-input, since there is hardly a possibility for an emergency to occur during the last sequence step.

Emergency-stop modification module for P.L.C.

For pneumatically controlled circuits it was recommended that a modification be used to produce a substitute for the missing preparation signal usually coming from the last keep relay in the sequence control. Because of the emer-

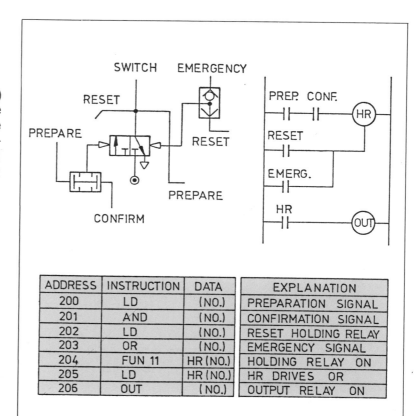

ADDRESS	INSTRUCTION	DATA	EXPLANATION
200	LD	(NO.)	PREPARATION SIGNAL
201	AND	(NO.)	CONFIRMATION SIGNAL
202	LD	(NO.)	RESET HOLDING RELAY
203	OR	(NO.)	EMERGENCY SIGNAL
204	FUN 11	HR (NO.)	HOLDING RELAY ON
205	LD	HR (NO.)	HR DRIVES OR
206	OUT	(NO.)	OUTPUT RELAY ON

Fig. 11–18 Step-counter module with emergency-reset provision.

Fig. 11–19 Emergency-stop module combined with preparation signal substitute module.

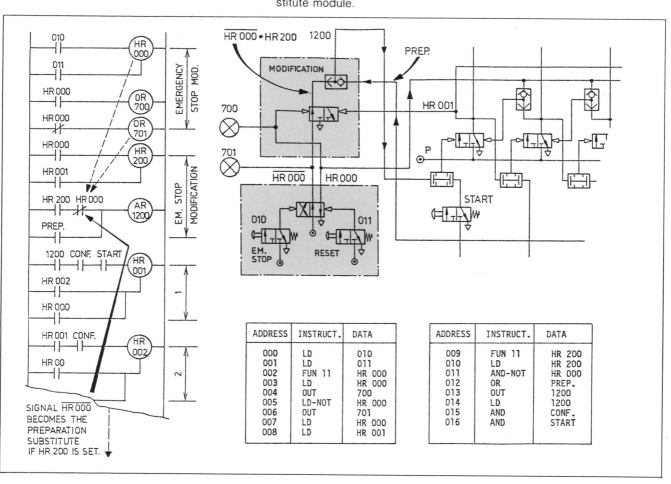

ADDRESS	INSTRUCT.	DATA
000	LD	010
001	LD	011
002	FUN 11	HR 000
003	LD	HR 000
004	OUT	700
005	LD-NOT	HR 000
006	OUT	701
007	LD	HR 000
008	LD	HR 001

ADDRESS	INSTRUCT.	DATA
009	FUN 11	HR 200
010	LD	HR 200
011	AND-NOT	HR 000
012	OR	PREP.
013	OUT	1200
014	LD	1200
015	AND	CONF.
016	AND	START

gency interruption, the last step-counter module does not get set, and therefore no preparation signal is available to start a new cycle. It is, therefore, also necessary for P.L.C. programming to use this modification. For programming instruction see fig 11-19.

Where the continuation mode demands that after the emergency command was given all actuators must return to their initial cycle position, one also has to use "OR" functions on the power valve signal inputs to achieve such emergency return. A typical control example including this continuation mode is given in fig. 11-20.

Programming the P.L.C.

As previously explained, no commercially available electronic programmable controller can be programmed in exactly the same way as another brand.

The "OMROM C20", however, has many features in common with the more widely used brands for fluid power control. To give the reader a real recipe for the programming of a P.L.C. the following illustrations and procedural steps are based on the "OMROM C20". The "OMROM" manual suggests that the following seven basic programming steps should be used.

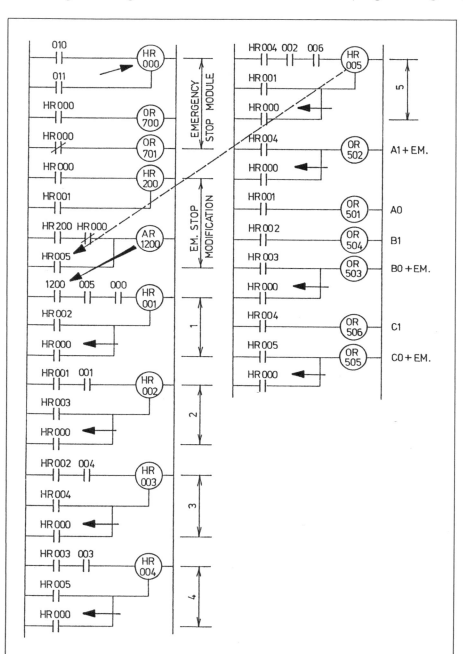

Fig. 11-20 Control example including an emergency stop module, a preparation substitute module, and a control program for the sequence: A0-B1-B0-C1/A1-C0. The emergency signal causes all actuators to return to their cycle start position at once.

Program Simplification

Programing can be very simple, or complicated, depending on the procedure used. Generally, a complicated circuit can be divided into several simple blocks. Program each block and then combine them.

For example, let's program the following circuit.

(1) First, divide the above circuit into small blocks. In this example, it is divided into six blocks: a, b, c, d, e, and f.

(2) Program each block from top to bottom, and left to right. Always combine blocks vertically, and then left to right. Therefore, the order in which the blocks are programmed and combined is from ① to ⑤.

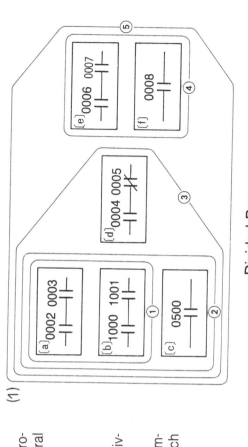

Divided Program

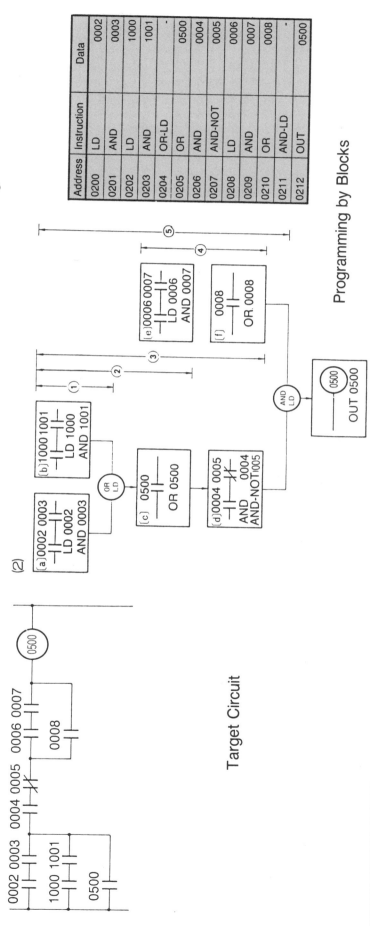

Address	Instruction	Data
0200	LD	0002
0201	AND	0003
0202	LD	1000
0203	AND	1001
0204	OR-LD	-
0205	OR	0500
0206	AND	0004
0207	AND-NOT	0005
0208	LD	0006
0209	AND	0007
0210	OR	0008
0211	AND-LD	-
0212	OUT	0500

Programming by Blocks

Target Circuit

Fig. 11-21 Program simplification (target circuit — divided program — programming by blocks).

- Determine what the system to be controlled must do, and in what order the sequence steps must occur (draw a traverse-time diagram).
- Assign labels to all input and output devices. That is, designate name or number to all limit switches, start switches, or other input switches and also to all solenoid valve commands which cause these valves to redirect or pass flow (A1, B0, C1, a_0, b_0, START).
- Draw a ladder diagram using relay and contact symbols as shown in the previous illustrations. This diagram must represent the correct sequence at which the program is to execute its switchings.
- Code the ladder diagram with appropriate relay and contact numbers as shown in figs. 11–5 and 11–16. The C.P.U. of the controller does not accept any other instructions.
- By means of the programming console transfer these coded numbers and instructions into the C.P.U. of the controller.

- Edit the program, and test the program for errors. These functions are accomplished with inbuilt "SEARCH" facilities which enable the programmer to locate, insert, delete and rectify any parts of the program before it is permanently saved onto a cassette tape.
- Save the program, after it has been tested, onto a cassette tape. The same tape may also be used to program other compatible P.L.C. controllers.

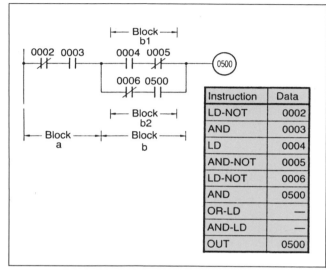

Fig. 11–24 The parallel circuit block of the series-parallel circuit shown below can also be divided into two branches. In this case, program block a, and then blocks b1 and b2, in this order. Then combine blocks b1 and b2 with OR-LD. Finally, combine block a and block b with AND-LD.

Fig. 11–25 Connecting parallel circuits in series. To program two or more parallel circuit blocks in series, first divide the entire circuit into the parallel circuit blocks. Then subdivide each parallel circuit block into the individual blocks. Program each of the parallel circuit blocks, and then combine them in series.

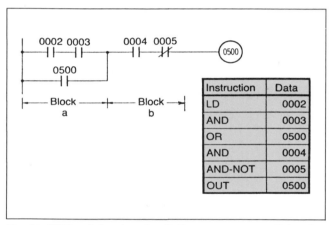

Fig. 11–22 Parallel-series circuit. To program a parallel-series circuit, simply program the parallel circuit blocks first, and then the series circuit blocks. In the following example, first program block a and then block b.

Fig. 11–23 Series-parallel circuit. To program a series-parallel circuit, divide the circuit into the series circuit blocks, and parallel circuit blocks. Program each block and then combine the blocks into one circuit. In the following example, divide the circuit into blocks a and b, and program each block. Then combine blocks a and b with AND-LD.

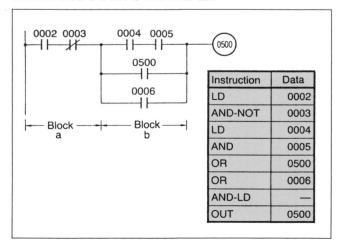

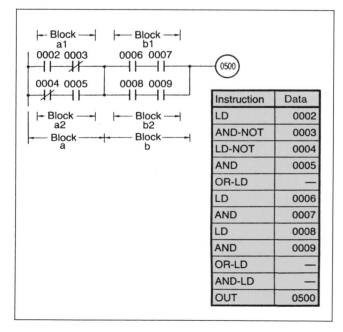

Name	No. of points	Relay number									
Input relay	80	0000 to 0415									
		00CH		01CH		02CH		03CH		04CH	
		00	08	00	08	00	08	00	08	00	08
		01	09	01	09	01	09	01	09	01	09
		02	10	02	10	02	10	02	10	02	10
		03	11	03	11	03	11	03	11	03	11
		04	12	04	12	04	12	04	12~	04	12
		05	13	05	13	05	13	05	13	05	13
		06	14	06	14	06	14	06	14	06	14
		07	15	07	15	07	15	07	15	07	15
Output relay	60	0500 to 0915									
		05CH		06CH		07CH		08CH		09CH	
		00	08	00	08	00	08	00	08	00	08
		01	09	01	09	01	09	01	09	01	09
		02	10	02	10	02	10	02	10	02	10
		03	11	03	11	03	11	03	11	03	11
		04	12	04	12	04	12	04	12	04	12
		05	13	05	13	05	13	05	13	05	13
		06	14	06	14	06	14	06	14	06	14
		07	15	07	15	07	15	07	15	07	15
Internal auxiliary relay	136	1000 to 1807									
		10CH		11CH		12CH		13CH		14CH	
		00	08	00	08	00	08	00	08	00	08
		01	09	01	09	01	09	01	09	01	09
		02	10	02	10	02	10	02	10	02	10
		03	11	03	11	03	11	03	11	03	11
		04	12	04	12	04	12	04	12	04	12
		05	13	05	13	05	13	05	13	05	13
		06	14	06	14	06	14	06	14	06	14
		07	15	07	15	07	15	07	15	07	15
		15CH		16CH		17CH		18CH			
		00	08	00	08	00	08	00			
		01	09	01	09	01	09	01			
		02	10	02	10	02	10	02			
		03	11	03	11	03	11	03			
		04	12	04	12	04	12	04			
		05	13	05	13	05	13	05			
		06	14	06	14	06	14	06			
		07	15	07	15	07	15	07			
Holding relay (retentive relay)	160	HR000 to 915									
		00CH		01CH		02CH		03CH		04CH	
		00	08	00	08	00	08	00	08	00	08
		01	09	01	09	01	09	01	09	01	09
		02	10	02	10	02	10	02	10	02	10
		03	11	03	11	03	11	03	11	03	11
		04	12	04	12	04	12	04	12	04	12
		05	13	05	13	05	13	05	13	05	13
		06	14	06	14	06	14	06	14	06	14
		07	15	07	15	07	15	07	15	07	15
		05CH		06CH		07CH		08CH		09CH	
		00	08	00	08	00	08	00	08	00	08
		01	09	01	09	01	09	01	09	01	09
		02	10	02	10	02	10	02	10	02	10
		03	11	03	11	03	11	03	11	03	11
		04	12	04	12	04	12	04	12	04	12
		05	13	05	13	05	13	05	13	05	13
		06	14	06	14	06	14	06	14	06	14
		07	15	07	15	07	15	07	15	07	15

Name	No. of points	Timer/counter number					
Timer/counter	48	TIM/CNT00 to 47					
		00	08	16	24	32	40
		01	09	17	25	33	41
		02	10	18	26	34	42
		03	11	19	27	35	43
		04	12	20	28	36	44
		05	13	21	29	37	45
		06	14	22	30	38	46
		07	15	23	31	39	47

Fig. 11–26 Assignment of input/output channels and relay numbers for an "OMRON C20" programmable electronic controller.

Allocation of I/O channels and relay numbers

The C20 comes in two versions, a basic unit and an expandable unit.

The capabilities of the basic unit include up to 1194 program statements, 28 I/O points, and 136 internal auxiliary relays. The expandable unit is functionally identical to the basic unit except that it can be expanded using expansion I/O units to include up to 140 I/O points.

The C20 offers the flexibility of either RAM- or EPROM-based operation. When a RAM chip is installed as the memory, programs written by the user can be modified and rewritten. EPROM provides a semi-permanent storage for complete programs.

The programming console can be detached and is upwardly compatible with the full line of SYSMAC C-Series programmable controllers.

Due to its compact design, the C20 incorporates the detachable I/O terminals and microprocessor functions in a single housing called the CPU. The detachable programming console functions as the programming device. Additionally, various optional peripheral devices are available to support system expansion.

The input channel of the CPU is fixed at channel 00 while the output channel is fixed at channel 05.

When an expansion I/O unit, I/O link unit, or both are connected to the CPU, the I/O channel of the connected units are automatically assigned and registered in the CPU.

For example, when a 28-point expansion I/O unit is connected, input channel 01 is automatically assigned to the expansion I/O unit and registered in the CPU. As the output channel number of the expansion unit, 06 is automatically assigned and registered.

The input/output signals (devices) are connected to the input/output terminals on the CPU. Because the CPU uses the numbers assigned to the input/output terminals when executing the program, assignment and management of the input/output terminal numbers are required and must be correctly performed.

The numbers assigned to each relay are listed in the following tables. Use these to keep a record of which relays have been used.

Input relays

The CPU has 16 input relay points (one input channel). The number of input points can be increased to a maximum of 80 by the addition of expansion I/O units. Because one channel equals 16 points, this means that a maximum of five channels (from channel 00 to 04) are available.

The data from the SYSBUS are received by the input channel relays assigned to the I/O link unit.

Output relays

As with the input relays, the CPU has one output channel consisting of 16 relay points. However, of these 16 points, numbers 12 to 15 are internal auxiliary relays used to carry out CPU internal processes. For this reason, the number of output relays the CPU actually possesses is 12.

When expansion I/O units are connected to the CPU, a maximum of 60 output relays are available for a total of five channels, 05 to 09. The statuses of all the output channel relays assigned to the I/O link unit are transferred to the SYSBUS when an I/O link unit is employed.

Internal auxiliary relays

The CPU has 136 internal auxiliary relays (No. 1000 to 1807) that constitute channels 10 to 18.

Holding relays (retentive relays)

The CPU has 160 holding relays (No. HR000 to 915) that constitute holding relay channels 0 to 9. The holding relay retains data during power failure.

Timers/counters

The CPU has 48 points of timers/counters, TIM/CNT00 to 47, that can be used for either timers or counters. Timers and counters cannot be assigned the same number.

Temporary memory relays

The CPU has eight temporary memory relay points (TR0 to 7).

Special auxiliary relays

The CPU has 16 special auxiliary relays, some of which operate or release according to internal conditions controlled by the hardware irrespective of the statuses of the I/O devices. Each of these special auxiliary relays functions as follows:

Relay 1808
This relay operates when a battery failure occurs. An alarm signal indicating a battery failure can be output to an external device by programming a circuit incorporating the contact of this relay.

Relay 1809
Normally OFF

Relay 1810
Normally OFF

Relay 1811
Normally OFF

Relay 1812
Normally OFF

Relay 1813
Normally ON

Relay 1814
Normally OFF

Relay 1815
This relay is turned ON for one scan time at the start of program execution.

A *scan time* is the time required for the PC to execute the user program once starting from address 0000 to the program's end instruction.

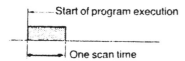

Relay 1900
This relay is used to generate a 0.1 s clock. When used in conjunction with a counter, it functions as a timer that can retain its present value during power failure.

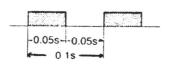

Note: The ON time of a 0.1 s clock is 50 ms. If a longer time is required for program execution, the CPU may fail to read the clock.

Relay 1901
This relay is used to generate a 0.2 s clock. When used in conjunction with a counter, it functions as a long-time timer that can retain its present value during power failure.

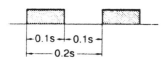

Relay 1902
This relay is used to generate a 1 s clock. When used in conjunction with a counter, it functions as a

long-time timer than can retain its present value during power failure. The relay output can be also used as a flicker signal.

Relay 1903
This relay turns ON when the result of an arithmetic operation is not output in BCD form.

Relay 1904
This relay serves as a carry flag and operates or releases according to the result of an arithmetic operation. It can be forcibly turned ON by the SET CARRY (STC) instruction and turned OFF by the CLEAR CARRY (CLR) instruction.

Relay 1905
This relay turns ON if the result of the COMPARE (CMP) instruction is ">" (more than).

Relay 1906
This relay turns ON if the result of the compare operation is "=" (equal to). It may also turn ON if the result of an arithmetic operation is 0.

Relay 1907
This relay turns ON if the result of the compare operation is "<" (less than).

Relays TR0 to 7
These are temporary memory relays and may not necessarily be assigned in sequence. The same coil number of these relays must not be used in duplicate within the same block of a program. However, the same coil number can be used in a different block. When using a temporary memory relay, the letters "TR" must be prefixed to the relay number (e.g., TR0).

Assigning I/Os

Points and channels

The programmable controllers of the SYSMAC C-Series all use the concept of I/O channels to identify individual I/O terminals, or points. Each of these channels consists of 16 points.

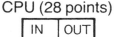

CPU (28 points)

The four-digit number used to identify an I/O point therefore can be broken down into the left-hand two digits, which identify the channel, and the right-hand two digits, which identify the point within the channel.

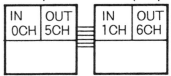

CPU (28 points)
+ I/O expansion unit (28 points)

For example, "0000" identifies the first point of the first channel and "0104" identifies the fifth point of the second channel. In the C20, the first five channels (00 to 04) are used for input and the next five (05 to 09) for output.

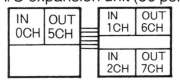

CPU (28 points)
+ I/O expansion unit (56 points)

Note: for simplicity reasons the author has omitted the first digit on all input relay numbers. Thus number 0002 appears as 002, etc.

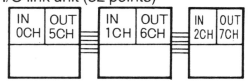

CPU (28 points)
+ I/O expansion unit (28 points)
+ I/O link unit (32 points)

The actual number of points used by the basic C20 is 16 inputs and 12 outputs.

Remember that the C20 can be expanded by adding expansion I/O units. Up to four new channels can be added for a total of 80 (5 × 16) input and 60 (5 × 12) output points.

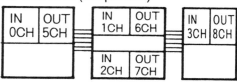

CPU (28 points)
+ I/O expansion unit (56 points)
+ I/O link unit (32 points)

I/O channel assignment

The assignment of I/O channels is as shown on the left.

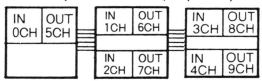

CPU (28 points)
+ two I/O expansion units (56 points)

Note: The total number of I/O points, including those of an I/O link unit, cannot exceed 140.

Programming console

This is the standard programming device used with the C20. The control programs written by the programming console are stored and run in the CPU.

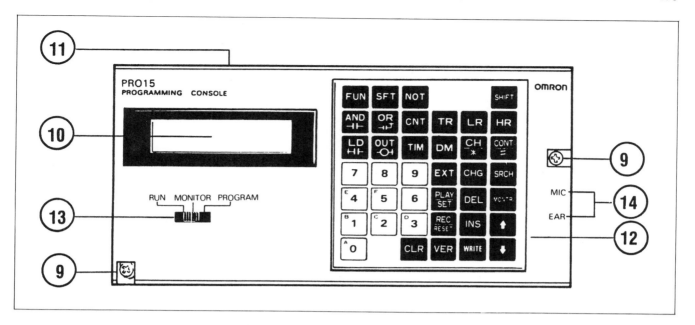

9 Mounting screws
These two screws secure the detachable programming console to the CPU.

10 LCD
This displays the program as it is being written and is used for checking and monitoring program operation. It also displays error messages.

11 Display light switch/contrast switch
This display switch controls the light that illuminates the LCD for use at night or in poorly lit environments. The contrast switch adjusts contrast level of display.

12 Keypad
The color-coded keypad is functionally divided into the following areas:

White   10 keys used to input program addresses, timing values, and other types of numeric entry

Red   One key used to clear the display

Yellow   12 keys used to provide editing functions while writing and correcting the control program

Grey   16 keys used to input instruction words used in the program

The function of each key is detailed later in this chapter.

13 Mode selector switch
This three-position switch selects one of three operation modes of the PC: program, monitor, and run. These modes are explained in detail in Chapters 4, and 5.

14 Jacks for connecting cassette tape recorder
Programs may be saved to a standard cassette tape recorder connected to the output (MIC) jack. Previously written programs can also be supplied to the CPU via the input (EAR) jack.

The keyboard

Numeric keys
These are the white keys numbered 0 to 9.

These keys are used to input numeric values used for program data. For instance, in a program as shown in figs. 11–13 and 11–21, these keys would be used to input the input/output numbers and timer/counter numbers and values.

These keys are also used in combination with

the function key (FUN) for special instructions such as FUN 11 (keep relay or holding relay).

CLR key

This red key is used to clear the display. It is also a key used when keying in the "password", which is used to foil unauthorized access to the PLC's program. Rather than using an actual password, though, one gains access using this two-keystroke entry:

When this is done, on the console display one will see written either PROGRAM, MONITOR, or RUN. Pressing the CLR key again makes the word disappear and prepares the PLC for the operation selected with the three-position mode switch.

Operation keys

These yellow keys are the ones used to carry out the editing functions of the programming console. These functions will be explained in more detail later but at this point it's important to know how three of these keys in particular are used.

The first two are the arrow keys. To move through the program a step at a time, press the bottom (down) arrow key. The displayed address of the program will increment once for each press. To go in the opposite direction, press the other (up) key. The program will then decrement one step at a time until it reaches its beginning.

The arrow keys are normally used for moving only a small number of steps in the program. Later several ways are given to move right to the program step you want.

The third key important in the yellow group is the WRITE key. During programming when one has written an instruction and its data, this key is used

to register the instruction in the PLC memory at the address desired.

Instruction keys

Except for the SHIFT key on the upper right, these grey keys are the ones used to place instructions in the program. The SHIFT key is similar to the shift key of a typewriter, and is used to obtain the second function of those keys which have two functions.

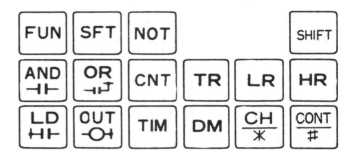

Each of the remaining grey keys has its function indicated by an abbreviation. The abbreviations mean:

FUN   Selects a special function. Used to key in special instructions. These special instructions are realized by pressing FUN and then the appropriate numerical values. The special instructions are listed in Appendix B.

SFT   Enters SHIFT REGISTER instruction.

NOT   Forms NC contact.

AND   Enters AND instruction used for ANDing two contacts.

OR   Enters OR instruction used for ORing two contacts.

CNT   Enters counter instruction. Must be followed by counter data.

LD   Enters LOAD instruction used for loading a specified input.

OUT   Enters OUTPUT instruction for outputting to a specified output point.

TIM   Enters timer instruction. Must be followed by timer data.

TR   Enters temporary memory relay instruction.

LR *   Enters link relay instruction.

HR   Enters holding relay instruction.

DM *   Enters data memory instruction.

CH *   Specifies a channel.

CONT   Used to search for a contact.

* Although these functions are not available on the C20, these keys are provided to ensure programming console compatibility with other SYSMAC C-Series PLCs.

A complete overview of all the keys on the programming console is given below.

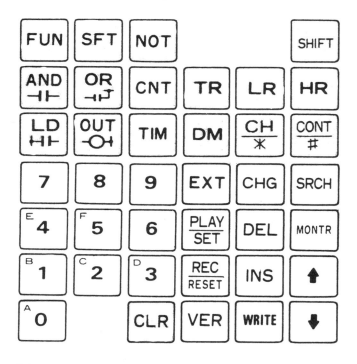

Mode switch

Now let's turn to the three-position mode switch one will use to select the operating mode of the PLC. Three modes can be selected.

The RUN mode is the one used to begin PLC operation. When this mode is turned on the PLC begins controlling the equipment using the program written into the PLC memory.

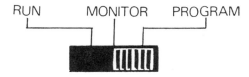

The MONITOR mode allows visual monitoring of the operation in progress. For instance, if one wants to check that a particular relay is in the correct state (either ON or OFF) at the proper time, one can move to the address (or step) which references that relay.

Using the PLC in the RUN and MONITOR modes is explained later.

The PROGRAM mode, which is fully explained below, is used during the programming operation.

Console display

The easy-to-read display is a practical window to look into the workings of the PLC. The display format changes depending on the mode selected. Let's first examine the PROGRAM mode.

PROGRAM mode

If the mode selector switch is turned to PROGRAM, this is what you'll see when you apply power to the PLC.

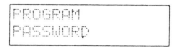

This is your electronic sentry that blocks unauthorized use of the PLC. To gain access, an actual password isn't necessary—just this key sequence:

The PLC responds with a beep each time you depress a key. This is what you now will see, informing you of the mode you are in.

To clear this, press

At this point you are ready to begin entering your program step-by-step using the code you wrote from your ladder diagram. The display you see is the starting address of the memory. This location is used in connection with the quick-search editing functions.

Let's skip this for now because we won't need to use it until after we have input our program.

Although it is possible to overwrite an old program with a new one, this is not recommended because it may cause confusion and lead to program-writing errors.

If the programming console has been used before and an old program currently exists in the memory, carry out the procedure given next.

To erase existing memory

Caution: The following procedure will entirely and permanently erase any program which currently exists in the CPU memory.

Anytime you want to erase the memory or start again when inputting your program, use this key sequence:

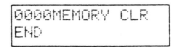

At this point, the display will be the one shown at left to allow you to reconsider. If you then press

you have finished the memory clearing operation and the display will change to this.

Now if you press

you can begin writing your program.

Summary
Putting all these steps together, here's a summary on what to do to begin programming the CPU:

1. Confirm whether a RAM chip is mounted and the DIP switches are set correctly.
2. Apply power to PLC
3. Turn mode selector switch to PROGRAM
4. Do these key sequences:

5. If you want to completely clear the old memory, press

Example program
For practice, let's key in the program code given below. This is the coding chart used as an example.

Address	Instruction	Data
0000	LD	0000
0001	LD	1000
0002	CNT	47
		#0005
0003	LD	CNT47
0004	OR	1000
0005	AND.NOT	TIM00
0006	OUT	1000
0007	LD	1000
0008	TIM	00
		#0020
0009	OUT	0500
0010	END(FUN01)	

Entering the program in the CPU

The first information to be keyed in is the LD instruction. To do this, press

and you see this display

The first set of four 0s at left is the beginning address position, and is where the LD instruction will be stored. The second set of four 0s at right is a numeric value representing the input point. Currently the input point is 0000. Since this is the value you assigned to the input of Counter 47, you can leave it unchanged. Then, to WRITE this LD instruction to memory address 0000, press

and this display appears.

```
0001READ
NOP (00)
```

Here, 001 is the address number; READ means you are reading the program; and NOP (00) means that no operation has yet been assigned to this address.

Next enter

again to specify the reset input of the counter. READ disappears because you are now *writing* to address 0001 instead of only *reading* its contents. Now key in the relay number and write it to the address with

In the next address we want to input the counter. This is done by pressing

Now we have to specify the coil number of the counter we want, which is 47. Do this by keying in

This second counter display

```
0002CNT    DATA
           #0000
```

indicates the set value of the counter. Currently it is 0000. We want to change this to 5, the number of signals that must be input by the optical sensor. So key in

Now we have to enter the LD instruction to specify the contact of the counter that corresponds to the coil of Counter 47. Do this by pressing

For address 0004, key in

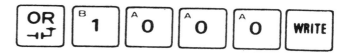

to enter the OR instruction and specify the internal auxiliary relay number. So that the timer acts to turn on the solenoid only 2 seconds, for address 0005 key in

This causes the contact of the timer instruction to act as an NC (normally closed) contact. Key in

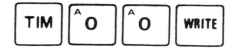

Now we have finished entering the timer contact. The timer coil for this contact must exist somewhere in the program for the timer instruction to be executed.

For address 0006 to designate the relay coil, key in

For address 0007, key in the contact corresponding to the internal auxiliary relay coil we have just entered.

Remember that the coil and the time value for the contact keyed in for address 0005 must still be specified. This is done by keying in

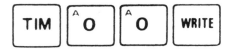

for address 0008.

This message is displayed asking for the timer data. Since the data includes a decimal point before the last digit, to set a value of 2 seconds, we enter

The only thing that remains is to complete the circuit by specifying the output. Do this by entering

The output relay number for this output is 0500, which is automatically selected by the PLC. (Remember that 05xx is the ouput channel designation, while 00xx is the input channel designation. The relay number is xx.)

The final instruction in our program is the END instruction, which tells the PC that the program is complete. Writing this requires the use of the FUN key and the value 01 which represents the END instruction. So key in

and you're finished.

Now check your program

To check whether the program has been correctly written, go back through it using the arrow keys to scan what you've done. If you see you need to make corrections, just overwrite the statements that are in error.

Deletion and insertion of instructions is explained next.

Deleting instructions

Let's return for a moment to Timer 00. For practice, let's delete it from the program using the following procedure.

First, go to address 0008. Then key in

This eliminates the TIM instruction and moves the next instruction (LD, in this case) into the 0008 address location.

This function is useful when you wish to modify or correct an existing program.

Inserting instructions

Inserting an instruction somewhere in a program is almost as easy. Let's assume that we now realize it was a mistake to eliminate Timer 00 from the program. To put it back into the program, we follow this six-step insert procedure.

1. First key in

This takes you to the first address of the memory.

2. Then key in the number of the address where you wish to make the insertion.

3. Press

4. And then

to move the program down one step, reserving address 0008 for the TIM instruction.

5. Key in the TIM data

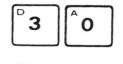

6. Then press

to complete the insertion.

Note: Steps 5 and 6 are necessary only to input the timer and counter data. These are not required for insertion of other types of instructions.

You can check that the insertion has been correctly done by moving the program using the arrow keys.

Spend some time becoming acquainted with these editing functions by inserting and deleting various instructions. In the next chapter you'll learn how to put these to use in debugging much larger programs.

Checking and running your program

Overview

The C20 has a number of helpful features to help you debug your program and fine-tune the operation before and after it is under way.

This instruction covers the final three steps for programming. We now explain how the quick-search editing functions make locating instructions quick and easy. Also explained is how to correct the program using the C20's debugging capabilities.

Once you're ready to begin a test run of the program, you'll learn how you can monitor the operation and make any necessary modifications based on the actual performance of the equipment.

Finally, we show you how to use a standard cassette tape recorder to save and load your program.

Editing the program

Quick-search editing functions

The C20 programming console has three features which greatly facilitate program editing and debugging. One allows you to go directly to specific program addresses. Another allows you to hunt for specific instructions. The last lets you speedily check each contact. These editing features eliminate the need for you to travel through the program a single step at a time—quite a tedious process when a large number of addresses are involved.

Going directly to a known address

This is a simple way of going straight to the address you want. Do this by first clearing the display by pressing

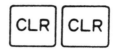

Then key in the number of the address you want.

Let's assume that we want to change the time value from 2.0 seconds to 3.0 seconds. We first key in its address

and follow that with

When you do this you will see that the display has gone directly to address 0008. Before you can change the time value, however, you must first go to the timer's data by pressing

to move the program down one step. This is the display.

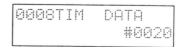

At this point you can make the change to 3.0 seconds by keying in

Check the display. It should now look like this.

When correcting or modifying a program, this method of address locating is particularly useful to go quickly to a new part of the program. Note, though, that you must always start at the first address in memory which is reached by pressing

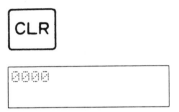

one or more times.

Searching for a specific instruction

This second editing technique gives you a convenient way to locate a particular instruction for which you do not have to know the program address. Assume, for example, that you wished to go to an unknown address containing Timer 00. To do this, press

several times until you see the first address.

Then key in

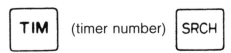

This takes you to address 0008 which contains Timer 00.

This editing feature has a wide range of uses. For instance, in a long program you may wish to deter-

mine whether you have inadvertently used the same timer or counter number more than once. This normally would be an error, since a particular timer or counter should be used only once in a program.

Let's imagine for a moment that our program is actually quite large and that there's a possibility that Timer 00 has been duplicated somewhere. We would search for the first reference to this timer by using the approach we just learned. Do this by keying in

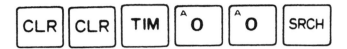

which moves us again to address 0008. To search for any illegal timers, simply press

again, which tells the PLC to scan the rest of the program for any other mention of a timer with the number 00. As you can see, there are no others in our program. (To test this further, go to any address, say 55, write in TIM00, and then repeat the search procedure.)

NO END INSTRUCTION message

In doing this you may have reached this display message.

The number 1193 here represents the total number of address locations available. Normally when you see this message, it means that you forgot to put an END instruction at the end of your program.

This is one time when you can ignore the error message, though. The search function has carried you into the unused region of the memory (address locations 0011 to 1193). No new END statement is actually required here.

Searching relay contacts

The quick-search operation can be performed still another way. If you wish to locate specific contacts, this can be accomplished easily in any of the three modes with this simple operation. Start from the first address of memory. Then press

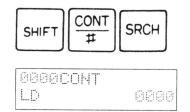

If you still have our sample program in the CPU memory, this should be the display. Then, to go to the next contact with the same number, again press

The display now changes to the next address containing an instruction with the same contact number (the four digits at the lower right of the display). This continues until the end of the program is reached.

If you want to search for a contact having a different contact number, follow the same steps as before. But this time, enter the numeric value of the contact you want to locate.

Using our sample program, if you enter 1000 and press the search key, the display will show the LD instruction in address 0001. This is how you would do it:

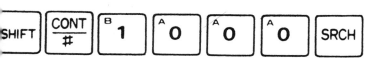

Search function summary

The C20 gives you three convenient ways to search through a long program for specific addresses, instructions, or contacts. The key entries for these are summarized here:

Address search

Specific instruction search

Specific contact search

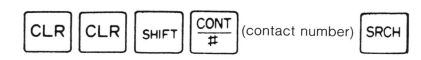

Testing for errors

The C20's debugging features can be used to catch many types of programming errors. For this you use the FUN and MONTR keys in any of the three modes.

To see if the program you entered in the PLC has a programming error, press

This is the display if no error is found.

If an error is found in the program, the corresponding error message number will be displayed. Refer to Appendix E, Maintenance and troubleshooting, and carry out the corrective action listed there.

If more than one error exists, continue pressing

MONTR

to display each error message one at a time.

There are basically two levels of errors—fatal and non-fatal. A fatal error, such as a memory error, prevents the PLC from operating. A non-fatal error, such as a battery failure, allows the operation to proceed but still must be corrected. In case both types of errors have occurred, the error messages for the fatal error take precedence over the other and must be corrected first.

Status check

At times you may want to scan the status (whether ON or OFF) of each relay contact before you start the control operation. This can be an effective troubleshooting technique and is available in both the RUN and MONITOR modes.

You would normally perform the status check after you have carried out the debugging step above. To begin, go to the first address and press

You'll see the address, the instruction, and the word OFF or ON displayed in the upper right hand corner of the LCD. OFF or ON indicates the current state of the relay contact. Continue pressing the DOWN arrow key to check the status of each of the relays that follow.

Forced set/reset

During the execution of the program, this operation is used to force set or reset (for one scan time) the operating status of each I/O relay, internal auxiliary relay, holding relay, timer, or counter. This operation is only meaningful while the PLC is in the MONITOR mode.

The most common use of this function is during a trial run of the controlled system. For instance, if a particular task (such as illuminating a heat lamp for drying) would normally take 30 minutes but you simply wish to test the contact, use this function to force reset the contact after a few moments of lamp operation. The program could then continue to the next task in the control sequence without delay.

Forced relay set

To do a forced set of a relay contact, first place the PLC in the MONITOR mode and press

Now specify the contact that you wish to force set. In this case, Relay 1000 is to be forced set. Press

The display shows the present status of the relay which in this case is OFF

Now, to force set the relay (to ON), press

This force sets the relay from OFF to ON.

Forced relay reset

To force reset a relay to the OFF state, press

This turns the currently ON relay to OFF.

Forced timer set

To force set a timer, put the PLC in MONITOR mode and press

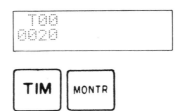

The display shows the first timer, Timer 00, and its preset time if the timer is not currently running. Then, to clear the set time for one scan time, press

Due to the short cycle period, however, you will not be able to see this clearing actually take place. If the timer is currently operating, depressing

displays the current value of the timer. If the timer has completely gone through its timing period, you will see this value.

Depressing

restarts the timing operation from the set time.

Note: Special auxiliary relays 1808 to 1907 cannot be forcibly set or reset. These relays serve as flags that are internally "raised" and "lowered" (set and reset) to enable monitoring of PLC operation. If an attempt is made to set or reset one of these relays, you'll hear a beep and no key input will be accepted by the programming console.

Changing set value of timer or counter
It's also possible to change the set value of a timer/counter during program execution. This is used for such purposes as slowing down or speeding up an assembly line operation without having to stop the program.

This operation can only be done in MONITOR mode. To reset the value of a timer, first key in

 (timer number)

In our sample program, the timer number is 00.

Now press

to move to the timer data display.

The timer currently is set for 3.0 seconds. To change this to 6.0 seconds, press

and the PLC display will query you about the new data.

0008DATA? informs you that the operation to be

performed is the setting of a value for the instruction address 0008; T00 is the number of the timer; #0030 is the timer's present set time; and #???? asks what you want the new set time to be.

In response, key in your new set time and write it to the PLC memory.

Follow the same procedure to change the set value of a counter.

Input/output monitor
While the PLC automatically controls the equipment, you can keep a watchful eye over the operation using the PLC's monitoring capabilities.

You may, for instance, have a particular unit of equipment that you want to check for correct operation. Or you may want to keep a running check on the value of a down counter.

The C20 allows you to easily check the status of any device connected to it by this procedure:

Place the PLC in the MONITOR mode during operation.

Locate the device you wish to check. We could do this for Counter 47 in our practice program, for instance, by keying in

which gives this display.

Press

and the display changes to this.

If the PLC were currently operating, we would be able to watch the counter as it is decremented. The same thing is also possible with timers.

Rapid check of counter/timer values

For a fast way to visually scan the set value of all timers and counters in a program, go to an address where a timer or counter is located. Then press

and the set value of the timer/counter is displayed as a four-digit number.

Then, each time you press

you will be able to see the set time of the next timer/counter.

Rapid check of relay status

You can check the status of relay contacts using the same technique. To do this, go to the address of a relay you want to check. Then press

to display the current status (whether ON or OFF) of the relay. Then, to move to the next relay, press either

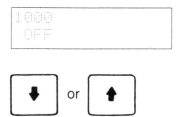

Saving your program to cassette tape

At this point it's assumed that you have completed program debugging and the trial run of the equipment and are satisfied with the operation. Now it's time to save the program to a cassette tape.

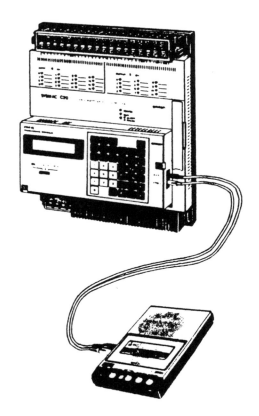

For this, you can use any reliable cassette tape recorder.

This provides a way to save your programs for later use. The same tape can also be used to program other C20s which control identical operations. Multiple copies of the tape can be made using a conventional tape copier, at either normal or high speed.

Store only one program to a tape (or side of a tape). The reason for this is that there is no way to identify individual programs if more than one have been stored on the same side of a cassette. The only requirement is that the tape be at least 7 minutes long. Either a standard or microcassette tape can be used.

First you'll learn how to save the program you have in the PLC's memory. Then you'll see how to load it back into the PLC and verify that it was loaded correctly.

To save the program, carry out this procedure:

1. Plug one end of a Type SCY-PLG01 cable into the MIC jack of the PLC and the other into the MIC jack of the tape recorder.*
2. Plug a second cable into the EAR jacks of both devices.**
3. Turn the volume and tone controls of the tape recorder to their maximum levels.
4. Switch the PLC to the PROGRAM mode.

5. Press

6. Press

A message is displayed to ask you to start the save operation.

 * On some tape recorders this may correspond to the LINE-IN jack.
** On some tape recorders this may correspond to the LINE-OUT jack.

7. Press the button or buttons on the tape recorder that begin the recording.
8. Then within 5 seconds press

on the programming console keyboard. A blinking rectangle appears in the right corner of the display. This means the program is being saved to the tape.

9. Wait about 7 minutes.

The program is being saved from the first program address to the end of the RAM/EPROM memory area. The PLC display increments as each address goes into the cassette recorder. During this period you can halt the save operation by pressing

10. When the program (plus vacant memory) has been completely recorded, the save operation stops with this message.

Refer to this flowchart for the correct procedure when saving your program to the cassette recorder.

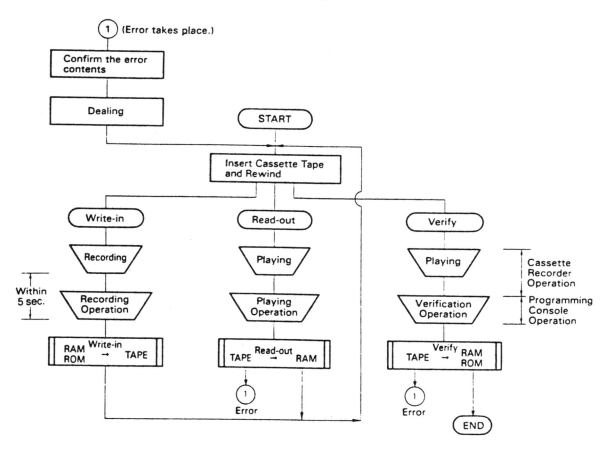

Loading and verifying the program

Before loading a program into the C20, determine that the tape is correctly positioned. The taped program should begin to load into the PLC within 5 seconds after the loading operation has been initiated. You can determine this by listening for the sound that marks the beginning of the program. Then rewind the tape slightly so a few seconds of blank tape precedes the beginning of the program. Then follow these steps:

1. Press PLAY on the tape recorder to start loading the tape into the PLC.
2. Press

on the PLC keyboard. When you do, this message appears and you'll see a blinking rectangle on the upper right portion of the display.

3. Wait about 7 minutes for the program to be completely loaded.
4. When the loading is complete, a message appears.

Verifying the program

It's always a good practice to check that the program was loaded correctly. This is done by performing the following procedure:

1. Rewind the tape until you reach the beginning. Provide about 5 seconds of blank tape leader before the taped program segment begins.
2. Turn on the PLAY button of the tape recorder.
3. On the C20 keyboard, press

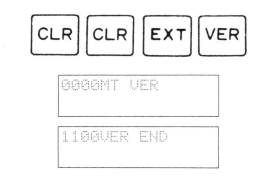

When you do, a message appears and a blinking rectangle indicates that program verification is taking place.

4. Wait about 7 minutes. A message will appear when verification has been successful.
5. If a program loading error has occurred, one of the messages described in Appendix E will be displayed. In this case, repeat the tape load operation.

12 Compressed Air Servicing

The final stage of compressed air servicing or conditioning is a process carried out on the air supply, directly before its point of usage. This process is accomplished by three devices, collectively called the *air service unit*, sometimes abbreviated F.R.L. or A.S.U. The air service unit consists of a filter, a pressure regulator, and in most cases also a lubricator. To read off the adjusted system pressure, a pressure gauge should also be attached to the pressure regulator (fig. 12–1). One of the three aims of the air service unit is to provide the air consuming equipment (actuators, motors, air tools and control circuits) with compressed air of suitable cleanliness: that is, without contaminants such as dirt and water. This aim is accomplished with the air filter.

The air service unit should also provide pressure stabilised air at no more than the required maximum pressure, which is normally about 50 kPa below the cut-in pressure of the air compressor capacity controller, but may be adjusted to less if required.

Lastly, it provides an air supply (flow), which carries lubricating oil in the correct adjusted quantities to lubricate the various components, such as valves, cylinders and motors.

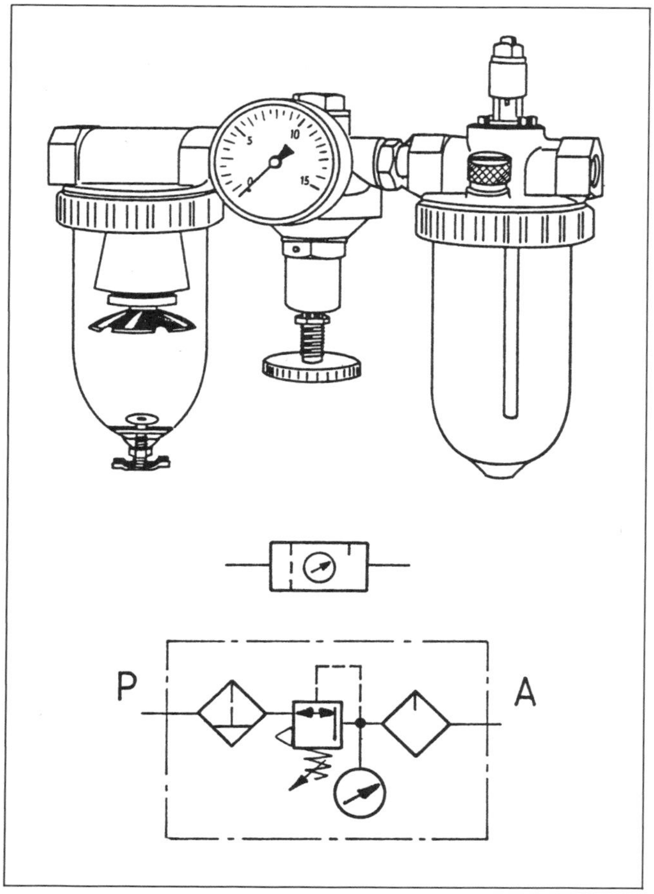

Air line filter (separator)

The air line filter is in fact more than just a filter, since it serves a twofold purpose. It is designed to filter out contaminant particles, but it also

Fig. 12–1 Air service unit consisting of filter, pressure reducing valve and lubricator.

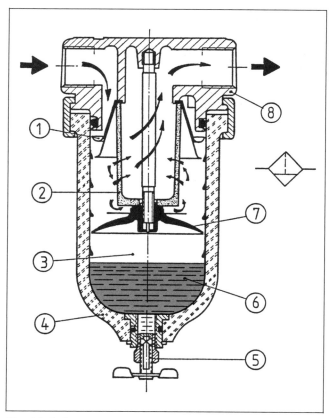

Fig. 12–2 Cross section through an air line filter (separator).

separates condensate from the passing air (fig. 12–2).

For most industrial applications, the air filter element is rated around 40 to 60 micrometres (microns). Some air control components and special tools require special filtration for maximum efficiency. Moving part logic components (see Chapter 8) and non-contact type air sensing equipment (see Chapter 6) are particularly sensitive to air contamination. It is therefore advisable to check with manufacturer's specifications to determine what filter rating is required.

To avoid undue pressure drop, correct sizing of the filter is important. The air flow in cubic metres/min F.A.D. through the air service unit must be ascertained, and an appropriate filter selected. It is recommended that, wherever possible, a filter with an automatic drain should be chosen (fig. 12–3).

With manually drained filters, if at any time, the condensate in the bowl accidently rises above its maximum level, it could then find access into the filter element, which would mean that all the contaminants and the water accumulated in the bowl would be forced through the filter element and would stream into the pneumatic system. This would not only reduce the flow capacity of the filter, but would also block the filter and contaminate the pneumatic system.

The filter cartridge should not need servicing for periods of less than twelve months. The filter bowl, however, must be drained whenever the maximum water level is reached. Servicing normally requires dismantling and cleaning of the bowl. The filter element on some filters may be reverse-flushed with air and cleaned with kerosene. Some filter cartridges, however, are of the disposable type (impregnated paper), and can therefore not be washed and cleaned.

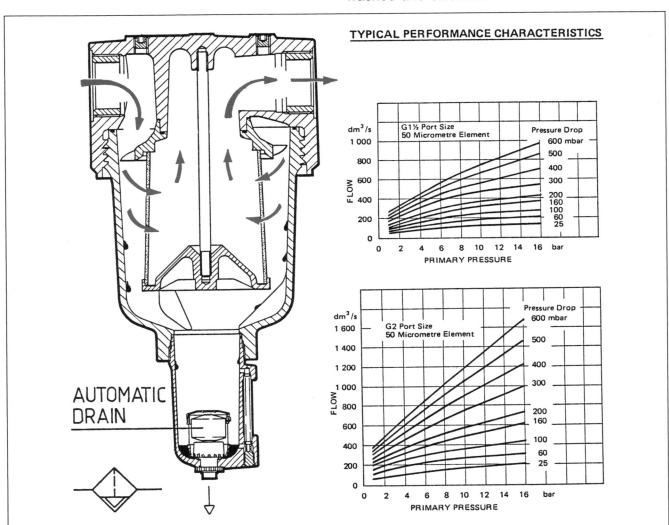

Fig. 12–3 Filter with automatic drain.

Pressurised air enters the air inlet port and passes through directional vanes or louvres ①, which force the air into a swirling motion (whirlpool effect). The centrifugal force caused by this swirling motion throws the larger contaminant particles and condensate droplets outwards, where they collect on the side of the bowl ④ and fall past the baffle into the "quiet zone" ③ (see fig. 12–2). From there they can be drained off ⑤, either manually or automatically. The baffle ⑦ prevents condensate and contaminants in the quiet zone ③ from being stirred up again and carried into the outgoing airstream. This baffle may also serve as a maximum condensate level indicator.

As the pressurised air leaves the bowl it passes through a filter element ②, which removes smaller solid contaminants. From there the air streams through the exit of the filter. Filters are supplied with a transparent bowl made from polycarbonate plastic ④, but where required, they can be supplied with metal bowls or with protective metal guards. Such guards may be necessary, since the polycarbonate bowl can deteriorate and become dangerous, as it could explode under pressure.

If the bowl through ageing or cleaning loses its transparency, it is advisable to have it exchanged for one made from metal or a new plastic bowl with metal guard. This applies to all other pressurised items in the air service unit, which use transparent polycarbonate plastic parts. Volatile liquids, such as thinner and some solvents, are not compatible with the polycarbonate plastic parts in the air service units, and cannot therefore be used for cleaning. If cleaning is required, one may use soapy water, dish washing detergents, or kerosene. Even momentary exposure to volatile chemicals or chemical fumes will start chemical deterioration of the plastic parts. A list of materials that will attack polycarbonate plastic bowls is given below.

Some materials which will attack polycarbonate plastic bowls:

Acetaldehyde	Ethylene glycol
Acetic acid (conc.)	Formic acid (conc.)
Acetone	Freon (refrigerant & propellant)
Acrylonitrile	
Ammonia	Gasoline (high aromatic)
Ammonium fluoride	Hydrazine
Ammonium hydroxide	Hydrochloric acid (conc.)
Ammonium sulfide	Lacquer thinner
Antifreeze	Methyl alcohol
Benzene	Methylene chloride
Benzoic acid	Methylene salicylate
Benzyl alcohol	Milk of lime (CaOH)
Brake fluids	Nitric acid (conc.)
Bromobenzene	Nitrobenzene

Butyric acid	Nitrocellulose lacquer
Carbolic acid	Phenol
Carbon disulfide	Phosphorous hydroxy chloride
Carbon tetrachloride	
Caustic potash solution	Phosphorous trichloride
Caustic soda solution	Propionic acid
Chlorobenzene	Pyridine
Chloroform	Sodium hydroxide
Cresol	Sodium sulfide
Cyclohexanol	Styrene
Cyclohexanone	Sulfuric acid (conc.)
Cyclohexene	Sulphural chloride
Dimethyl formamide	Tetrahydronaphthalene
Dioxane	Thiophene
Ethane tetrachloride	Toluene
Ethyl acetate	Turpentine
Ethyl ether	Xylene
Ethylamine	Perchlorethylene and others
Ethylene chlorohydrin	
Ethylene dichlroide	

Pressure reducing valve (pressure regulator)

Because the distribution system pressure must always be somewhat greater than the pressure required for any particular air operated device or system, some means of pressure reduction is required, preferably at each individual point of air consumption. The pressure reducing valve (regulator) is used for this purpose. Although the term "pressure reducing valve" more correctly describes the function of this device, the commonly used term is "pressure regulator". If air pressure driving an air operated device is greater than actually required, then some compressed air is wasted (and excessive air pressure means compression power unused). Since force output of an actuator or air tool depends on the system pressure, it can be said that the pressure reducing valve is a force regulator (force = pressure × area).

Pressure reducing valve without relief

Pressure reducing valves come in two versions, those with secondary system relief function (down stream relief) and those without. Providing the primary pressure is always somewhat higher than the adjusted secondary pressure, the pressure reducing valve will provide two important functions which may be summed up as follows:

- It maintains a nearly constant secondary pressure (adjusted pressure) independent of pressure fluctuations on the primary side (inlet side).

• It maintains a nearly constant secondary pressure (adjusted pressure) independent of flow demand fluctuations on the secondary side (outlet side).

A simplified cross-sectional illustration of a pressure reducing valve without secondary system relief function is given in fig. 12–4. This illustration is used to explain the somewhat complex operation of this valve.

F_S = Spring force (adjust regulation force)
$p2$ = Secondary pressure (adjusted pressure)
$p1$ = Primary pressure (inlet pressure)
A = Diaphragm area
F_D = Force from $p2$ on diaphragm
Q = Flow rate through valve
Δp = Pressure drop $p1$ to $p2$

In a condition, where the primary pressure $p1$ does not fluctuate and the flow demand on the secondary side is stable, the valve will also remain stable and the forces F_S and F_D will be equal ($F_S = F_D$). During such a condition the valve disc assumes a position (regulated by the diaphragm to which it is attached) which causes a pressure drop (Δp) exactly to the amount required to reduce the pressure from $p1$ down to $p2$ (fig. 12–4).

Where the previously explained force balance is disturbed due to a reduction of $p2$ (caused by an increased flow demand [Q] in the system), the spring must force the diaphragm up, and the valve disc increases the flow opening (orifice). By this, the pressure drop (Δp) through the orifice decreases and $p2$ increases until the adjusted pressure and force balance is reached again ($F_S = F_D$).

Where $p1$ fluctuates, the pressure $p2$ in the secondary chamber will also tend to fluctuate, but the spring loaded diaphragm reacts to these fluctuations and constantly adjusts the valve disc. Using this it opens or closes the orifice to maintain a constant $p2$.

Where $p2$ is on the increase, so that the force balance ($F_S = F_D$) becomes disturbed, the diaphragm will move down. This reduces the valve orifice and increases the pressure drop across the orifice (Δp). By this the secondary pressure will decrease again to find its equilibrium with the

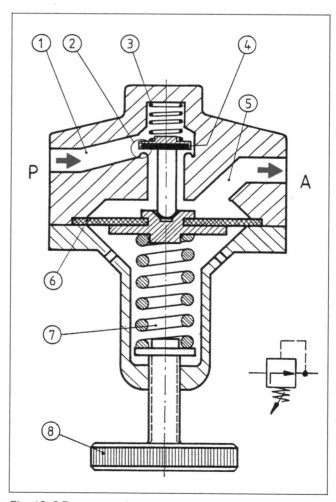

Fig. 12–5 Pressure reducing valve without secondary system relief failure.

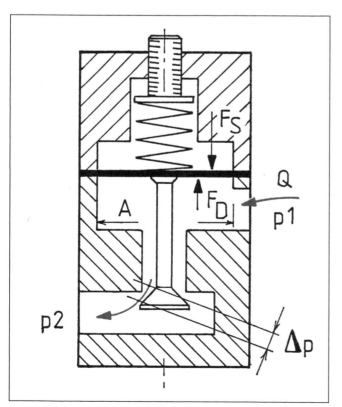

Fig. 12–4 Function principle of a pressure reducing valve.

spring force, and therefore, the force balance on the diaphragm is restored.

In a no-flow demand condition, the pressure in the secondary chamber (p2) will increase until it reaches the level of the primary pressure (p1). Pascal's Law may now be applied, for this is a static condition. The diaphragm is now forced down until the spring force $F_S = p1 \times A$. The valve disc now closes the orifice off completely until such time that flow from the system is demanded again (figs. 12–4 and 12–5).

Pressure reducing valve with relief function

Most pressure reducing valves are equipped with an inbuilt secondary pressure relief function. Their construction has a relief orifice in the centre of the diaphragm hub (fig. 12–6). During non-relief operating conditions the relief orifice is closed by the valve stem. If an overpressure situation occurs in the downstream line, the diaphragm will move up and the valve disc closes the main orifice completely. This uncovers the relief orifice in the centre of the diaphragm hub and excess air from the secondary side (p2) can bleed off through the hub orifice into the spring housing and then via vent holes to atmosphere.

When the secondary system over-pressure is relieved, the diaphragm reseats onto the valve stem and closes the orifice completely off. Thus all the forces within the valve are again in balance ($F_S = F_D$). The pressure relieving feature is a great advantage in closed or "dead end" circuits (cylinder applications).

Applications of pressure reducing valves

The pressure reducing valve must have a flow capacity greater than the sum of all the compo-

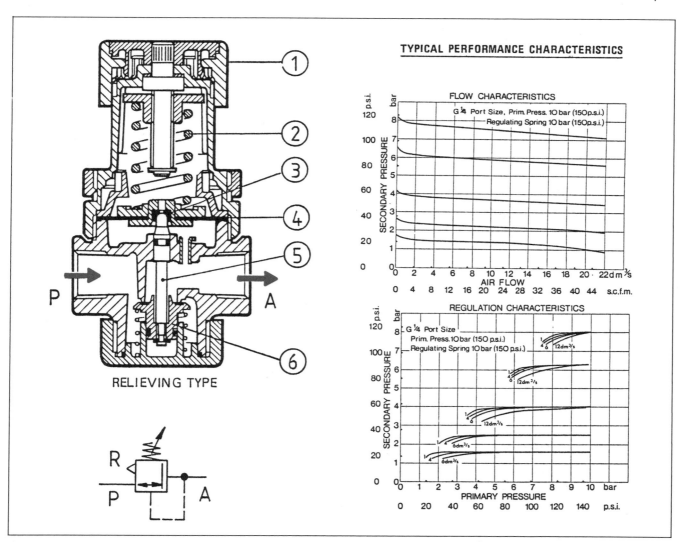

Fig. 12–6 Pressure reducing valve including secondary system relief feature.

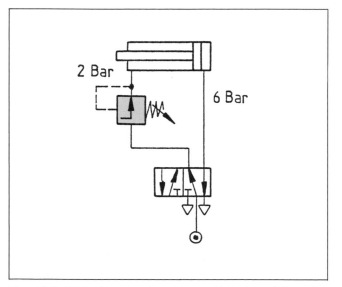

Fig. 12-7 Typical application circuit with reduced force for actuator retraction.

Similarly, one may use maximum system pressure only on the extension stroke of a linear actuator, but greatly reduced pressure (force) for its retraction stroke (if that retraction stroke is not a power stroke). The savings achieved by such pressure reduction will soon pay for the pressure reducing valve (fig. 12-7).

Further compressor savings may be achieved if the power circuit (see fig. 12-8) is operated on the adjusted pressure of the pressure reducing valve in the air service unit (say at 700 kPa or 7 bar). An additional pressure reducing valve may now be used to supply compressed air of only 250 kPa (2.5 bar) to all limit valves. The pneumatic pilot signals emanating from these limit valves require a pressure of no more than 250 kPa (2.5 bar) to actuate the power valves or internal logic valves (cascade valves or step counter modules). Figure 12-8 shows a typical sequential circuit where reduced pilot pressure is applied.

nents downstream of the valve. To provide individual control of actuator force in multi-actuator circuits, it is quite common to use a pressure reducing valve for each individual actuator, thus controlling the force and power consumption of each individual actuator.

Air lubrication

Air lubrication is still widely used today to reduce friction and corrosion in air valves, linear actuators and air motors. Physically speaking, lubricating oil is mixed into the air stream and this oil mist (fog) is in theory carried by the flow of compressed air into

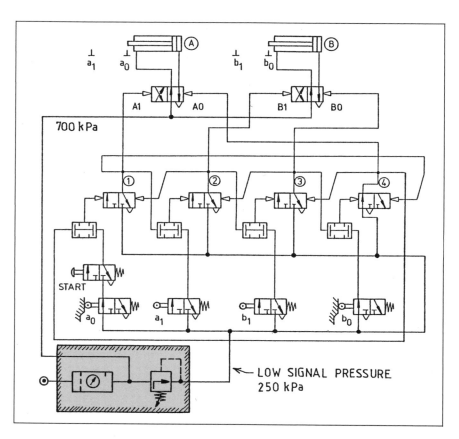

Fig. 12-8 Sequential circuit where reduced pilot pressure is applied.

every crevice, clearance gap and bearing, where sliding movements cause friction. When air exhausts through the exhaust ports of valves and air motors, the same quantity of lubricating oil which was previously injected into the air stream, must eventually emerge from these exhaust ports! In recent years, a number of reports have pointed out that pneumatic exhaust air, carrying oil mist has, in fact, found its way into the lungs of factory workers operating machines in close proximity to pneumatic circuitry or air tools.

For this reason, manufacturers have made extreme efforts to develop pneumatic valves and actuators, which no longer need any form of air lubrication. Numerous modern type logic control valves, better known as moving part logic control equipment, may even start to malfunction, if air lubrication were used for their supply. With these thoughts in mind, air line lubrication may for many applications be regarded as on its way out and air lubrication should, therefore, only be used where absolutely necessary!!

Air lubrication must be applied correctly, if the pneumatic components in need of such lubrication, are to receive the lubricant consistently and adequately, whilst they are in operation. For this

reason, great care must be applied to the selection, installation and adjustment of a suitable lubricator especially when used on complex pneumatic control systems. Many such systems are composed of actuators demanding lubrication, but these actuators may be controlled and sequenced by components (valves), which must not receive any lubrication at all. For controls of this nature, special provisions must be made to keep the lubricated and the unlubricated systems separated. A typical circuit, showing limited air lubrication, is shown in fig. 12-9.

Venturi action to transfer lubricant into the air stream

Air lubricators utilise the pressure difference caused by the compressed air as it streams through a *venturi tube* (fig. 12-10). Transfer of lubricant is based on the Venturi principle, which states that different flow velocities produce different amounts of suction. The velocity increase that results from a restriction in a flow line (see point B in fig. 12-10) will have associated with it a decrease in static pressure. This decrease is directly used to draw the lubricant into the air

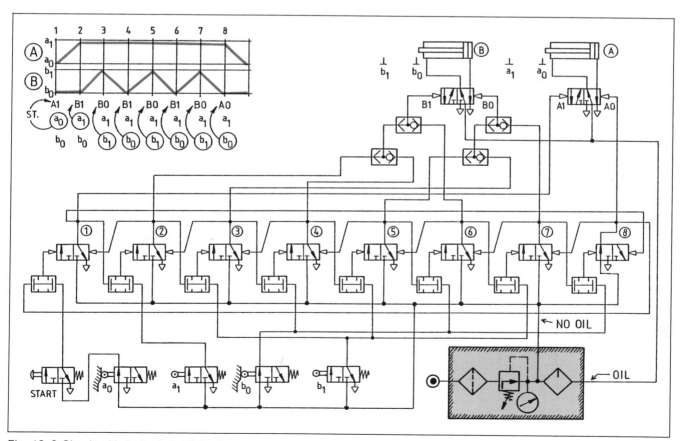

Fig. 12-9 Circuit with limited air lubrication.

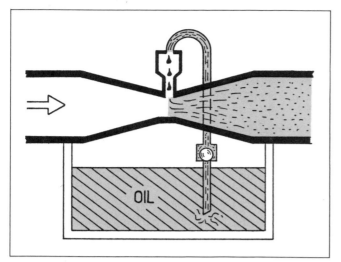

Fig. 12–10 Venturi principle used on air line lubricators to transfer lubricant into the air stream.

stream. The pressure measured at point A of the venturi acts onto the surface of the lubricant in the reservoir. When the compressed air leaves the venturi "throat" and enters section C at the end of the venturi tube, it has regained its original pressure, but its velocity has also returned to the lower velocity it had at point A. The velocity at point B, however, is much higher than at point A and point C.

Macro lubricator (oil fog lubricator)

Compressed air enters the lubricator at the inlet port ④ and streams through and around the venturi assembly ③ and past a flow sensor ⑨ to the outlet port ⑧. Inlet air pressure acts through hole ⑤ to pressurise the lubricant bowl (fig. 12–11). Differential pressure between the lubricant bowl and the sight feed dome ⑪ causes oil to flow up

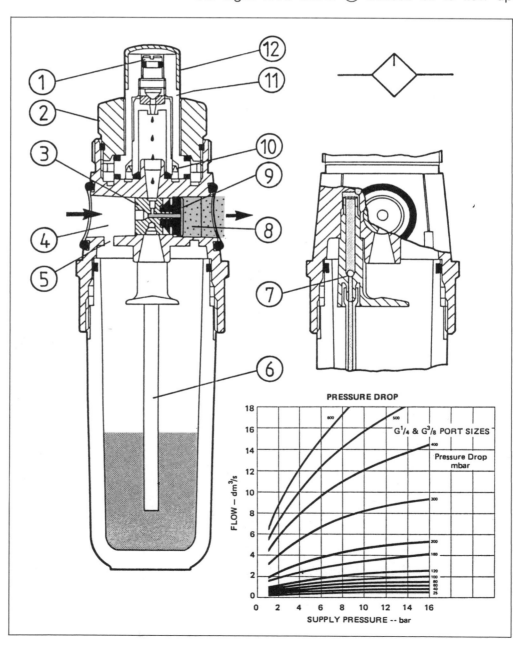

Fig. 12–11 Oil fog lubricator.

through the siphon tube ⑥ past a check valve ⑦. From there it flows through the filter screen ⑩ into passages of the sight feed dome. The oil drip rate is controlled by adjusting screw ① from where the oil drips into the venturi assembly, where it eventually is "atomised" by the passing air stream and becomes an airborne oil fog. This oil fog is then carried to the pneumatic devices or control circuit. The automatic flow sensor (elastomer plug) flexes as necessary to allow more or less air to flow past it, thus permitting changes in flow demand ⑨.

To refill the lubricant bowl, the air must be shut off (on some lubricators), and the system vented, before lubricant can be refilled through the filler cap ②. To adjust the drip rate for this particular lubricator the plastic cap ⑫ must be removed and a screw driver is used to rotate the adjusting screw ①.

With oil-fog lubricators all the oil dripping through the sight feed dome will enter the air stream and will be "atomised" into the air stream. This means that small or larger droplets are carried through the air stream, whereby the larger droplets tend to fall out of the air stream, which means they are no longer airborne. For this reason, oil fog lubricators should only be used for systems where the components to be lubricated are close to the lubricator.

Micro lubricator (oil mist lubricator)

Oil mist lubricators use a totally different principle to "atomise" the lubricant (compare with oil fog lubricator fig. 12–12). Compressed air enters the lubricator at port p and flows through its main passage to the outlet port A (fig. 12–13). A small portion of the passing air is diverted to the venturi nozzle ⑥, which is located inside a "mist generator" (atomiser) ①. As the lubricant drips into the high velocity air stream it becomes "atomised". The air-oil mixture is then sprayed onto a baffle plate ③, which causes further oil fragmentation and most importantly redirects the air stream in vertical direction (90° turn). Leaving the baffle, the oil-laden air streams across the bowl to a large exit opening ⑤. Oil-laden air flowing through this opening then rejoins the air stream in the main passage, mixes thoroughly with it, and finally leaves the lubricator on its outlet port A.

Whilst the air-oil mixture is redirected on the baffle plate and travels to the exit opening, all large oil particles greater than 2 μm drop out of the air stream and fall back into the lubricator reservoir ④. Thus, only the smallest oil-mist particles find entry into the air stream and it is generally assumed that only about 5% of the injected oil quantity (of the oil drops counted in the drip dome), will find its way into the pneumatic system.

Most oil mist lubricators use a self-adjustable air diverter valve ② (usually a neoprene flap), to divert more or less flow to the mist generator, depending on the flow rate through the lubricator.

Since these oil particles from this type of lubricator are so small they stay airborne much longer, and tend not to coalesce into larger droplets. As

Fig. 12–12 Different atomising principles as used on oil fog and oil mist lubricators (macro and micro).

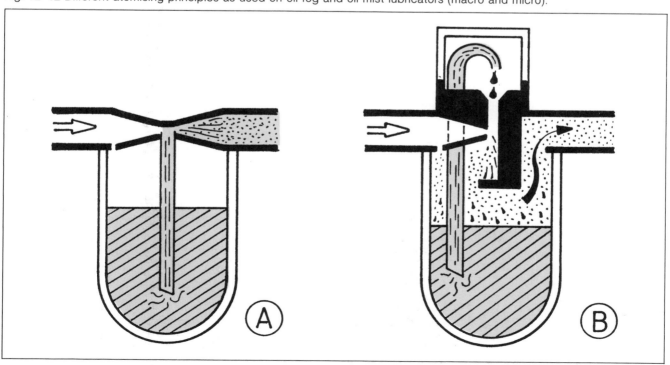

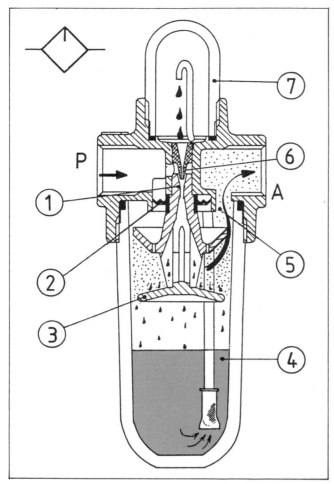

Fig. 12–13 Oil mist lubricator.

This is not the case with oil fog lubrication! For this reason it is advisable to use predominantly oil mist lubrication for complex and diversified pneumatic systems.

Oil feed rate

Suggested oil feed rates are given in fig. 12–14, but as explained previously, these rates are to be strictly monitored and experimented with, to avoid overlubrication, which might cause hazardous situations to machine operators and factory personnel.

Suggested lubricants for air line lubrication

Manufacturers of lubricators and air equipment suggest that only light mineral oils should be used for air line lubrication. Oil viscosity should be approx. 9–11 mm^2/s (9–11 cSt) at 40°C. This corresponds to ISO-CLASS VG 10 as per ISO 3448.

Some suggested oil brands are: Avia Avilub RSL 10, BP Energol HLP 10, Esso Spinesso 10, Festo Special oil, Mobil DTE 21, Shell Tellus Oil C 10.

Installation of air service units

The air service unit is usually installed as close as possible to the pneumatic components supplied by its airflow. It should also be located in a position where it is easily accessible (for maintenance purposes) and where possible, the refill opening as

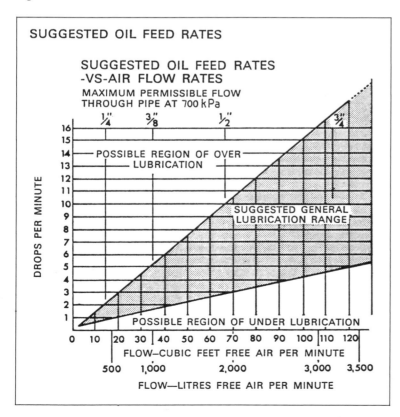

Fig. 12–14 Suggested oil feed rates.

well as the drain opening should have ready access for servicing.

Most modern type air service units come in modular form, which makes them easy to install and service, since every item (filter, regulator and lubricator) can individually be removed, whilst the other items need not be removed from the assembly. It is advisable that each air service unit is equipped with a shut off valve on its upstream side. This permits isolation of the machine being serviced by the air service unit and also permits maintenance on the air service unit, whilst other machines and the compressor stay on stream.

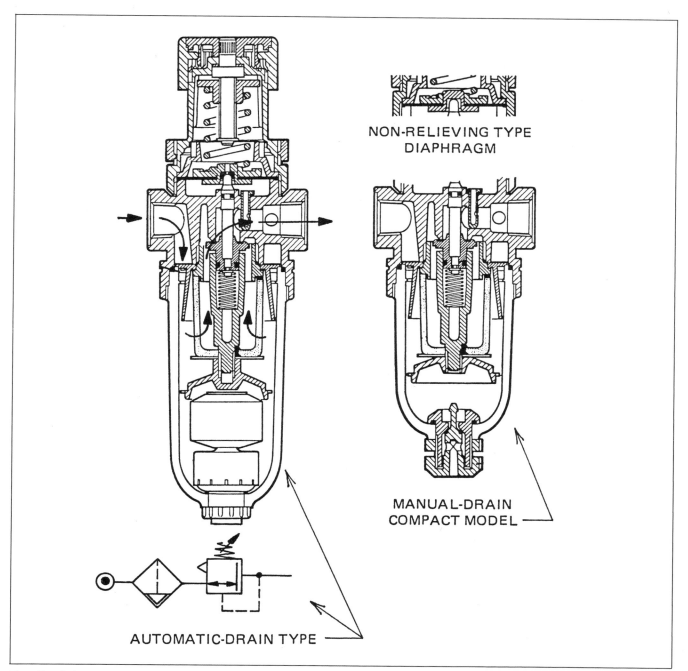

NON-RELIEVING TYPE
DIAPHRAGM

MANUAL-DRAIN
COMPACT MODEL

AUTOMATIC-DRAIN TYPE

Fig. 12–15 Air service unit without lubricator.

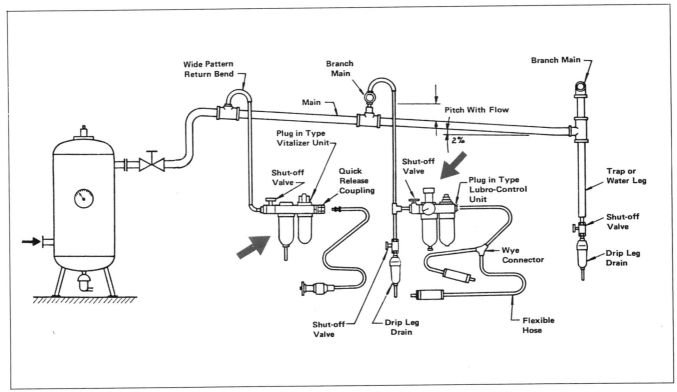

Fig. 12–16 Typical air service unit installations.

13 Air Compression

The preceding chapters explained how useful work and complex logic control may be obtained from compressed air by the design of pneumatic systems and control circuits for use in industry. But so far it has been assumed that the compressed air used to operate these actuators and systems is clean enough and at a sufficient pressure level and flow rate to enable the intended control and work functions to be performed efficiently for many thousands or indeed millions of cycles. How these already assumed requirements on the compressed air supply are accomplished and the methods and equipment used to bring this about, will be described in the following chapters.

Free Air

An understanding of the term "free air" and its significance in relation to compressed air is important, as it is the term used for measuring the flow rate of all devices producing, consuming, processing, distributing and storing compressed air.

Free air (or ambient air) is defined as air at atmospheric conditions at any specific location and time. It should be noted, however, that because the altitude, barometric air pressure and air humidity may vary at different geographical localities and times, the term "free air" does not mean air under identical or standard conditions. In view of such variations and in order that stated air flows relating to pneumatic equipment can be calibrated and compared, a standard is essential. There are unfortunately many standards depending on the industry or country where equipment is produced, but the two most common are the European Standard and the Imperial Standard. The European standard for free air is based on a pressure of 101.3 kPa absolute, at a temperature of 0°C, whereas the Imperial standard is based on 14.7 p.s.i. absolute at 15.6°C.

Hence, the term "free air" based on the previously mentioned standards is quite a simple and useful rating to size and compare delivery rates of air compressors. If the compressor, however, is installed in an environment which has a free air pressure and temperature other than either of the standards above, then its flow rate will alter from the manufacturer's standard specification.

Compressors, for example, are always rated for flow output based on the flow rate (ambient air) they take directly from the atmosphere through their air intake. To illustrate: If a compressor had a rated output of 4 m^3/min F.A.D. (free air delivery), for each minute the compressor was operating, eight cubic metres of "free air" would be drawn into the intake from the atmosphere. This would be comparable to a volume of air 2 metres long by 2 metres wide by 1 metre deep, being drawn in each minute (see fig. 13–1 (a) and (b)). The output rate from the compressor, however, would be far less, since the air would now be compressed to a much smaller volume, but to a much higher pressure than at the original ambient pressure prevailing at the compressor intake (see also Chapter 1, fig. 1–12).

Diagrams (a) and (b) of fig. 13–1 will assist in understanding this important concept of "free air". The gas laws explained in Chapter 1 can be used to calculate the relationship between the volumes and pressures given in these diagrams.

Altitude effects on air compression

In some locations, the effects of altitude on the output of air compressors and air consuming devices must be considered. At altitudes of for example 500 and 1000 metres, there is a corresponding reduction in atmospheric pressure of approx. 5 kPa and 10 kPa respectively.

For compressors, these reduced intake pressures cause a decrease in the compressor power requirement. However, the lower atmospheric intake pressure also means that the compression ratio of the compressor is higher. This results in an increased power requirement. The net result for a compressor operating at an altitude above sea level would mean a decrease in the power required to drive the compressor, and a decrease in the free air delivery.

Air consuming devices such as air tools and linear actuators, if operated at sea level, would consume more air than if operated above sea level.

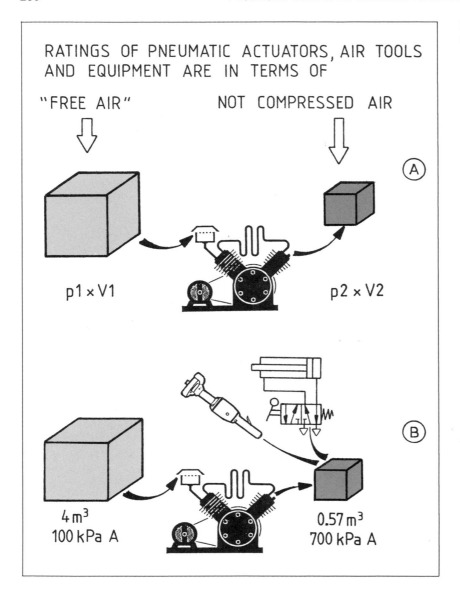

Fig. 13–1 (a) and (b) Diagram of the "free air" concept.

RATINGS OF PNEUMATIC ACTUATORS, AIR TOOLS AND EQUIPMENT ARE IN TERMS OF

"FREE AIR"   NOT COMPRESSED AIR

(A)

$p1 \times V1$ $p2 \times V2$

(B)

$4\,m^3$ $0.57\,m^3$
$100\,kPa\ A$ $700\,kPa\ A$

Fig. 13–1 (a) and (b) Diagram of the "free air" concept.

Air flow rates

The S.I. standard unit for flow rate is cubic metres per second (m^3/s). This unit is not often used in air compressors and pneumatic component specifications, because the flow rate numbers involved are usually small decimal fractions and, therefore, would be difficult to comprehend. Hence, the units more commonly in use are: litres per minute (L/min), and cubic feet per minute. A conversion to these units can be made as follows:

To convert from:	Operation
$m^3/s \rightarrow m^3/min$	$\times 60$
$L/min \rightarrow m^3/min$	$\div 1000$
c.f.m. $\rightarrow m^3/min$	$\div 35.3$
$m^3/min \rightarrow m^3/s$	$\div 60$
$m^3/min \rightarrow L/min$	$\times 1000$
$m^3/min \rightarrow$ c.f.m.	$\times 35.3$

Air flow rates used in the following chapters will be given in m^3/min, unless otherwise stated.

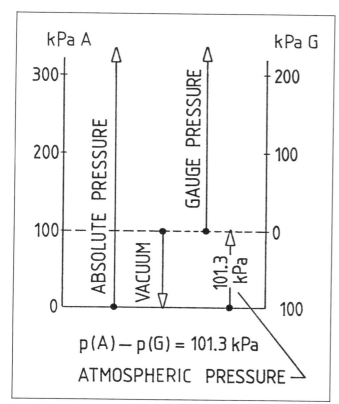

Fig. 13-2 Various methods of expressing pressure.

Air pressure

Pressures will be quoted in kilopascals gauge (kPa) and bar unless otherwise stated (see also Chapter 1).

An atmospheric pressure of 101 kPa absolute has been adopted for the conversion of gauge pressures to absolute pressures for use in calculations shown in the text. The relationship between gauge and absolute pressure is shown in fig. 13-2.

Most pressure gauges used on pneumatic equipment are calibrated in gauge pressure. Thus, if a pressure gauge is calibrated for kPa A it is capable of measuring pressures in the vacuum range ("A" stands for absolute pressure calibration). But if it is calibrated just for kPa then the pressure gauge shows pressures above atmospheric pressure only (see fig. 13-2).

Air compression process

Before studying the various types of compressors and their characteristics, it is necessary to look at some of the principles governing the compression process. Chapter 1 dealt with the gas laws, which define the relationship between volume, pressure and temperature of a gas. Comprehension of these laws is essential for understanding the design of compressors and their function.

Whenever air is compressed its temperature rises. This temperature rise may be experienced, for example, when a bicycle tyre is pumped up, and the pump and its connection become hot. The rise in temperature is due to an increase in the average velocity of the molecules, bombarding the walls of the compression space: the volume of which is being rapidly reduced. This causes a transference of some of the kinetic energy of the molecules into heat energy. In an air compressor, if this temperature rise is not controlled by some cooling mechanism, it can lead to a very inefficient compression process, or worse still to excessive wear and final break down of the air compressor. Furthermore, the compressed air at the outlet of the compressor would be too hot for direct usage.

Air can theoretically be compressed by various processes, and these are described in the following definitions:

Adiabatic compression is that process in which the air under compression retains all the heat gained during the compression process. This purely theoretical process would require perfect insulation of the air under compression, and there would be a very large amount of power required for its compression. The increase in power required is a direct consequence of the expansion of the air, caused by the heat generated during compression. The compression element reduces the volume of air, but also has to compress an additional volume brought about by this rise in temperature. Hence, power is wasted (see Chapter 1, fig. 1-12).

Isothermal compression is also a purely theoretical process in which all the heat gained during the compression process is removed, and thus the air under compression is kept at a constant temperature. This process would require perfect cooling, in order to immediately carry away the heat generated during compression. The power required for isothermal compression is only about 36% of that required for the adiabatic process. It can therefore be understood, why compressor designers strive to design compressors which compress air as close to isothermal as possible (fig. 1-12).

Polytropic compression can be considered as that compression process which lies between adiabatic and isothermal. It is far more practical and is in fact the only commercially feasible process. Figure 13-3 shows theoretical performance

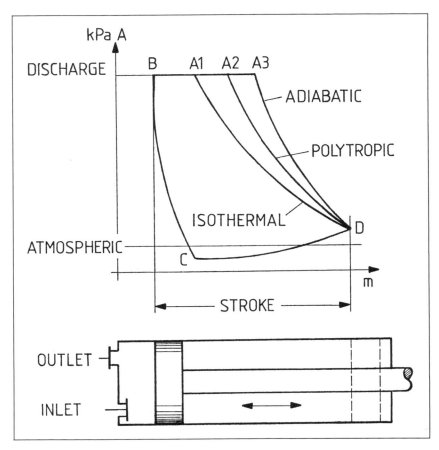

Fig. 13-3 *Left:* Theoretical compression curves for piston compressors. The areas under each curve represent the work required for each method.

A1 B C D — Isothermal
A2 B C D — Polytropic
A3 B D C — Adiabatic

Fig. 13-4 *Below:* Compressor design types.

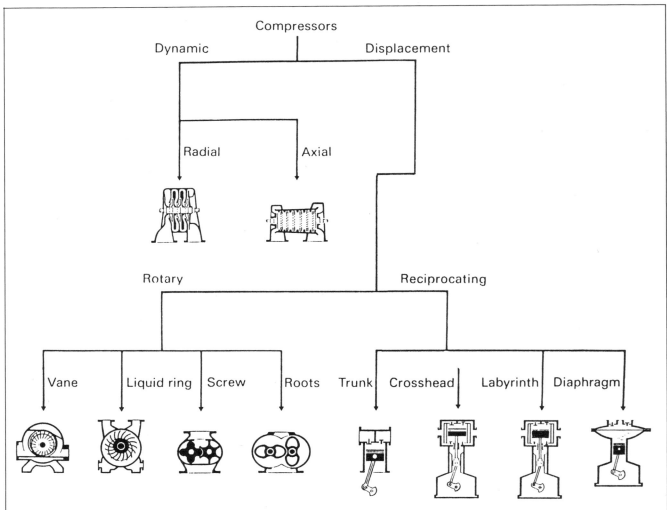

curves for a reciprocating piston compressor for various compression methods. The areas under each curve, and plotted within the points A, B, C, D, represent the work requirements for each method.

Compressor design principles

There are many different types and designs of compressors operating in industry, from the simple type used in every garage for pumping up tyres and performing other garage jobs, to medium and mammoth compressors used in the manufacturing and mining industry. Compressed air is used in many different applications, and to suit these and still compress air efficiently requires compressors of different types and sizes.

There are basically two principles of compressor design; "positive displacement" and "dynamic". Emphasis will be placed on *positive displacement compressors* in this book, because these are most commonly used in industry. However, dynamic compressors will be discussed briefly.

Positive displacement compressors may be defined as those air compressors in which successive volumes of air are enclosed and then elevated to a higher pressure by reducing the volume of the enclosure holding the air.

Dynamic compressors are rotary continuous flow machines, in which a high speed rotating element accelerates the air, converting a velocity head into pressure. Figure 13–4 shows the various types of compressors in these two design methods.

Compressor Staging

Compressors of all types are made up of one or more individual compression spaces, called compression elements.

In order to obtain better efficiency and thus cut air compression power costs, larger compressors are designed to compress air in stages. By this method, two or more compression elements are used to raise the air pressure to the final delivery pressure. These compression elements are all driven from a single common source of power e.g. (electric motor).

Cooling of the air between each stage is essential, and is called intercooling (it cools the air whilst it passes from one stage to the next). The intercooler is an integral part of the staged compressor. Intercooling results in volume reduction of the compressed air which is directly responsible for the power reduction gained; thus intercooling combined with staging is a step towards the ideal

isothermal compression process described previously.

Piston Compressors

The piston compressor (sometimes also called reciprocating compressor) is by far still the most common type of air compressor in use in industry today, and is available in a wide range of types and sizes. It is very versatile with respect to capacity and pressure level attained, since it is capable of compressing air up to 1200 kPa (12 bar) as a single stage compressor and up to 14000 kPa (140 bar) as a multi-stage compressor. It provides air flow deliveries from around 0.02 m³/min F.A.D. to 600 m³/min F.A.D.

The average motor service station would have a compressor of around 0.8 m³/min F.A.D.

Reciprocating compressors tend to be noisy and cause vibration, and for this reason are anchored to their workplace. This makes them less desirable as a mobile unit (mounted on a trailer), especially for larger sizes. When compared with rotary type compressors, such as screw and vane compressors, these drawbacks are significant. Modern designs, however, have reduced the noise and vibration remarkably and depending on the application, such noise and vibration can sometimes be tolerated.

Because the air is compressed in pulses, pulsation damping is required to smooth out these pulses. Some compressors utilise a pulse tank especially matched to the compressor's pulsing characteristics and fitted directly after the outlet port. However, a receiver is generally used to smoothe pulsations. This makes the overall compressor package bulky. However, a receiver is required in any case, for compressed air storage and cooling.

The operation of a piston compressor is described below, in conjunction with fig. 13–5. On the intake stroke, a partial vacuum is created above the descending piston and air is therefore forced into the compression chamber through the inlet or suction valve. This valve opens automatically due to the pressure differential caused by the partial vacuum above the piston. When the piston reaches bottom dead centre (the lowest possible position) and starts to rise again, the trapped air due to a pressure imbalance above and below the inlet valve will close the inlet valve and thus compression of the trapped air begins to occur. Once the air pressure inside the compressor has risen to a level where it can force the inlet valve to open, the partially compressed air will then flow through the outlet valve into the compressed air system.

Both inlet and outlet valves are usually spring loaded, and are made to open and close automatically by a pressure differential across each valve. After the piston reaches top dead centre and begins to descend again, the outlet valve closes and the cycle of compression and air intake is repeated. Compressor cooling is of course essential to carry away the heat generated by the compression of the air, and either air or water cooling may be used.

Some piston compressors are double acting, meaning that compression takes place during both strokes. Figure 13–6 shows such a compressor, which requires the use of a crosshead to link the connecting rod to the piston rod. The crank end of the cylinder is closed and fitted with additional inlet and outlet valves to allow compression also to take place on the return stroke.

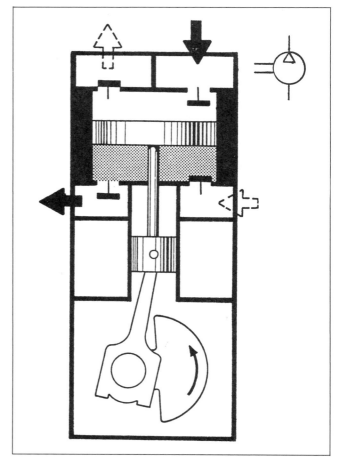

Fig. 13–6 Horizontal double acting piston compressor.

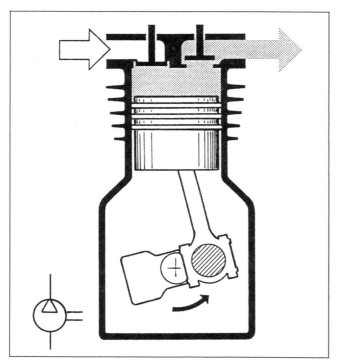

Fig. 13–5 Piston compressor operating principle.

Although making the compressor arrangement mechanically more complicated, double acting compression is beneficial for larger compressors, resulting in larger air delivery and greater efficiency whilst keeping the compressor size relatively small.

Staging of piston compressors

Most large piston compressors (single or double acting) above about 3 m^3/min F.A.D. employ two or more stages, but two stage types are the most common. Referring to fig. 13–7, there is a compression chamber (compression element) for each stage and both pistons are driven from the same power source.

The first stage, the low pressure stage, is comparatively large and takes in air directly from the atmosphere. It then raises the pressure to some portion of the total compressor output pressure. If 700 kPa is required, the first stage may raise the pressure to 500 kPa. The first stage cylinder discharges the partially compressed air into a heat exchanger (an intercooler), which may be air or water cooled. Whilst passing through the intercooler, the air is cooled and its volume is reduced, and whenever air can be cooled during compression, work savings can be made due to the reduction in volume and temperature (see fig. 13–8).

Whilst the partially compressed air passes through the intercooler, some condensate will drop out and can be removed. (Condensate is an important consideration in the compression of air, and is discussed more fully in Chapter 14.)

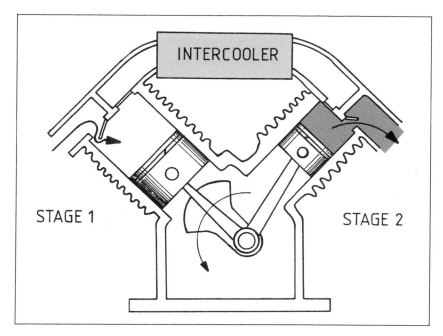

Fig. 13–7 Two stages, single acting piston compressor arrangement.

Leaving the intercooler, the air then passes into the second stage or high pressure cylinder, which is smaller in diameter than the first stage. The second stage cylinder compresses the air to the final output pressure. If extremely high pressures are required, more than two stages must be used. The following calculations, based on typical compression specifications, serve as an example to show how the gas laws can be used to solve a calculation problem relating to compressor staging.

Calculation example:
A two stage compressor with a capacity of $0.16 \, m^3$/min F.A.D. takes in atmospheric air at a pressure of 101 kPa A (absolute) and a temperature of 20°C.

The first stage discharges compressed air at 124°C and 450 kPa gauge.

Whilst passing through the intercooler, this air is cooled to 40°C, and thus its pressure drops to 420 kPa gauge. With this reduced pressure and temperature the air then enters the second stage, where it is further compressed to 700 kPa.

Assuming 100% compressor efficiency, calculate the actual reduction in volume of the air across the intercooler per minute (m^3/min), and also the percentage reduction.

Conditions at the inlet and outlet ports of the first compression stage after one minute of operation.

At the inlet, these conditions are:
$p_1 = 101 \, kPa \, A$
$V_1 = 0.16 \, m^3$
$T_1 = 20 + 273 = 293 \, K \, (20°C) \, (1°C = 1K)$

At the outlet these conditions are:
$p_2 = 450 + 101 = 551 \, kPa \, A$
$V_2 = ? \, m^3$
$T_2 = 124 + 273 = 397 \, K \, (124°C)$

Solving for V_2, the reduced volume after first stage compression is:

$$V_2 = \frac{p_1 \times V_1 \times T_2}{p_2 \times T_1}$$

$$V_2 = \frac{101 \times 0.16 \times 397}{551 \times 293} = 0.04 \, m^3$$

Conditions at the inter-cooler inlet and outlet ports after one minute of operation.

At the inlet (i.e. as for first stage outlet):
$p_2 = 551 \, kPa \, A$
$V_2 = 0.04 \, m^3$
$T_2 = 397 \, K \, (124°C)$

At the outlet these conditions are:
$p_3 = 420 + 101 = 521 \, kPa \, A$
$V_3 = ? \, m^3$
$T_3 = 40 + 273 = 313 \, K \, (40°C)$

Solving for V_3, the volume after intercooling:

$$V_3 = \frac{p_2 \times V_2 \times T_3}{p_3 \times T_2}$$

$$V_3 = \frac{551 \times 0.04 \times 313}{521 \times 397} = 0.033 \text{ m}^3$$

This is the reduced volume after intecooling.

The volume reduction per minute is:

$$= 0.04 - 0.033 \text{ m}^3/\text{min}$$
$$= 0.007 \text{ m}^3/\text{min}$$

Therefore the percentage reduction is:

$$\frac{0.007 \times 100}{0.04 \times 1} = 17.5\%$$

This sample calculation shows that the volume reduction is significant and amounts to 17.5%, but when all factors are taken into consideration, it can result in practical power savings of up to 15% for a two-stage compressor compared with a single-stage compressor (see fig. 13–18).

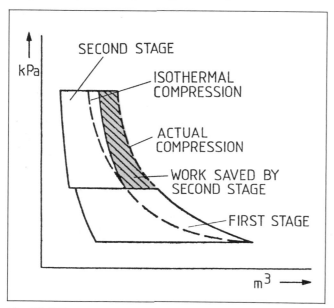

Fig. 13–8 Two stage piston compressor curves showing work savings due to staging.

Theoretical compression curves for a two stage positive displacement compressor are shown in fig. 13–8, where the work saving is the result of intercooling. This graph plots the pressure rise as a direct result of the volume reduction of each compression stage, and therefore the shaded area represents work savings (pressure × volume). The shaded area, however, can also represent power savings if the cylinder volume is considered as volume per unit time or flow rate.

Oil free piston compressors

Oil inevitably gains entry into the air compressed in a standard compressor which has an oil filled crankcase and lubricating oil fed to valves, bearing and other points. Industries such as the food and chemical industries and hospitals cannot tolerate this oil contamination, and for these applications oil free piston compressors are made. These compressors have dry crankcases, and the piston rings, bearings and other sliding components are made of teflon, graphite or similar substances. This overcomes the need for oil lubrication of the compressor which delivers air free of oil contamination.

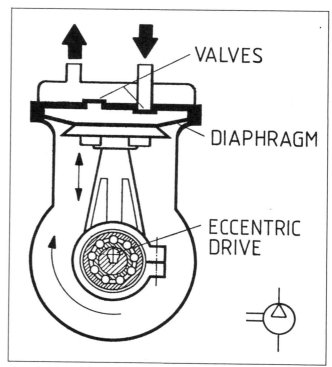

Fig. 13–9 Diaphragm compressor.

Diaphragm Compressors

The diaphragm compressor (fig. 13–9) is a reciprocating compressor, but the piston is replaced by a disc and diaphragm assembly. The diaphragm is attached to the reciprocating disc and to the cylinder wall and acts like a piston in a piston compressor but positively isolates the compression chamber from the oil lubricated crank case, so that no oil comes in contact with the air being compressed. The air intake and outlet delivery is controlled by inlet and outlet valves. Diaphragm compressors are mechanically simple, and are available for a wide range of delivery volumes and pressures suitable for industrial shop air. They are easy to maintain and very reliable.

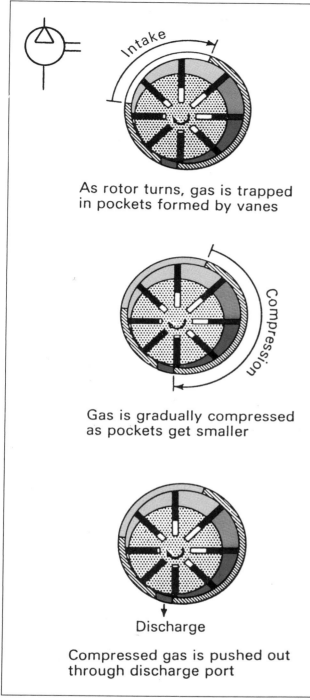

As rotor turns, gas is trapped in pockets formed by vanes

Gas is gradually compressed as pockets get smaller

Discharge

Compressed gas is pushed out through discharge port

Fig. 13-10 Compression steps for a rotary vane compressor.

Rotary Vane Compressors

In rotary vane compressors an eccentrically arranged slotted rotor with radially movable vanes inserted into the slots rotates within a circular stator or housing (fig. 13-10). During rotation, the centrifugal force drives the vanes outward against the circular housing where they make contact and follow the contour of the housing. These vanes, in conjunction with the rotor and the circular housing, form compression spaces. As the rotor turns, these spaces are reduced in volume as they move towards the air outlet, and increased as they move towards the air inlet. Air is drawn into the inlet, and while it moves through the compressor, it is compressed by the reducing air spaces and then discharged into the pneumatic system. A non-return valve (check valve) is usually installed at the outlet port to prevent the compressor from motorising (being driven in reverse) when compression is stopped.

The rotary vane compressor is predominantly an oil flooded type in which oil is sprayed into the compression chamber where it performs three functions. It lubricates all the moving parts; it assists in sealing the clearance gaps between the vanes and the housing, and between the vanes and the rotor; and most importantly, acts as a heat exchange medium. The heat of compression is picked up by the oil and dissipated through an oil cooler (fig. 13-11).

Before the oil can be cooled, however, it must be separated from the compressed air by an oil separator. Two types are used; the felt pad type, and the replaceable cartridge type. When operating correctly, these separators remove virtually all the oil from the compressed air. This oil then passes through the oil cooler, before being returned to the compression chamber for reuse. Oil flooding provides relatively cool air (80–90°C) directly from the compressor which in many applications overcomes the need for further cooling.

Other beneficial features of oil flooded rotary vane compressors are a non-pulsating air flow and low control pressure differential (see compressor control). These features make a receiver unnecessary which produces a compact compressor package, and this coupled with their low noise level and lack of vibration makes them the ideal choice for mobile "tow along" compressors and for general use in industry.

Rotary vane compressors are, however, most efficient when they are operated continuously at the top end of their air delivery ratings. These compressors are manufactured with more than one stage and their ratings are: 800 kPa for a single stage compressor, or 1000 kPa for a two stage compressor with capacities from .3 m³/min to 30 m³/min F.A.D.

Although oil free rotary vane compressors are manufactured, many of the benefits resulting from oil flooding are absent.

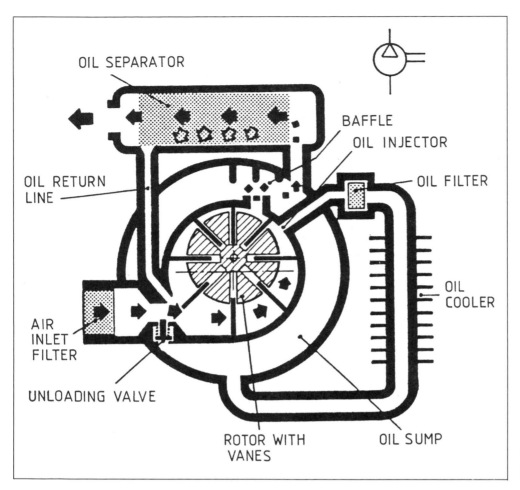

OIL SEPARATOR

BAFFLE

OIL INJECTOR

OIL FILTER

OIL RETURN
LINE

OIL
COOLER

AIR
INLET
FILTER

UNLOADING VALVE

ROTOR WITH
VANES

OIL SUMP

Fig. 13–11 Schematic diagram of an oil flooded rotary vane compressor.

For industrial applications, one important operating aspect of oil flooded compressors should be mentioned, particularly those types employing felt pad oil separators (fig. 13–11). Strict attention must be paid to correct periodical maintenance and supervision of the oil separator! If this is lacking, and the separator becomes ineffective, oil can be discharged from the compressor and enter the compressed air lines. This would cause malfunctioning of pneumatic control circuits due to clogging with oil.

Most recent designs of vane compressors have dispensed with the felt pad oil separator in favour of the replaceable cartridge type separator, thus making maintenance easier, and virtually overcoming the problem of ''oil carry over''.

Rotary screw compressors
(helical lobe compressors)

Rotary screw compressors, which came into general use around 1950, have captured a large market for many applications. The rotary screw compressor consists essentially of a casing, which houses two screws or rotors, one male and one female (fig. 13–12). The male screw usually has one or two less lobes than the female has flutes, into which the lobes mesh. The two screw shafts, in some designs (notably on oil free designs), are

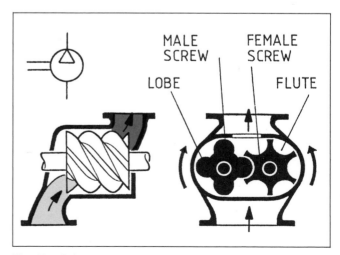

MALE
SCREW

FEMALE
SCREW

LOBE

FLUTE

Fig. 13–12 Arrangement of a screw compressor showing the two screws and casing.

synchronized by gears and so kept from contact with each other. The screws also do not make contact with the casing which encloses them. The clearances between the screws and the housing, however, are very small.

As the screws revolve rapidly, air is drawn into the casing through the inlet port and into spaces between the screws, where the air is compressed between the lobes and flutes as the space becomes progressively smaller (see fig. 13–13). Compressed air then emerges from the outlet port into the casing. Both inlet and outlet ports are automatically covered and uncovered by the shaped end of the screws as they turn. The amount of compression (compression ratio) is predetermined by the shape of the lobes and flutes. A check valve is fitted at the outlet port of the compressor to prevent motorising when compression is stopped.

Rotary screw compressors can be supplied in either oil flooded or oil free versions. The oil flooded type has oil sprayed into the compression spaces in exactly the same manner as the rotary vane compressor and the same advantages are gained. Oil separation in rotary screw compressors, for virtually all designs, is of a replaceable cartridge type of oil separator.

The oil free screw compressor is used extensively in industry for those applications which require oil free air. Since oil is not injected, no internal air cooling occurs so the air has to be cooled by after cooling after it exits the compressor. Additional cooling is also required for the compressor itself.

Screw compressors have many of the same features and operating characteristics as vane compressors, such as steady and pulsation free air delivery, lack of vibration and low noise level. They are most efficient when operating to near full capacity.

Flow capacities start at 1.4 m³/min and extend to 60 m³/min F.A.D. Pressure ratings are up to 1000 kPa single stage and 2500 kPa staged.

Screw compressors are widely used in industry, and on civil and mining sites in the form of large mobile compressors. Many are also made as "packaged" units requiring only the connection to a source of electrical power to produce compressed air.

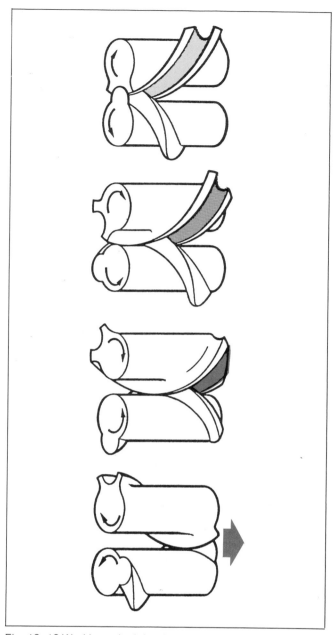

Fig. 13–13 Working principle of a screw compressor, showing only one lobe and one flute.

Dynamic Compressors

Dynamic compressors fall into two main categories. These are: centrifugal flow and axial flow. They are basically low pressure, high capacity machines, driven by electric motors, internal combustion engines or steam turbines. High pressure can however be obtained by multi-staging.

Advantages of dynamic compressors include oil free air delivery, high air delivery capacities relative to their physical size, pulsation free air delivery, and operation without vibration.

Some of the principal applications to which both the centrifugal and axial flow compressors are well suited are: air delivery to wind tunnels, blast furnace air delivery, tunnel ventilation, sewage agitation, and gas distribution.

The axial flow compressor is generally not suitable for industrial shop air, but there is on the market a centrifugal compressor designed for shop air application, providing pressures up to 700 kPa. Its capacity is around 60 m^3/min, normally too large for general usage.

Centrifugal Compressors

In centrifugal compressors each stage consists of a casing, an impeller, a diffuser and a volute (fig. 13–14).

When the drive rotates the multi-bladed impeller at high speed, air is trapped between the impeller and the casing and thrown outwards, leaving the blade tips with pressure and high velocity. This outward thrown air then enters the stationary diffusor ring, where the air can expand and therefore reduce its velocity, but most importantly can increase substantially in pressure. From the diffusor, the air enters the stationary volute, where it

expands further and increases in pressure, and is directed to the outlet or next stage.

In this type of compressor, unlike the axial flow type, air is thrown to the outside of the casing and so cooling and staging are relatively simple. By multistaging, pressures as high as 25 000 kPa can be achieved, but generally the design pressures are below 700 kPa. However, large capacities of from 3000 m^3/min to 20 000 are possible.

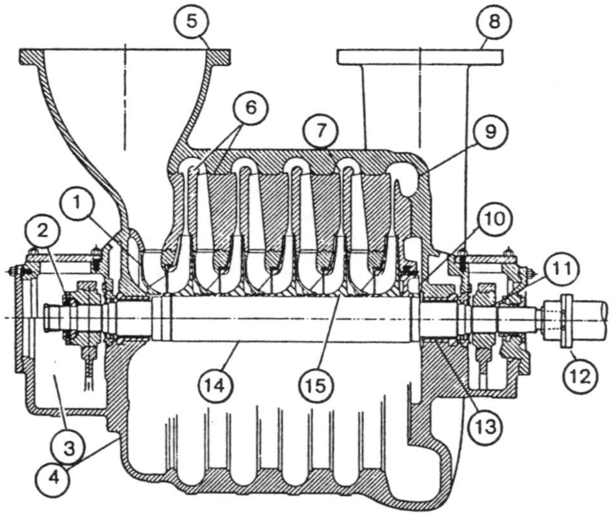

Axial Flow Compressors

In axial flow compressors the air is moved through a series of rows of alternating, rotating and fixed blades, with the direction of flow parallel to the axis of rotation. The rapidly rotating blades impart velocity to the air. As the rotor turns, the air velocity is converted to pressure as it passes through the adjacent stationary blades, which act as diffusors (fig. 13–15). The fixed and moving blades are curved opposite to each other, increasing the sectional area through which the air is flowing. Thus velocity is reduced, but pressure increased. Axial flow compressors are typically low pressure (up to 500 kPa) high flow machines with capacities from 280 m^3/min to 3 300 m^3/min F.A.D. Higher pressures of up to 30 000 kPa are possible with special designs.

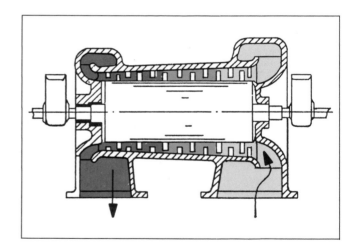

Fig. 13–15 Multi-stage axial flow compressor.

Compressor capacity control

Because a compressor installation must be capable of supplying more compressed air flow than is required at peak demand, the output of a compressor must be controlled or regulated to match the system demand. This is usually accomplished by monitoring the system pressure at some point where the air pressure is relatively stable and free from pulsation, such as at the receiver. The rise and fall of pressure at this point is used to load

1. guide vane
2. thrust bearing
3. bearing housing
4. casing
5. inlet flange
6. diaphragm
7. diffuser
8. discharge flange
9. volute
10. thrust balancing drum
11. shaft bearing
12. coupling
13. seal
14. shaft
15. impeller

Fig. 13–14 Five stage centrifugal compressor.

and unload the compressor by a valve or other device. The pressure range between the high and low points is called the *pressure differential*.

An important operating factor in relation to any type of compressor control, is to determine which process or device in the system requires the highest pressure to operate satisfactorily. If, for example, a cylinder must operate at 600 kPa to provide a specified output force, or an air tool to do its work efficiently, then the monitored pressure must never be allowed to drop below about 700 kPa before the compressor again commences to feed compressed air into the system, otherwise the system pressure may drop below the required 600 kPa before the compressor has been able to recharge the system. Many pneumatic circuits and tools malfunction because this factor is ignored.

There are numerous methods of capacity control used in compressor systems, but the following three basic methods are used predominantly:

- start-stop control,
- constant speed control,
- variable speed control.

Start-stop control

The start-stop method uses the monitored pressure to operate an electrical pressure switch. This switch stops the electrical motor driving the compressor when enough pressure (that is, compressed air volume) has built up in the receiver or at the pressure sensing point. "Start-stop" control is best suited for use on electrically driven piston compressors up to about 1 m^3/min F.A.D. (fig. 13–16).

A piston compressor using this form of unloading, requires a valve to automatically exhaust the compression cylinder after it has stopped, thus preventing restarting under load.

The start-stop method may also be used on compressors other than piston types as long as certain conditions are satisfied, especially with respect to the storage volume of the air receiver.

The cut-in/cut-out pressure differential for this method of control is around 140 kPa, and the system must be designed so that no more than about 10 starts per hour are required, otherwise the electric motor, due to the starting current, may overheat.

Rotary vane and screw compressors work most effectively and efficiently on constant load, and are not suited to "start-stop" capacity regulation. When vane compressors are stopped, the compression spaces must be exhausted to atmosphere. This may either be done manually or automatically, prior to restarting, otherwise scuffing between the vanes and cylinder walls will occur, which would lead to rapid wear.

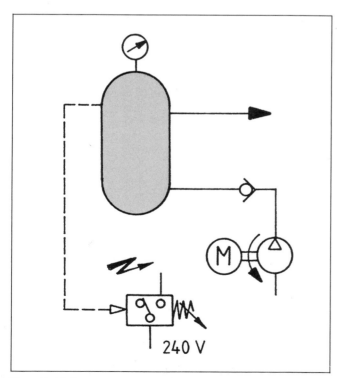

Fig. 13–16 Start-stop control circuit.

Constant speed control

Constant speed control means that the compressor and its drive motor maintain for all practical purposes a constant speed. Capacity control (pressure unloading) is carried out by various means which still depend on the monitoring of the system pressure. The control once again is applied mainly to electrical drives, and is suitable for all types of compressors. Three design methods are used to achieve constant speed control.

Inlet valve unloading (constant speed control)
Inlet valve unloading is applied only to piston compressors and is sometimes also called "grip arm" control. The inlet valve is held open by an actuating device which grips the valve and prevents air from being compressed in the compression chamber. It is the most efficient method of constant speed control. Figure 13–17 shows a diagram of this arrangement and fig. 13–18 shows a cross-sectional drawing of an unloading actuator.

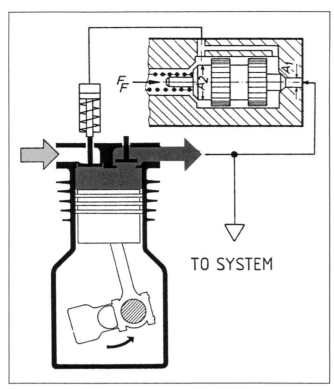

Fig. 13–17 Inlet valve unloading (grip arm constant speed control).

Fig. 13–18 Unloading actuator (grip arm design).

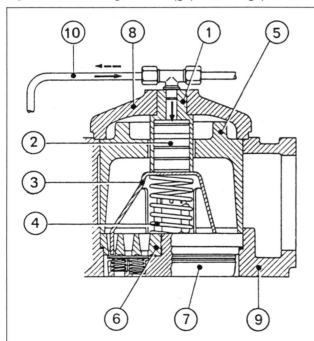

1 Unloading cylinder
2 Unloading piston
3 Unloading claw (depress the valve discs and keeps the suction valve open during the unloading period)
4 Unloading spring (returns unloading claw at start of loading)
5 Valve retainer
6 Seat of suction valve
7 Suction valve guard
8 Suction valve cover
9 Cylinder head
10 Compressed air to and from unloading device

The monitored pressure operates an unloading valve (sequence valve) which, when the system pressure has risen to a preset value, causes the unloading actuator with its claw to hold the inlet valve open. In some designs, this unloading actuator is simply an air cylinder built on to the inlet valve (fig. 13–17).

The only load that the drive motor has to overcome, when unloaded, is the friction of the free running compressor itself, and the movement of atmospheric air, in and out of the cylinder on each stroke. When the monitored pressure has dropped to the loading point, the unloading valve exhausts the actuator and the inlet valve is allowed to operate again, thus compression recommences.

The cut-in/cut-out differential for this method can be as low as 35 kPa, but is usually around 50 to 70 kPa.

Intake throttling (constant speed control)

Intake throttling is mainly used on rotary vanes, screws, and some dynamic compressors. The monitored pressure is used to throttle the intake valve either by a simple open-shut control (see fig. 13–19), or by varying the throttle opening to match system demand. A pressure operated servo valve as a pilot is used to control the opening and closing of the throttle valve. (See also fig. 13–11.)

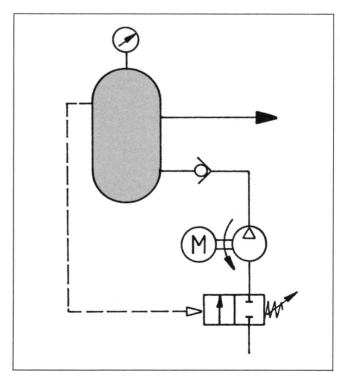

Fig. 13–19 Intake throttling control.

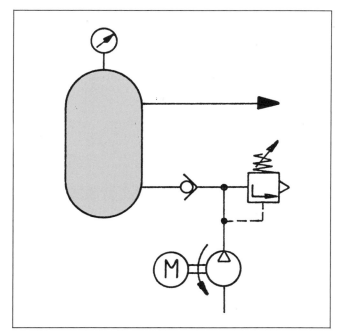

Fig. 13–20 Exhaust regulation control circuit

Intake throttling is not as efficient as "grip arm" unloading, as compression and expansion is still occurring within the compressor whilst the intake is throttled, and therefore power is being used unnecessarily. Overheating can also occur in the compressor if the throttle is fully closed for long periods. One advantage of intake throttling is its very narrow pressure differential, which may only be 7 kPa.

Exhaust regulation (constant speed control)

Exhaust regulation is used on very small reciprocating compressors. It blows the excess compressed air off to the atmosphere through a relief valve. This is extremely wasteful, as no reduction of power is achieved whilst the compressor is exhausting (fig. 13–20). A variation of this method, which also brings no power reduction, is external by-pass control, where the excess air is returned (by-passed) to the compressor inlet. This method is used on some vane and screw compressors.

Variable speed control

Variable speed control, as a means of capacity control, can be used on all types of compressors, but is most commonly used on internal combustion engine driven compressors. The speed of the engine and compressor is controlled, once again, by monitoring the system pressure. The controller then operates a pressure operated engine throttle governor. A minimum idle speed must, however, be maintained to prevent overheating of the compressor, and with vane type compressors, due to the lack of centrifugal force at low speed, the vanes may not fully extend.

Compressor site (location of installation)

The compressor installation site has a great influence on the quality of compressed air delivered into a distribution system and therefore deserves special consideration. The compressor should be sited in a location or building that is:

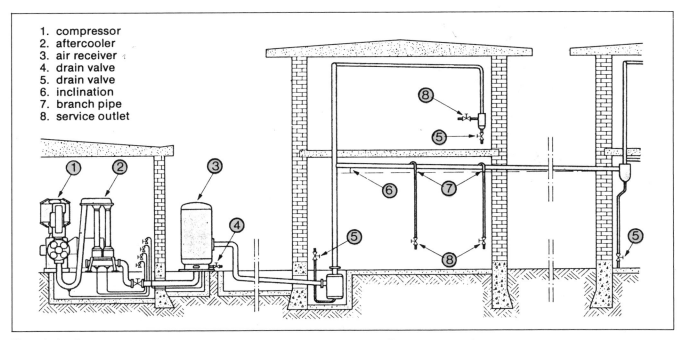

1. compressor
2. aftercooler
3. air receiver
4. drain valve
5. drain valve
6. inclination
7. branch pipe
8. service outlet

Fig. 13–21 Compressor installation showing some commonly used ancillary components.

- Cool and dry, to ensure that the minimum amount of water vapour is drawn into the compressor inlet (well away from steam boilers and other high humidity equipment).
- Dust free, to prevent blockage of the air cleaner, and the compressor cooling system.
- Adequately ventilated, to allow completely unrestricted entry of ambient air to the compressor inlet and ample air flow around the compressor to assist in its cooling.
- Adequately sized, to permit free and unrestricted access for servicing and maintenance of the compressor and its ancillary equipment.

Compressor ancillary equipment

A compressor performs the basic function of compressing air. However, to accomplish this function efficiently and effectively, it also requires accessory or ancillary equipment. A typical compressor installation is shown in fig. 13–21 and fig. 13–28, where such ancillary items are included. Some are essential to the compressor, and are ordered with it. One of these is the prime mover—an electric motor, or an internal combustion engine. Together with the compressor, it must be installed correctly to ensure efficient and reliable operation. Compressor installations, particularly for piston com-

pressors, may require a solid foundation and it is therefore advisable to follow manufacturer's specifications.

The following ancillary items are not usually supplied with the compressor, but are optional:

- Intake silencer
- Air receiver (air storage tank)
- After cooler (compressed air cooler)
- Moisture separator

Intake filter/silencer

Each compressor needs an intake filter to remove particles of dirt before the air enters the intake port (fig. 13–23). These are either oil bath or paper type element filters which require periodic servicing or replacement.

An air silencer is sometimes required to silence the noise of the air rushing into the compressor. It can be fitted before or after the air filter, depending on the silencing effect required (figs. 13–23 and 13–24).

Air receiver

A compressor plant is normally equipped with an air receiver which must be large enough to suit the compressor flow rate, and must be matched to the selected compressor regulating system (com-

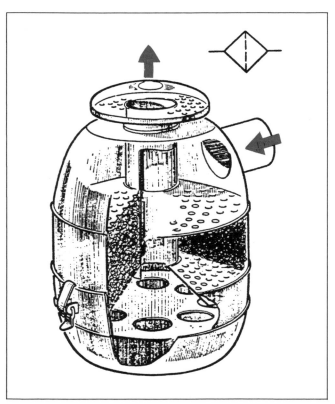

Fig. 13–22 Oil bath filter.

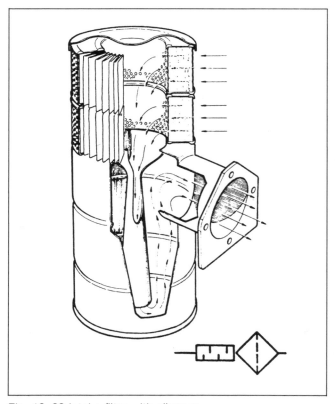

Fig. 13–23 Intake filter with silencer.

1. compressor
2. silencer with filter
3. venturi tube
4. filter
5. flexible connection
6. suction pipe
7. air intake

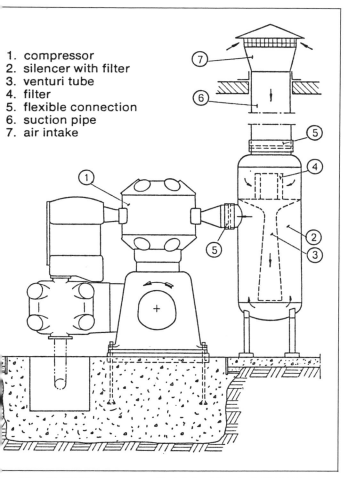

Fig. 13-24 Compressor with intake filter and silencer.

pressor capacity control). For some systems it may also be the only means of air cooling to reduce the moisture in the system (fig. 13-25). An air receiver serves four main functions:

• storage of compressed air; thus eliminating the need for the compressor to run continuously;
• pulsation damping to smooth the pulsing flow of air from the compressor;
• heat exchange to assist air cooling and thus produce condensate drop out before the air enters the distribution system;
• collection and drop out point for dirt and condensate accumulating in the air after compression.

Receivers are usually fitted with an automatic condensate drain, or if manually operated must be periodically drained. Air receivers must also be fitted with a safety air relief valve! Since air receivers are pressure vessels, they must be registered with the appropriate Government Authority, which controls and checks that the air receiver (pressure vessel) is properly constructed, installed, maintained and remains in a safe working condition. Figure 13-25 shows a receiver and its accessories.

A piston compressor must always work in conjunction with an air receiver and up to 2 m³/min F.A.D. compressors are usually supplied with a small air receiver which is part of the compressor package.

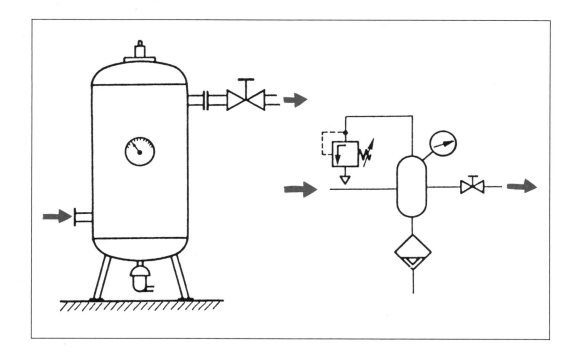

Fig. 13-25 Air receiver and its accessories.

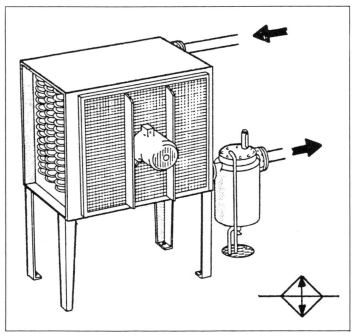

Fig. 13–26 Air cooled after cooler.

Receiver volumes for compressors with start-stop regulation can be calculated by the following formula:

$$V_R = \frac{15 \times Q \times p_1}{\Delta p_2 \times c}$$

Where:

V_R = Receiver volume (m³)

Q = Compressor delivery (m³/min F.A.D.)

p_1 = Compressor intake pressure (kPa A)

p_2 = Pressure differential (cut-in/cut-out in kPa A)

c = Number of starts per hour

Air receivers may also be sized by a simple rule of thumb method which states:

Receiver volume (m³) = capacity control factor × compressor F.A.D.

or: V = K × Q

The factor K depends on the type of capacity control used on the air compressor and is given as follows:

    for start-stop control: K = 0.9–1.0
    for constant speed control: K = 0.5

After cooler

An after cooler is required to cool the air after it has left the compressor. The after cooler is a heat exchanger similar to the intercooler and can be either air or water cooled (figs. 13–26 and 13–27). Ideally, the compressed air temperature on leaving the compressor installation prior to use, should be cooled to no higher than 15°C above that measured at the compressor intake. The use of an after cooler is one method of achieving this. It is usually used in conjunction with larger capacity piston compressors and non-oil flooded rotary compressors. These compressors have no inbuilt oil cooling mechanism after the final compression element and, therefore, require after cooling. As the name implies, the after cooler cools the air, after it leaves the compressor and helps to remove much of the condensate. This removal of condensate is of prime importance, before the compressed air enters the distribution system. An after cooler is usually equipped with an automatic drain and requires frequent servicing much the same as an intercooler. There are many design types of after coolers and they are always installed directly after the compressor delivery outlet, prior to the receiver.

Moisture separator

A moisture separator is used to help rid the airline of condensate, oil and dirt. It is usually fitted before the receiver, but can be fitted virtually anywhere in the distribution system where condensate collects. Moisture separators are simple mechanical devices, designed to impose a sudden change of velocity and direction on the airstream, thus causing water and oil droplets to be flung out of the air stream, where they can be collected and drained off at the bottom of the separator. In some separator types, the air also passes through a filter element, which removes dirt particles.

Intercooler

The intercooler is a heat exchanger, required between any two stages of a multi-stage compressor and is an integral part of a staged compressor (fig. 13–7). Whether it be air or water cooled, it requires periodic servicing to ensure the cooling mechanism is operating efficiently. Automatic condensate drainage must be fitted and its function and operation is described in the next chapter. This drain must never become blocked as otherwise one of the advantages of staging (the removal of condensate) would be negated.

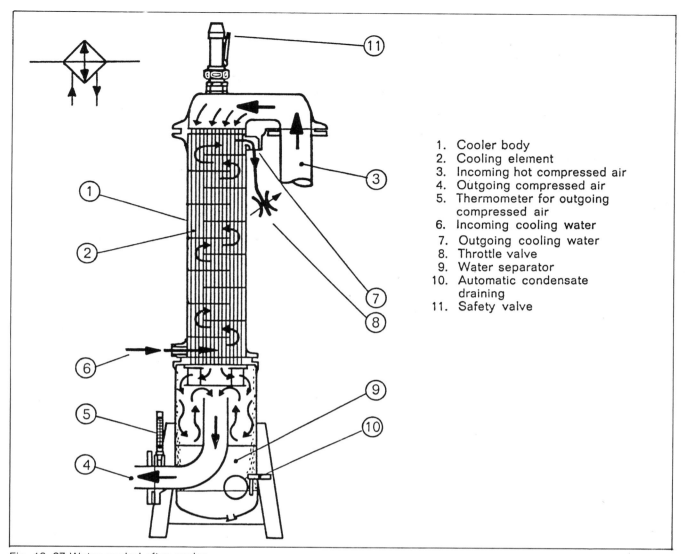

1. Cooler body
2. Cooling element
3. Incoming hot compressed air
4. Outgoing compressed air
5. Thermometer for outgoing compressed air
6. Incoming cooling water
7. Outgoing cooling water
8. Throttle valve
9. Water separator
10. Automatic condensate draining
11. Safety valve

Fig. 13–27 Water cooled after cooler.

Fig. 13–28 Compressor and ancillary equipment test points.

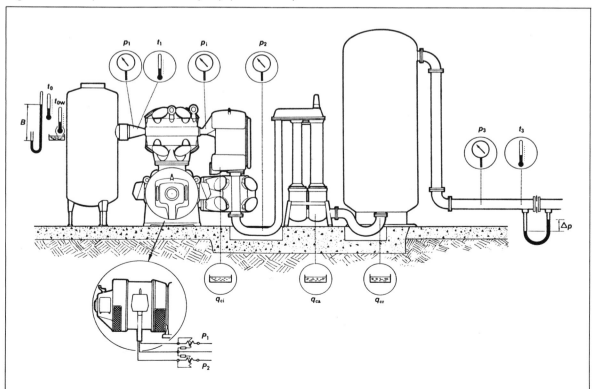

14 Compressed Air Drying

Water condensate is the greatest single source of contamination of compressed air. If this water condensate is not removed, it can cause the malfunction of components and the spoiling of manufacturing processes, especially where the compressed air comes in direct contact with the manufactured product. It may also cause corrosion of the pneumatic components and discomfort for operators of air tools.

The quantity of water vapour in ambient air, and consequently the quantity of condensate (water) that can be released under pressure, can be considerable. A typical small 30 kW air compressor (6 m^3/min F.A.D.), taking in ambient air at 20°C with a relative humidity of 60%, will deposit 20 litres of water in the air line during each 8 hour shift. In other words, for every cubic metre/min F.A.D. of compressor capacity 0.4 litres of condensate is produced!

Through correct sizing, design, installation and maintenance of the compressed air system (this includes the air service unit), the problem of water condensate drop out can be reduced or made to occur before the compressed air enters the components to perform work.

However, many industrial automation and manufacturing processes cannot tolerate *any* moisture in the compressed air. To meet such stringent requirements a compressed air dryer must be used. Drying the compressed air is a complex procedure, and to understand it fully and make the correct decision when a suitable dryer must be selected one must understand the physical laws governing the transition of air moisture from water vapour to condensate.

Concepts and terminology

The atmosphere is composed of a mixture of various gases, such as oxygen 21%, nitrogen 78% and percentage fractions of carbon dioxide, hydrogen and helium. This mixture of gases called "air" also includes water vapour which is invisible, together with the other constituents.

The air can suspend specific amounts of water vapour, depending on air temperature. If the air temperature is lowered so that saturation occurs,

the water vapour changes its state and becomes visible, in the form of fog or mist, which are forms of water condensate. Conversely, if the air temperature is raised sufficiently, the condensate will change its state once again and return to invisible water vapour. Air is said to be saturated when it has taken up all the water it can suspend (fig. 14–1) at that temperature.

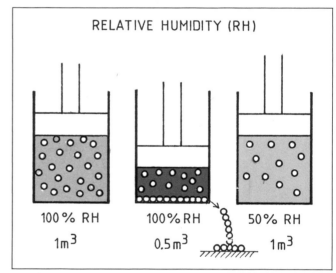

Fig. 14–1 The effect of water vapour in air under isothermal compression. RH = relative humidity.

The following terms and concepts are frequently used in compressed air drying:

- "*Saturation quantity*" is the maximum amount of water vapour by weight, which a sample of air can suspend at a particular temperature.
- "*Absolute humidity*" is that amount of water vapour by weight which a sample of air is suspending when not fully saturated. This humidity depends on geographical and atmospheric conditions. For example, on a rainy day the absolute humidity is much higher than on a dry sunny day. Absolute humitidy is expressed in grams of water vapour per cubic metre of air.
- "*Relative humidity*" is the ratio of the absolute humidity at a particular temperature, to the saturation quantity, at the same temperature. Rela-

tive humidity expresses the water vapour content of the air as a percentage (%) relative to the vapour concentration that would be required to reach saturation quantity. Relative humidity is measured with a "hygrometer". Relative humidity is an important factor for calculating the condensate drop out of a compressed air system (see calculation later in this chapter).

• "*Dew point*" is the *temperature* at which a given sample of moist air becomes saturated. This dew point definition is general and is usually applied to atmospheric or free air. It is that temperature at which a sample of atmospheric air in a closed vessel, when cooled sufficiently, will drop out condensate in one of its forms. Con-

versely, if the same sample is reheated, the dew point is that temperature at which all of the previously shed condensate reverts to vapour form. Dew point is sometimes referred to as "atmospheric dew point" or A.D.P. (see fig. 14–2).

• "*Pressure dew point*" is the term used to indicate the dryness of the compressed air when it leaves the air dryer. It is mainly used in connection with industrial compressed air dryer performance specifications. The pressure dew point (P.D.P.) is that *temperature* at which water vapour contained in compressed air of a specified pressure (other than atmospheric pressure) will condense to form liquid water. This P.D.P. is usually given for 700 kPa (7 bar) pres-

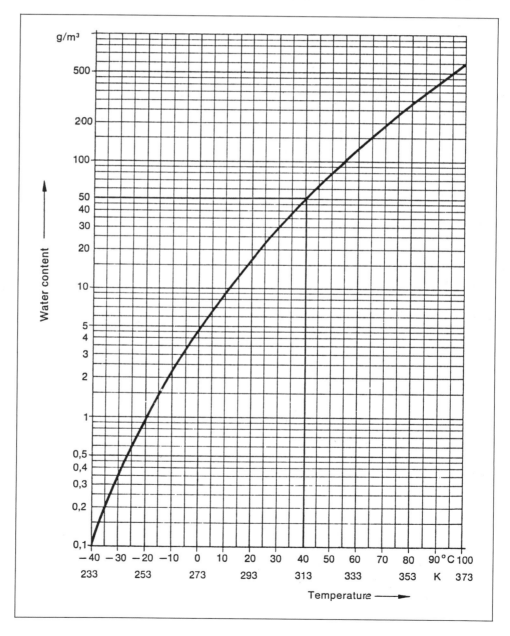

Fig. 14–2 Dew point chart for all pressures.

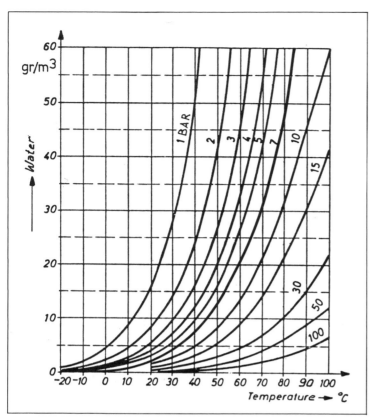

Fig. 14-3 Chart showing pressure dew points (P.D.P.) for various pressures and operating temperatures.

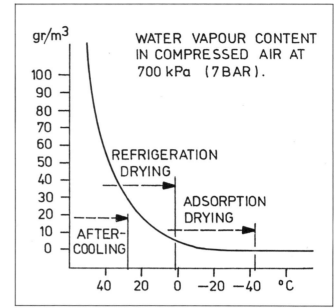

Fig. 14-4 Drying pressure dew point for various drying methods at a pressure of 700 kPa.

sure, the pressure predominantly used in industrial compressed air systems.

When comparing the performance of air dryers it is essential that the difference between atmospheric and dew points (A.D.P. and P.D.P.) is understood, as suppliers of dryers vary as to the form of dew point quoted.

The dew point capability of the dryer at the air pressure at which the compressed dryer is to be used must be known if a suitable dryer is to be chosen, because if the temperature of the compressed air at the point of use falls below the dew point, condensate drop out will occur.

To illustrate the difference between dew points of air at different pressures, refer to chart Fig. 14-3. This chart shows the effect that compressing a sample of air can have on the dew point. For example, consider a sample volume of air containing 10 gr/m³ of water at atmospheric pressure (1 BAR). The dew point (A.D.P.) will be 12°C. If this is compressed to 5 BAR its dew point (P.D.P.) will be 40° C and if compressed to 7 BAR its dew point will have risen further to 45°C. It can therefore be seen that the correct interpretation of dew point figures is essential in assessing an air dryer's performance.

When a compressor operates, it takes in large volumes of atmospheric air, which always contains some moisture in the form of water vapour. The compression process reduces the volume of the air, but the volume of water cannot be reduced. For this reason it must be noted that the compressed air emerging from a compressor (at a temperature slightly higher than the ambient intake temperature) is always 100% saturated!

The following calculation, using the gas laws and the dew point graph (fig. 14-2), is a typical example of the amount of condensate generated during compression.

Example calculation:
A compressor is delivering 4 m³/min F.A.D. at 700 kPa gauge, and the air being drawn in at the compressor inlet has an ambient temperature of 25°C. The relative humidity of the drawn in air is 70% and its pressure is 101 kPa absolute. Calculate the amount of condensate per hour that would drop out at a point after the compressor, where the compressed air temperature has cooled down to 35°C.

Calculations over 1 hour:
4 m³/min F.A.D. = 240 m³/h F.A.D.

Conditions at the inlet port of compressor:
$p_1 = 101$ kPa A
$V_1 = 240$ m³
$T_1 = 25 + 273 = 298$ K

Conditions at the delivery port of compressor:

$p_2 = 700 + 101 = 801$ kPa A

$V_2 =$ to be solved (m³)

$T_2 = 35 + 273 = 308$ K

Using the Gas Law:

$$\frac{p_1 \times V_1}{T_1} = \frac{p_2 \times V_2}{T_2}$$

$$V_2 = \frac{p_1 \times V_1 \times T_2}{T_1 \times p_2}$$

The volume after compression is:

$$V_2 = \frac{101 \times 240 \times 308}{298 \times 801} = 31.28 \text{ m}^3$$

From the dew point graph (fig. 13–2) 1 m³ of air at 25°C can hold 22 g of moisture at 100% R.H.

At 70% R.H. 1 m³ holds $\dfrac{22 \times 70}{100} = 15.4$ g

Moisture taken in at the compressor inlet:

$= 15.4 \times 240 = 3696$ g

Referring again to the dew point graph 1 m³ of air at 35°C can hold 38 g of moisture at 100% R.H.

Moisture suspended by the air after compression is therefore:

$38 \times 31.28 = 1188.64$ g

Condensate drop out = moisture taken in minus the moisture the compressed air can suspend:

$3696 - 1188.64$ g $= 2507.36$ g

or 2.5 litres of condensate would be dropped out each hour.

(1 litre = 1000 grams water)

The calculation shows that 2.5 litres of condensate was dropped out for each hour the compressor was operating under these conditions. If by way of comparison the inlet temperature (ambient) was 30°C and then after compression the air temperature was dropped to 40°C (assuming relative humidity would remain the same), the previously calculated amount of 2.5 litres would increase to 3.7 litres per hour. Figure 14–10, at the end of this chapter, further illustrates how temperature and pressure volume variations affect the moisture as it travels through a two stage compressor down the distribution line.

Aim of compressed air drying

The aim of compressed air drying is to dry the compressed air shortly after it leaves the compressor to a P.D.P. lower than that which is likely to occur in the distribution system or at the points of use. In doing so, all the moisture over and above that critical point, which produces saturation or P.D.P., will not occur again. Therefore, condensate will no longer drop out, whilst the air is being distributed to the various points where it is used. The lowering of the pressure dew point does not necessarily require cooling of the air, as physical and chemical processes can also be used to achieve the same effect.

Dryer selection criteria

If pneumatically controlled and operated equipment or a manufacturing process demands dry air, certain factors have to be determined before a suitable air dryer can be selected.

These factors are:

- total free air flow (m³/min F.A.),
- maximum system pressure (kPa max. cut out),
- minimum system temperature (°C at point of use),
- air inlet temperature (°C at dryer inlet),
- pressure dew point (P.D.P.) of the dryer,
- drying energy costs (dollars),
- pressure drop through the dryer (Δp).

The first four factors are pertinent to the system to be dried, whilst the last three are important dryer characteristics, usually obtained from manufacturer's specifications.

- *Total free air flow* is the flow rate in cubic metres/min F.A. (Q), which the dryer must be capable of passing and drying. If all compressed air from the compressor is to be dried, then the flow capacity of the dryer (Q of dryer) must be not less than the flow rate of the air compressor (Q of air compressor in F.A.D.). But to save energy and reduce capital costs, for some systems it may prove economical to dry only those branches of the system for which dry air is critical.

- *Maximum system pressure* is the pressure level of the compressor when it cuts out (capacity control). The dryer must have a pressure rating of no less than this maximum pressure.

- *Minimum system temperature* is the lowest possible temperature ever likely to be encountered at any point in the distribution system or place of air usage. For a typical industrial manufacturing plant, this requires a study of the ambient temperature conditions likely to be encountered over a full year. This temperature is particularly important to ascertain for installations in cold countries, or places with extreme fluctuating temperatures. When an air pipe must cross from one building to another and sub-freezing temperatures may affect the air temperature in the pipe, this must also be taken into consideration (see fig. 13–21). The pressure dew point (P.D.P.) achieved by the selected dryer must therefore always be lower than the lowest expected temperature in the system.

- *Air inlet temperature* is another factor to be considered when an air dryer is to be selected. A dryer treating compressed air at 45°C inlet temperature would have to extract 70% more water vapour than if the air entered at only 35°C.

  The compressed air temperature at the inlet of absorption and adsorption dryers must never exceed 38°C, as this can damage and render ineffective the drying material (desiccant) used in these dryers. To satisfy such temperature requirements it is common to place the dryer downstream of the after cooler and the receiver (see fig. 14–5).

- *Pressure dew point* (P.D.P.) capability of a dryer is a major criterion. Even if the term P.D.P is understood, it can often be difficult to appreciate and judge the dryness between one P.D.P. and another. Figure 14–6 shows the variations in water content of compressed air starting with a P.D.P. of 20°C at a pressure of 700 kPa (7 bar), assuming 100% saturation. A refrigeration dryer with a P.D.P of 6°C would remove 57% of the moisture (water) contained in the air at 20°C. However, a refrigeration dryer with a P.D.P. of 2°C would remove up to 67% of the water. In other words, a small reduction of the P.D.P. (4°C reduction), achieves a sizeable reduction of the moisture content (10% less moisture).

  To compare these figures, an adsorption dryer with a P.D.P. of −20°C would remove up to 95% of the water (moisture) contained in the compressed air at +20°C inlet temperature. Thus it can be seen that adsorption dryers are considerably more effective than refrigeration dryers and absorption dryers.

- *Drying energy costs* are a most important factor in the selection of air dryers. The following fig-

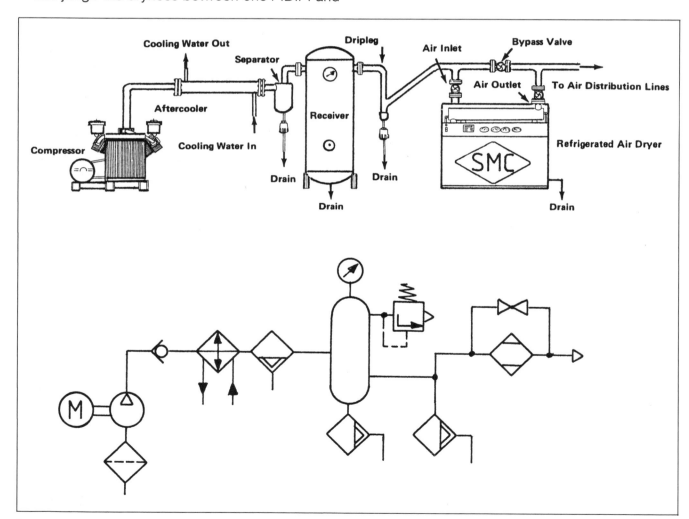

Fig. 14–5 Typical compressor air receiver and dryer arrangement.

ures serve as a comparison for refrigeration and adsorption drying.

REFRIGERATION:
+2°C P.D.P.   4 kWh/1000 m³ (F.A.)

ADSORPTION: (compressed air regeneration)
−20°C P.D.P.   18 kWh/1000 m³ (F.A.)

(electrical filament regeneration)
−20°C P.D.P.   8 kWh/1000 m³ (F.A.)

(heated fan-flow air regeneration)
−20°C P.D.P.   7 kWh/1000 m³ (F.A.)

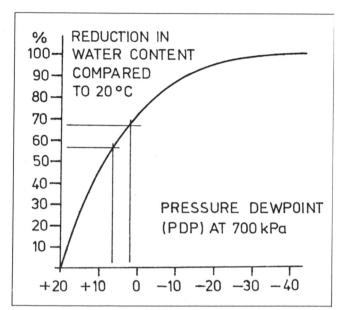

Fig. 14–6 Percentage reduction in moisture from compressed air at differing dew points, but relative to 700 kPa (7 bar) system pressure.

- *Pressure drop through dryers* must also be considered. Based on the energy required by the compressor to produce compressed air in the first place, it can be calculated that each 10 kPa (0.1 bar) pressure drop through the dryer requires a power input to the compressor of 0.8 kWh/1000 m³ (F.A.D.). This power figure in fact applies also to all other equipment producing pressure drops. The following figures serve as a comparison for refrigeration and adsorption dryers. adsorption dryers.

Refrigeration Dryers
Δp 30 kPa (0.3 bar) = 2.5 kWh/1000 m³ (F.A.)

Adsorption Dryers
Δp 70 kPa (0.7 bar) = 5.6 kWh/1000 m³ (F.A.)

There are basically three drying methods used for industrial systems:

- refrigeration drying (low temperature)
- adsorption drying
- absorption drying.

Refrigeration dryers

Only refrigeration dryers actually lower the temperature of the compressed air in order to lower the pressure dew point. They are sometimes referred to as "low temperature dryers". Refrigeration air drying is closely related to domestic refrigerators (fig. 14–7).

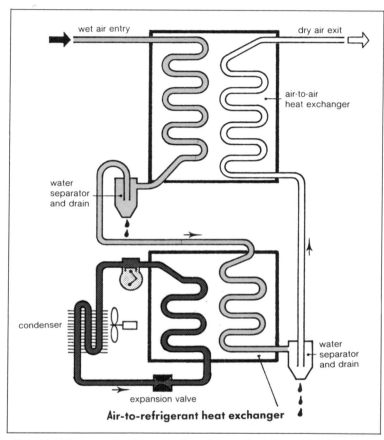

Fig. 14–7 Schematic diagram of a refrigeration dryer.

The compressed air to be dried enters at the inlet and flows through a "hot inlet air to a cold outlet air" heat exchanger. This "air to air" heat exchanger causes a temperature exchange between the two pipes passing through it, causing the incoming hot and saturated air to be cooled and thus releasing some of its moisture as condensate, whilst the outgoing cold and saturated air is warmed. In this way the relative humidity of the outlet air drops away from the saturation point, and therefore will stay dry during its flow through the distribution pipes.

After leaving the air to air heat exchanger, the air passes into the cooling unit (refrigeration unit), which further lowers the temperature of the air to about +2°C. This final cooling process removes most of the moisture (see P.D.P. curve fig. 14-4). The extracted condensate is removed by automatic drains as it is generated. Air still in a saturated condition, but now at a much lower temperature, flows back into the air to air heat exchanger, where it is warmed again.

Refrigeration dryers, although only capable of drying compressed air to a pressure dew point of just above 0°C, are quite adequate for use in many general applications, where moderate, minimum temperatures are experienced. Pressure drops of around 30 kPa are typical of such dryers, and their air inlet temperature (although not advisable) can be as high as 44°C. A certain quantity of oil in the air stream can be tolerated and removed with the moisture. To ensure that the specified dew point is maintained, and freezing does not occur, correct sizing of the dryer and correct adjustment of the refrigeration unit is important. Periodic maintenance is minimal and is virtually limited to checking the automatic drains.

Adsorption dryers

The adsorption dryer does not cool the air in order to lower the pressure dew point, but achieves this by a physical process called adsorption. Adsorption uses the properties of materials called desiccants. Such desiccants have the ability to attract and retain water vapour, whereby the moisture is deposited on the surface of the desiccant material. The moisture can be removed by various means, such as heating the desiccants or by purging it with dry air. This purging process is called regeneration (reactivation), and after completion the desiccant can be reused.

Adsorption dryers consist of two pressure vessels, which store the desiccant in pellet or granule form. Alternatively, one vessel is constantly in use, whilst the other is being "dried out" (regenerated).

Compressed air enters the drying vessel at the bottom and passes upwards through the desiccant material, where the process of adsorption occurs (fig. 14-8). The air then exits at the top of the vessel, having lost most of its moisture. In the meantime, the other vessel is being regenerated. The changeover from drying compressed air to regeneration is controlled automatically and reverses usually after a set period of time. The drying process, however, is continuous.

Adsorption dryers can achieve pressure dew points to −20°C and some may even reach a P.D.P. of −50°C, depending on desiccant material used. Pressure drop through such dryers may be as high as 70 kPa, but is generally about 20 kPa. Common desiccant materials are silica gel

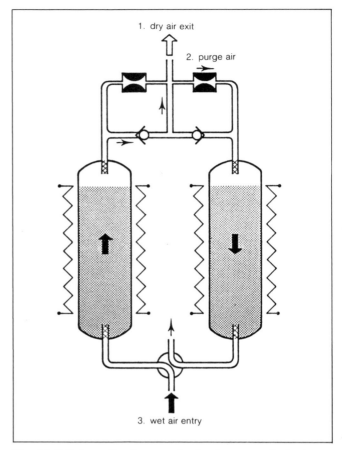

Fig. 14-8 Schematic diagram of a heat regenerated adsorption dryer.

or activated alumina. In both adsorption and absorption dryers, the actual composition of the commmercial desiccant is a very closely guarded secret, as each company claims lower dew point depressions for its dryers and drying agents. For maximum efficiency, an inlet temperature of 38°C should not be exceeded. For adsorption dryers, oil must first be removed to avoid damage to the desiccant material.

Small single unit or cartridge adsorption dryers are also available for low air flow applications. Regeneration is carried out manually by removing the cartridge and applying heat or dry air.

Absorption dryers

The absorption dryer uses a special type of desiccant, called a deliquescent. This deliquescent, when it comes in contact with water vapour from the compressed air, reacts chemically and liquefies. By this process, water vapour is absorbed from the air.

The pressure vessel is constructed to hold a large amount of deliquescent material in pellet or granule form (fig. 14-9). Typical deliquescent materials in use are: urea, lithium, and calcium

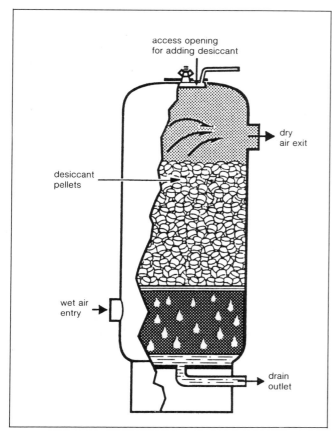

Fig 14–9 Cutaway diagram of an absorption dryer.

Labels in figure:
access opening for adding desiccant
dry air exit
desiccant pellets
wet air entry
drain outlet

chloride. In operation, the compressed air enters at the bottom of the vessel and passes upwards through the desiccant bed, where most of the moisture is absorbed, before it passes through the outlet port at the top end of the vessel. During absorption, the deliquescent desiccant slowly liquefies and seeps to the bottom of the vessel, where it is automatically drained off by a special automatic drain which can handle the fluid residue.

The pressure dew point or dew point "depression", as it is called in this type of dryer, is not a fixed temperature, but a function of the inlet temperature and can be as high as 15°C (Δ°C). If, for example, the temperature of the compressed air entering the dryer is 25°C, then the pressure dew point at the outlet will be 10°C. This low P.D.P. depression (Δ°C) is in many applications not a disadvantage, provided that the compressed air temperature at the dryer inlet has been lowered by aftercooling or an air receiver, to a temperature close to that at the point of usage.

Pressure drop through an absorption dryer is very low, typically less than 15 kPa. Any oil present in the air stream should, however, be removed before the air enters the dryer. Oil in the air can damage the desiccant material.

Absorption dryers are relatively economical, as

they require no electrical power supply. The deliquescent desiccant, however, must be continually replaced as it is used (liquefied). With a correctly sized and operated dryer, refilling on a monthly basis should be sufficient. Some deliquescent drying agents are strongly corrosive and there is also the added risk that they might be carried along with the dried air, thus causing corrosion to pipe systems and pneumatic components. A further drawback is that the drying agents may soften and bake together, if the inlet temperature is not precisely controlled (exceeds 30°C). This may lead to increased pressure drop and added power costs.

The liquid residue (a hydrous solution) from most absorption dryers on the market may be directly discharged into storm water or sewage drains, as apparently there are no constituents in the liquid which do not meet the waste liquid discharge requirements of relevant statutory authorities?! This should be positively ascertained, if pollution is to be avoided. If treatment of the liquid is required before discharge into drains can occur, this would then affect the operation and economy of absorption dryers.

Oil removal

Virtually all pneumatic processes which require dry air also require that this air is free of oil contamination.

Since no type of dryer can completely remove all oil particles present in the air stream, and desiccant material used in adsorption and absorption dryers can be damaged by oil contamination, oil must be removed from the air before it enters a dryer.

If the system does not warrant an oil free air compressor (see Chapter 13), oil removal filters are available with air flow capacities of up to 20 m³/min F.A.

There are two types of oil removal filter — coalescing (impingment) and adsorption. The coalescing filter uses a cartridge filled with a network of fibres. As air is passed through the filter any liquid oil particles present collide with the fibres (and each other), thus gradually forming larger droplets which gravitate to the bottom of the cartridge to be drained away.

The cartridge does not require changing until the air flow is restricted by dirt build-up, indicated by a pressure drop across the unit.

The adsorption oil filter uses a cartridge filled with particles which have a great affinity for oil, both in the liquid and vapour form, and adsorb this oil as air is passed through the filter.

When the particles can no longer adsorb any oil present in the air stream, a colour change occurs. The cartridge must then be discarded and replaced, as regeneration is not possible.

15 Compressed Air Distribution

Designing the system

After compression the air must be distributed to the points of usage, ideally with the least loss of pressure and flow. Pressure loss is caused by air turbulence in pipes and fittings, and flow loss is the result of external leakage. To achieve these goals proper sizing and designing of the air distribution system is thus of great importance and economical significance (fig. 15–1).

Sizing the system begins with calculation of the main pipe size required to carry the flow of compressed air from the compressor to the points of use (compressed air consumers). But first, a number of system operating factors or parameters must be determined:

- pressure drop through the system (Δp),
- calculated maximum system pressure (kPa max.),
- flow rate (Q) in free air,
- pipe length (m),
- pipe material (steel, copper, plastic),
- moisture drop out.

Pressure drop

In a compressed air system pressure drop is unavoidable. It is the result of turbulences and friction in the compressed air whilst it flows through pipes, fittings and valves. The sharper the bends in fittings and the rougher the inside surface of pipes, the greater the pressure drop will be.

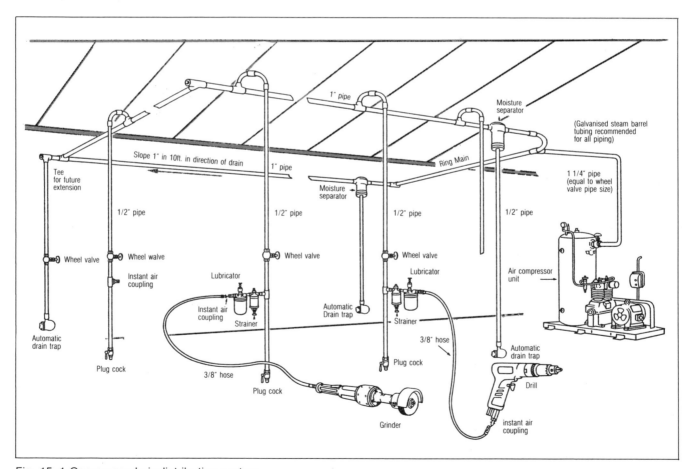

Fig. 15–1 Compressed air distribution system.

Tool	c.fm	m³/min
Adhesive guns (average)	10	0.28
Air hoists		
up to 1000 kg	1 cu.ft./ft.lift	0.09 m³/m lift
1000 kg to 5000 kg	5 cu.ft./ft.lift	0.45 m³/m lift
5000 kg and larger	15 cu.ft./ft.lift	1.38 m³/m lift
Air winches (1000 kg)	123	3.5
Blow guns (average)	18	0.5
Chipping hammers		
1 kg to 2 kg weight	12	0.34
2 kg to 7 kg weight	28	0.8
Concrete vibrators	30	0.85
Drills		
6 mm chuck capacity (1/4")	6	0.17
10 mm chuck capacity (3/8")	20	0.57
12.5 mm chuck capacity (1/2")	35	1.0
38 mm chuck capacity (1.1/2")	95	2.7
Air feed drill and tappers		
6 mm capacity (1/4")	21	0.6
12.5 mm capacity (1/2")	33	0.93
Grinders		
die 6 mm collet (1/4") (average)	10	0.28
disc 100 mm wheel (4")	20	0.57
175 mm wheel (7")	40	1.13
Jack hammers, pavement breakers etc.	180	5.1
Paint sprays (average)	7	0.2
varies from	2 to 20	0.06 to 0.57
Polishers		
175 mm diameter buff (7")	20	0.57
Sand-blasting units		
small	67	1.9
medium	113	3.2
large	300	8.5
Sanders		
disc 125 mm disc (5")	14	0.4
175 mm disc (7")	20	0.57
orbital average	10	0.28
Screw drivers (clutch type)		
bit size 4 mm (3/16")	5	0.14
bit size 6 mm (1/4")	10	0.28
bit size 10 mm (3/8")	21	0.6
Sheet metal shears and nibblers	10	0.28
Sump pumps (at 30 m head)		
2 l/s Diaphragm actuation	60	1.7
7 l/s Air motor actuation	109	3.1
13 l/s Air motor actuation	205	5.8
Wrenches		
ratchet 1/4" drive	7	0.2
3/8" drive	10	0.28
1/2" drive	12	0.34
impact 3/8" drive	9	0.25
1/2" drive	17	0.48
3/4" drive	25	0.71
1" drive	28	0.79

Fig. 15–2 Free air consumptions of tools.

Air speed through the pipe system is also important. Fast air flows will invariably lead to large pressure drops, particularly if these speeds exceed 10 m/s. A pressure drop of more than 50 kPa (for large systems) is uneconomical (see fig. 15–5).

Flow rate

The flow rate (Q) calculation of compressed air for a system with many actuators and processes consuming compressed air may seem to be straightforward if one simply adds together the free air consumptions of all compressed air actuators, tools and processes. This flow would however be uneconomically high, unless all the air consuming devices were operating together and continuously all day long. Therefore, in determining the real total flow required for such multi-use systems, one must assess their duty cycles. This assessment invariably brings a substantial reduction of the total flow requirement, with a consequent reduction of the distribution pipe diameter, and compressor size.

A typical chart of various tools and their relevant air consumptions is given in fig. 15–2.

System pressure

The *total system pressure* is made up from the *minimum operating pressure* plus the *system pressure drop* plus the *cut-in/cut-out* pressure differential plus a *safety margin pressure*.

The minimum operating pressure is that pressure which is required to efficiently operate all the actuators in a pneumatic system (air motors, air tools, linear actuators and air consuming processes—see Chapter 3, fig. 3–21). That pressure is usually about 600 to 700 kPa (6 to 7 bar), and must be based on that actuator in the system which requires the highest pressure to operate to its designed force or torque output.

The system pressure drop is an assumed pressure drop caused by turbulence and friction, whilst the compressed air flows through the distribution pipes. It includes pressure drops caused by ancillary equipment such as after coolers and dryers. For economy, this pressure drop should not exceed 50 kPa (0.5 bar).

The cut-in/cut-out pressure differential is governed by the compressor capacity controller and its purpose was explained in the previous chapter. The cut-in/cut-out differential is in the range of 50–100 kPa (0.5–1 bar).

The safety margin pressure or contingency allowance is an assumed pressure of approx. 10% based on the minimum system pressure.

Example calculation:

Minimum operating pressure	650 kPa
System pressure drop	30 kPa
Cut-in/cut-out pressure differential	100 kPa
Safety margin pressure	65 kPa
Total system pressure	845 kPa

From this calculation the total system pressure required for the calculation of the pipe size would be approx. 850 kPa.

Air leakage

Ideally the leakage factor should always be zero, but practically about 10% must be allowed for system leakage and must be added to the total flow rate required. The importance of such air leakage prevention cannot be underestimated. Figure 15–3 shows a chart for leakage air flow rates through various size holes, together with the associated power loss.

Compressed air leakage is power wastage, therefore measures should be taken for its prevention. One method of assessing and rectifying compressed air leakage is to periodically run leakage tests. Such leakage tests are carried out when the distribution system is not in use.

To carry out this test, the compressor is started and after the system pressure has been reached and the compressor has unloaded (cut-out) the system pressure will then begin to fall. The period of time (T) for the compressor to come on load (cut-in) is recorded and also the period of time (t) when unloading again occurs (cut-out). For greater accuracy the average time of a number of such cycles must be taken.

The leakage rate (Q_L) of the system can then be calculated as follows:

Q_L = System leakage rate (m³/min F.A.)
Q_C = Compressor free air delivery (m³/min)
T = Time taken between cut-out and cut-in (leakage time in minutes)
t = Time taken between cut-in and cut-out (charging time in minutes)

Formula

$$Q_L = \frac{Q_C \times t}{(T + t)} \text{ m}^3/\text{min}$$

Steps can then be taken to locate and rectify the leakage.

Hole diameter		Air leakage at 6 bar	Power required for compression	
True size	mm	m³/min	hp	kW
•	1	0.06	0.4	0.3
●	3	0.6	4.2	3.1
⬤	5	1.6	11.2	8.3
⬤	10	6.3	44	33

Fig. 15-3 Table showing actual hole size, equivalent leakage and power loss.

Air leakages, therefore, represent money wasted. A calculation carried out on a compressed air installation producing 15 m³/min F.A.D. for a large electrical appliance manufacturer in 1977, when the cost of electrical power at that time was 5 cents/kW hour, showed that the air leakage was in fact 25% of the total flow rate. The cost to that company at that time was $5,695 per year! The system was to all outside appearances well maintained and leak free. This unfortunately would be a typical leakage figure for many compressed air installations.

Air leakages are often unseen and in many cases also unheard, and therefore require periodical checks and stringent system maintenance, if they are to be avoided.

Pipe Length

The calculated pipe length can be ascertained by calculations or nomograms, and is made up of the actual pipe length and the equivalent pipe length.

The actual pipe length is that distance of the main pipe which is actually measured from the compressor to the closed off end of the pipe for a single line system, and for a ring main system the total pipe loop length is measured and then divided by two (fig. 15-4). The equivalent pipe length is an imaginary length of pipe which would have the same resistance to flow, and therefore cause the same pressure drop, as all the pipe fittings or valves used in the main pipe system (fig. 15-6).

Sizing can be carried out by mathematical formulae and calculation, but the most commonly used and easier method is by the application of nomograms and tables (figs. 15-5 and 15-6).

The calculation procedure for an air distribution system using both the nomogram (fig. 15-5) and

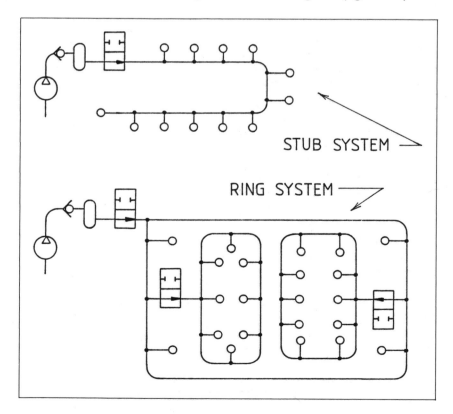

Fig. 15-4 Typical air distribution systems.

the pressure drop table (fig. 15–6), is presented below.

Example calculation:

Determine the diameter of a main distribution pipe to carry air at a flow rate of 10 m³/min F.A. The maximum system pressure is 700 kPa and the total actual length of the main pipe line is 200 metres. A maximum pressure drop of 10 kPa is permitted. The distribution main must include the following fittings:

6 bends with a radius of twice the pipe diameter,
2 elbow fittings,
4 tee connection fittings with 90° turn and 2 gate valves.

Step 1
Disregarding for the time being the equivalent pipe length, the previously given values must now be plotted into the nomogram shown in fig. 15–5 (200 m actual pipe length, 10 kPa permitted pressure drop, 10 m³/min F.A. flow rate and 700 kPa

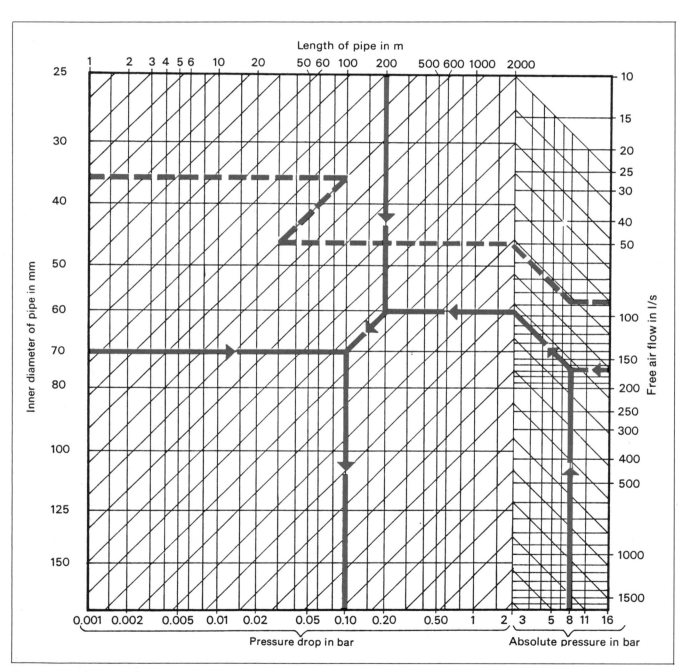

Fig. 15–5 Nomogram for calculation of pipe line dimensions.

Valves, etc.		Equivalent pipe length in m						
		Inner pipe diameter in mm						
		25	40	50	80	100	125	150
Seat valve		3–6	5–10	7–15	10–25	15–30	20–50	25–60
Diaphragm valve		1.2	2.0	3.0	4.5	6	8	10
Gate valve		0.3	0.5	0.7	1.0	1.5	2.0	2.5
Elbow		1.5	2.5	3.5	5	7	10	15
Bend R = d		0.3	0.5	0.6	1.0	1.5	2.0	2.5
Bend R = 2d		0.15	0.25	0.3	0.5	0.8	1.0	1.5
Hose connection T-piece		2	3	4	7	10	15	20
Reducer		0.5	0.7	1.0	2.0	2.5	3.5	4.0

Fig. 15–6 Table of equivalent pipe lengths for valves and fittings as used on air distribution systems.

maximum system pressure). From this nomogram, a pipe internal diameter of approx. 70 mm can be derived. This, however, is not the final pipe diameter since no consideration has been given to the pressure drop caused by fittings and valves. Therefore, this internal pipe diameter must be regarded as an interim result.

Step 2
The closest (next largest) commercially available pipe diameter to the 70 mm interim diameter derived in step 1 is 75 mm. This is the size now used in conjunction with values shown in table fig. 15–6 to calculate the total equivalent pipe length for all the fittings.

6 Bends (R = 2d)	6 × 0.5 m =	3 m
2 Elbow fittings	2 × 5 m =	10 m
4 T-connections (90° turn)	4 × 7 m =	28 m
2 Gate valves	2 × 1 m =	2 m
Total equivalent pipe length		= 43 m

Step 3
Total equivalent pipe length is now added to the actual pipe length of 200 m, to give 243 m the calculated pipe length. The nomogram is now again worked through exactly as before, but this time

243 m is used and a final pipe diameter of approx. 73 mm can be found. Since a 75 mm diameter pipe has already been selected, the additional pipe internal diameter required to account for the pressure drop caused by the fittings has been amply covered. The second plotting is not shown, for reasons of clarity.

Step 4
If at a later stage the system network had to be expanded, requiring the insertion of more fittings, and thus increasing the calculated pipe length to 550 metres, then an internal pipe diameter of 85 mm would be derived from the nomogram. This, however, is not a standard size and a pipe diameter of 100 mm would have to be bought (see fig. 15–5).

There are significant advantages in increasing the diameter wherever possible, as an increase for example of 1/3 (75 to 100 mm) increases the flow capabilities by the square of the diameter, or nearly twice in this case. This does not substantially increase the price of pipe and installation space required. A larger pipe diameter adds volume to the air distribution system, which can be regarded as additional air receiver volume. This volume brings advantages such as a less pressure drop, better cooling of the compressed air, and additional storage volume—which all help to increase the system economy and air quality.

Choice of pipe material
The usual choice for tubing for distribution systems is galvanised steel pipe screwed together with galvanised fittings. This choice is very popular, as the system can be built up relatively easily from pipe and fittings "off the shelf". It can be used in both large and small installations. Its main disadvantage is the connections, and since the pipe sections are screwed together, they require some form of thread sealing, which usually takes the form of hemp and sealing compound or thread sealing tapes (Teflon). If such sealing materials are not applied correctly, they can enter the pipe and cause air contamination and component failure further down the line. All such connections are also sources of turbulence, which lead to excessive pressure drop and often also become potential air leaks.

Galvanised pipe also tends to corrode at the joints due to the moisture condensation in the pipe system, and therefore becomes a contamination source (rust particles and scale), which is carried to the point of usage. When internal surfaces of the galvanised pipe become rusty, they tend to increase the pressure drop (turbulence) and for this reason alone long term system economy may be enhanced by the use of copper piping.

Galvanised steel pipe properly installed and fitted can, however, be an economical method of distributing compressed air for small to medium sized systems. For large systems galvanised steel pipe is not ideal and can prove extremely expensive, if long term economy is calculated.

Copper pipe is gaining popularity on compressed air distribution systems. Contrary to expectation, copper piping can be much more economical than a galvanised steel pipe system, has excellent pressure drop characteristics, and is almost corrosion-free.

Copper pipe systems are usually constructed by welding or brazing together all the joints and fittings, thus the problems associated with the connection and installation of galvanised tubing are automatically eliminated.

Wherever a medium to large distribution system is being designed (more than 10 m³/min F.A.D.), careful comparison of total costs including materials, labour and system economy should be made. The long term benefits of copper tubing should be carefully considered.

Moisture condensation and system design

The design of the distribution system should be such that the continually generated condensate can be drained away together with contaminants before the air reaches the pneumatically controlled machinery or air consuming processes.

Where take-off pipes leading to individual machines or compressed air tools are connected to the main air distribution line, they must be taken off with a half loop and connected into the top of the main air distribution line (see fig. 12–16). With such a connection, water condensate cannot flow into the take-off line. Arrows indicate that the piping of the ring system (main system line) falls or is graded down in the direction of flow. This grading is usually about 1–2 degrees.

Such grading permits the condensate to drain (flow) towards the drain valves, which are installed at the end of the air distribution lines at drop legs (fig. 12–16).

At the drain points (drop legs), it is advisable to fit automatic condensate drains. Such automatic condensate drains are also fitted to the intercooler, after cooler and air receiver. It is advantageous to automatically drain away the condensate as soon as it forms.

An automatic drain is a very important and valuable device used in compressed air distribution systems, since it eliminates the need for manual draining of the condensate at its collection points (fig. 15–7).

An automatic drain consists essentially of a small reservoir with a float. The float responds to the level of collected condensate in the reservoir and when the condensate reaches a set level, the float lifts and the condensate drain valve opens, thus discharging the condensate together with some compressed air. When the condensate level drops in the reservoir to the minimum level, the valve closes again.

Automatic drains can become ''air locked''. When this happens, a cushion of air builds up and prevents the condensate from flowing into the reservoir. Some automatic drains overcome this problem by fitting an anti-locking vent to the reservoir, which is connected to the drop leg by a small vent pipe (shown by the dotted line in figure 15–8).

An isolating valve should always be fitted ahead of an automatic drain to allow maintenance of the drain while the pipe system is under pressure.

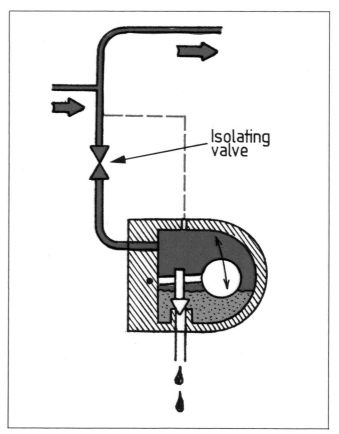

Fig. 15–7 Distribution pipe arrangement for controlling condensate drop out.

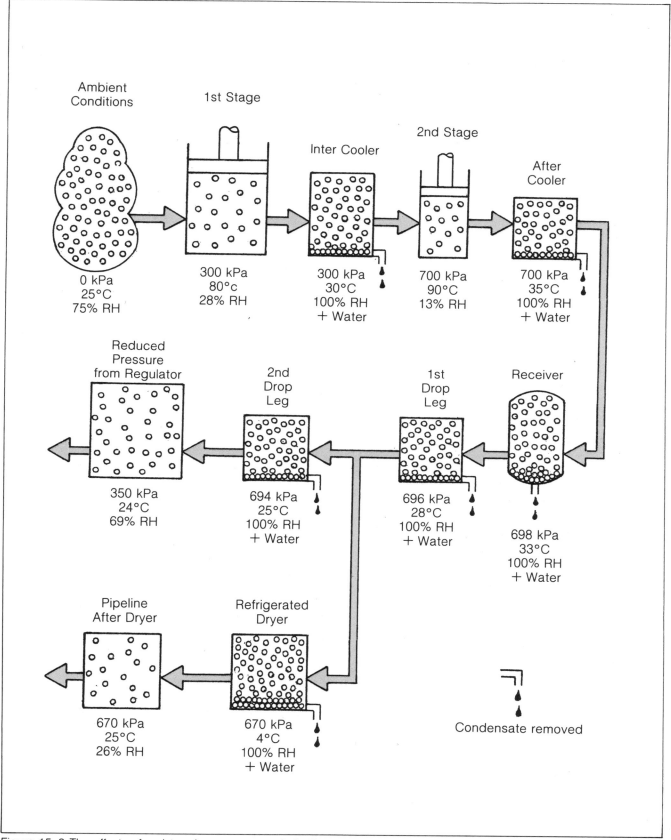

Figure 15–8 The effects of moisture in a compressed air distribution system from the compressor intake to points of usage. (Diagram by Gordon Smith.)

16 Circuit Documentation and Circuit Construction

The machine and its fluid power control functions should be fully documented. Documentation helps to translate the designed circuit into actual componentry such as cylinders, valves, fittings and tubing, and serves as an important reference for the operation and maintenance of the machine to help ensure trouble free operation over a long period of time. Such documentation must therefore include:

- complete and neatly drawn circuit drawing showing all fluid power components,
- traverse-time diagram,
- parts list of all fluid power components,
- servicing, maintenance and fault finding instructions,
- other relevant data.

Circuit drawing and documentation

The circuit drawing must conform to the internationally agreed symbols (see Appendices) and must show all components involved in the fluid power circuit. The power circuit, the logic control circuit (step-counter circuit, cascade circuit, limit valves) and the fringe condition modules should be clearly separated (see figs. 8–31, 8–47 and 9–22).

Time and pressure control valves must show their relevant time and pressure setting values (see fig. 2–52) and *all* signal lines must be labelled (START, b_1, A0 etc.).

Traverse-time diagram

The traverse-time diagram describes the sequence of the circuit step by step. This step by step description will assist a serviceman, for example in understanding the operation of the circuit for fault finding or other maintenance purposes. It should also include the emergency sequence if the circuit is equipped with an emergency start-stop provision (see fig. 8–36).

Parts List

Should the parts list not be included on the circuit drawing, then it must be added as a separate item in the documentation. Parts lists should carry the following suggested headings, but these can be varied, as required. Although the headings are self-explanatory, a few words of clarification are given (see fig. 16–2).

- The *item number* is referenced to the circuit diagram (fig. 16–1). Each component on the circuit diagram should carry an item number (preferably circled) which matches with a corresponding number in the parts list (fig. 16–2).
- The *part description* of the component (together with the port size) should be such that by referring to this description it can be easily identified on the machine without reference to a brand name, or a part number.

  For example: 4/2 Double pilot operated, flat slide valve.
  3/2 Roller operated poppet valve, normally open.

- *Port size* and thread type of the component. This may also include the type of tube fitting used on this component (push-in, push-on, etc.).
- The *brand* and *part number* which positively identify the valve or pneumatic component.
- *Comments* should state any noteworthy feature of the component such as pressure or time setting, or any deviation or modification from the original standard component.

Servicing, maintenance and fault-finding instructions

A detailed listing of the periodic servicing frequency which the machine requires, the type of oils used, and any other pertinent servicing information should be provided. This entails air service unit refill, draining and adjustment.

The servicing instruction must also include the compressor and other ancillary equipment such as a dryer if the operation of the machine is dependent on such items.

Other relevant data

Such data must include a general but brief *description* of the purpose of the machine and its method of operation.

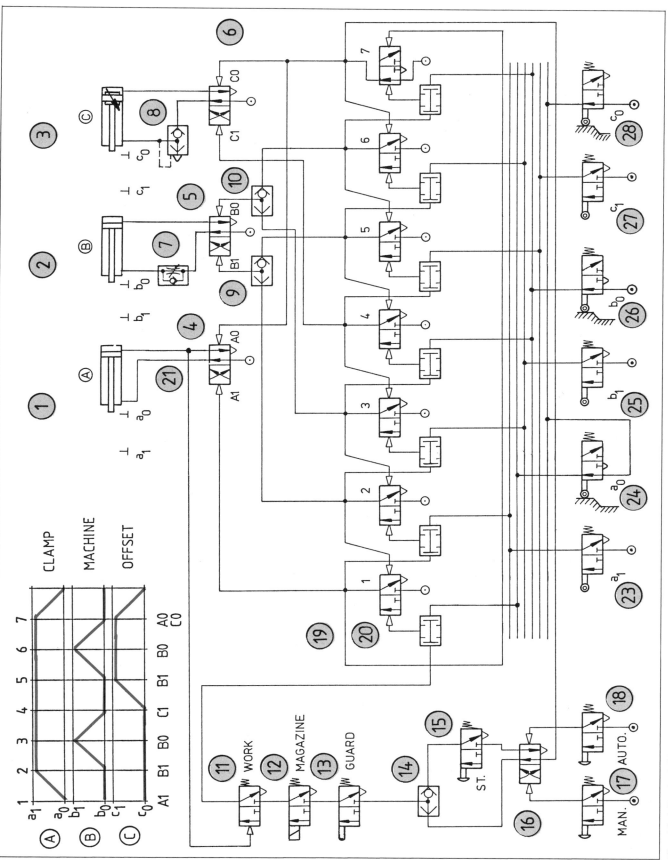

Fig. 16–1 Typical Circuit Diagram.

PARTS LIST

Item No	Part Description	Port Size	Brand	Part Number
1	25 mm bore × 25 mm stroke double acting cylinder	⅛″ B.S.P.	Festo	DGS-25-25
2	50 mm bore × 70 mm stroke double acting cylinder	¼″ B.S.P.	Festo	HC-50-70
3	50 mm bore × 50 mm stroke double acting cushioned cylinder	¼″ B.S.P.	Festo	HC-50-50-100771
4 & 16	4/2 Flat slide valve	⅛″ B.S.P.	Festo	JP-4⅛
5 & 6	4/2 Flat slide valves	¼″ B.S.P.	Festo	JP-4¼
7	Throttle relief valve	¼″ B.S.P.	Festo	GR-¼
8	Quick exhaust valve	¼″ B.S.P.	Festo	SE-¼
9, 10 & 14	"OR" element	⅛″ B.S.P.	Festo	OS-⅛
11	3/2 pilot operated valve	⅛″ B.S.P.	Bosch	0820212001
12	3/2 solenoid operated valve	⅛″ B.S.P.	Bosch	0820005104
13	3/2 plunger operated valve	⅛″ B.S.P.	Bosch	0820402001
15, 17 & 18	3/2 palm operated valve	⅛″ B.S.P.	S.M.C.	VM430-01-30
20	7 off step counter modules complete with AND elements bases and end blocks	4 mm push in	Crouzet	81-550-401 81-522-501 81-551-101 81-552-001
23 to 28	3/2 Roller operated valves	⅛″ B.S.P.	Bosch	0820402002
19	Air service unit module consisting of filter regulator & lubricator	½″ B.S.P.	Norgren	M4A-408 -A3EB -18-013-0111

Fig. 16–2 Typical Parts List.

The *method of operation* should describe the loading and operation of the machine, particularly from the operator's point of view. It should describe all connections which have to be made (power, air, water, etc.) to enable the machine to be installed, commissioned and operated.

Initial settings. These settings are very important for start up and machine commissioning.

Settings listed should be: regulator pressures, pressure sequence valve settings, time delays, valve settings, lubricator drip rates etc. Further-more, these listings should be updated whenever any of these inital settings are altered.

Circuit Construction

The components for the sequence control of the actuators should be laid out and attached to a circuit board or preferably mounted into a control cabinet. Electrical control boxes make ideal control cabinets for pneumatic control components. They are reasonably priced, are available in various sizes and are fitted with a detachable door for ease of access. The door may also have a compartment for the storage of the circuit diagram, and its control documentation.

Control cabinets provide considerable advantages, which increase in importance with the size and complexity of a control system. These include:

- clear layout of all control elements
- short line connections for fast signal processing
- rapid access to all control elements
- simpler fault finding
- no unauthorised access to control
- less, or no, soiling of control elements
- no tangle of lines on the machine
- only signal lines from limit valves and signal lines leading to positioning or operating elements emerge from the control cabinet; they can be combined by means of multi-connectors and protective hoses (flexible conduit)
- manual signal input devices can be grouped together in a small but accessible area
- modular design allows the use of standard size mounting frames for rapid component mounting.

Installation of Components

Pneumatically actuated machinery inherently vibrates, due to the action of the reciprocating cylinders, and although in many instances this vibration may be minimal, all components should therefore be secured by means of shakeproof fasteners (such as spring washers).

When correctly installed and efficiently maintained, pneumatic controls provide many hours of trouble-free operation. However, to achieve optimum performance, a few points should be noted to assist all users of pneumatic equipment.

- Mount pneumatic cylinders securely, ensuring that the piston rod moves freely through its full stroke.
- Check that no undue side stress is placed on the piston rod.
- Where piston rod or other machine components actuate signalling devices, ensure that this actuation is correct and that no excessive load is applied onto such signalling devices.
- Mounting brackets for cylinders and limit valves should have sufficient support and should not flex during system operation.
- For easier identification, label all components as per the circuit diagram (fig. 16–1).
- Make certain that components waiting to be assembled are covered and/or capped to eliminate the ingress of foreign particles.

- Tubing and fittings require leakproof joints. If thread or gasket sealants are used, apply to male thread only. This minimises the possibility of the sealant being forced into cylinder/valve ports or tube.
- Install air-line filters as close as practical to the actual point of use. If quick-disconnect couplings fall into dirt, chips etc. these contaminants could inadvertently be introduced into the circuit when the couplings are reconnected (see fig. 16–3). Filtration close to point of use removes such contaminants.
- When mounting pneumatic equipment on hot surfaces, check the manufacturer's specifications on temperature range for the equipment.
- Mount valves so they are protected and cannot be damaged by heavy falling parts, fork-lift trucks, etc.
- Make sure that the air supply is not restricted by undersized hose, kinked tubing and restrictive fittings.
- Assume that dirt, dust and water will settle on all components. Protect therefore exhaust ports facing up with silencers to prevent dirt from entering the component, but preferably mount all valves with exhaust ports facing down.
- Rapidly exhausting air is likely to be noisy. It may therefore be necessary to muffle the exhaust noise. Quick exhaust valves are usually noisiest! Plan to use silencers on all valves.
- Plan to mount all control components inside a cabinet, and where possible collect all exhaust air from the components into a common manifold to exit through a side of the cabinet.

Tubing

The layout of the interconnecting tubing between valves, cylinders, and other components is important if the circuit is to be pleasing in appearance and efficient in distributing the air signals to the components with minimal loss of pressure. Tubing size is usually set by the port sizes of the components, and if unrestricted flow (fast cylinder movement) is required, these sizes should be adhered to since they are designed to give optimal performance.

Where possible for ease of identification, colour coding of the tubing should be adopted, such as black for air supplies, white for cylinder port connections (working lines) and blue for valve pilot signal lines (these are only suggestions).

The material used for tubing is generally plastic. Plastic tubing is easy to cut, bend and fit and is inexpensive. However, in areas where excessive heat or physical damage may occur, copper is usually the preferred choice. It must be remembered that even though copper is stronger than plastic, physical damage may flatten copper tube, and cause a permanent flow restriction.

There is also a range of flexible tubing available which has an outer metal mesh covering (metal braiding), and this can be used to advantage for applications where abrasion damage could occur.

Fig. 16–3 Quick connect coupling.

Air line Fittings

Pneumatic fittings are made from materials such as brass, aluminium and plastic, and have various ways of making the connection to the tubing and to the valves or actuators.

For plastic tubing one may use one of the many brands of "push-in" fittings now available on the market. Such fittings are made for metric and imperial thread and tube sizes. Most brands of moving part logic components provide porting plates with permanently embedded push-in connections (see fig. 16–4(A)).

Push-in connectors are re-usable, inter-changeable, and easy to connect and disconnect. To make the connection, the tube is simply pushed into the fitting. To disconnect, either a screwdriver, fingernail, or special tool applied to a release ring on the fitting allows the tube to be removed. "Push on" fittings on which the tubing is pushed over a bulge on the fitting and secured with a sleeve nut are also widely used. They are also easy to connect and disconnect, however not as rapidly as the "push in" fitting. See fig. 16–4(B).

Plastic tubing used together with push-in fittings has made the piping up of the pneumatic system much easier than with the nut and olive compression fitting shown in fig. 16–5. The nut and olive fitting is, however, still used on copper or plastic-tubing.

Silencers

Compressed air devices have an inherent tendency to be noisy because pressurised air is allowed to exhaust to atmosphere on completion of its work. The pressure ratio between the exhaust

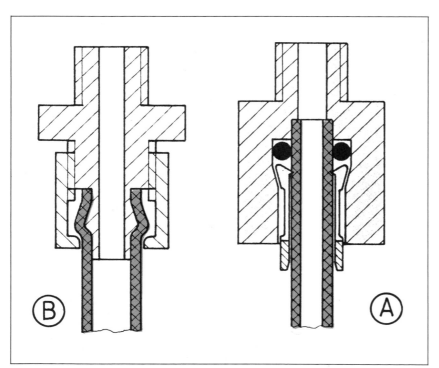

Fig. 16–4(A) Push in tube fitting (B) Push on tube fitting.

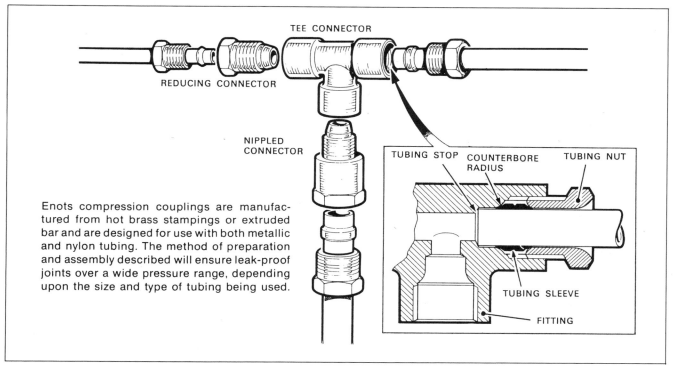

Enots compression couplings are manufactured from hot brass stampings or extruded bar and are designed for use with both metallic and nylon tubing. The method of preparation and assembly described will ensure leak-proof joints over a wide pressure range, depending upon the size and type of tubing being used.

Fig. 16–5 Compression fittings.

and the surrounding static air is normally high. As a result the velocity of the exhaust jet is supersonic, and thus generates shock waves and a considerable amount of turbulence as it mixes with the static atmosphere.

Much of this energy change is in the form of audible shock waves, capable of producing severe discomfort and damage to the human ear. Noise reduction is now required by health regulations and can be achieved by:

- eliminating shock wave formation, and
- reducing the exit air velocity to minimise turbulence.

The only effective way to reduce exhaust air noise is to allow the air to escape into an expansion chamber (silencer). Often, piping back the exhaust pipe through a long, larger pipe to a reservoir or atmosphere can effect considerable dampening.

The silencer acts as a filter to oil as well as other contaminants, and can gradually become saturated. A saturated silencer is indicative of an incorrectly adjusted or maintained air lubricator, and corrections must be applied. However there is a limit, and some fine oil mist will prevail, requiring consideration of the operator and the system's application/function positioning during design and installation.

Sintered bronze silencers

These silencers are made from sintered bronze, and some incorporate all the key requisites of the ideal silencer:

- high strength; withstands up to 100 MPa (1000 bar)
- does not deform and withstands high mechanical shock; unlike plastic or fabric types, can be attached to moving devices
- can quickly and easily be cleaned with common industrial solvents such as trichlorethylene, perchlorethylene, benzol, etc. and does not require replacement spares
- does not swell or deform in water or oil, as do some plastics and fabrics, thus preventing change of characteristics
- readily obtained in different grain sizes
- double chamber principle uses coarse size grains virtually unaffected by dirt, yet provides excellent damping ensuring high, predictable efficiency
- competitively priced
- provides deflector cap with 360° adjustment
- does not retain fluid contaminant or readily become saturated
- highly predictable pressure drop characteristic, and a wide range of use.

Appendices

1 Industrial fluid power symbols

These symbols are based upon the international I.S.O. 1219 fluid power symbols. Only the most common symbols have been included.

Composite symbols can be devised for any fluid power components by combining the relevant basic symbols.

Pumps, motors, and drives

	Fixed	Variable
Single direction pump		
Double direction pump		
Single direction motor		
Double direction motor		
Single direction pump/motor with reversal of flow direction		
Single direction pump/motor with single flow direction		
Double direction pump/motor with two directions of flow		
Hydrostatic drive, split system type		
Hydrostatic drive, compact, reversible output		
Semi rotary actuator		
Air compressor		
Vacuum pump		

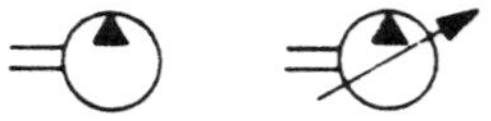

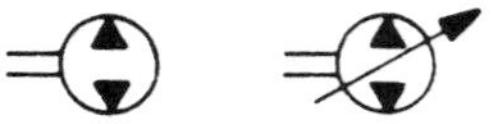

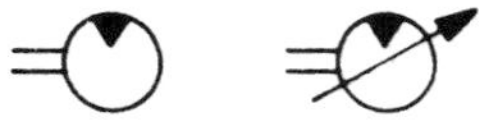

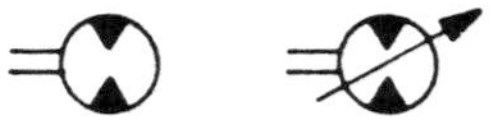

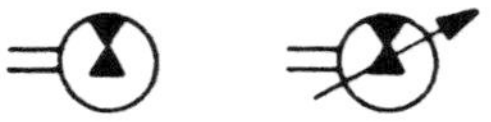

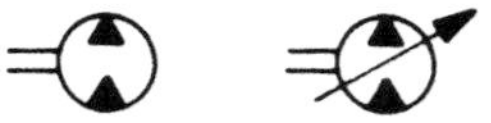

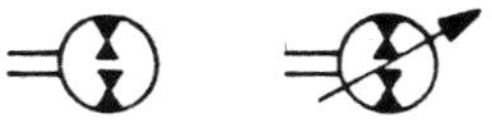

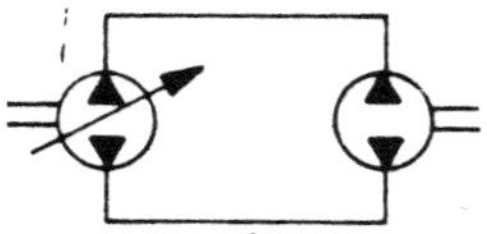

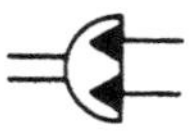

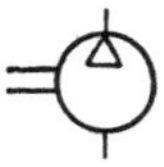

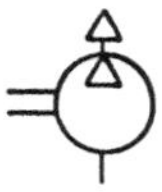

Air motor, single direction

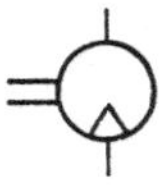

Air motor, double direction

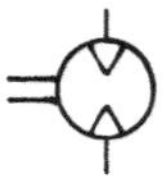

Semi rotary actuator

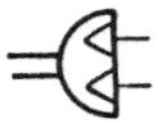

Linear actuators

Single acting ram
(load returns the ram)

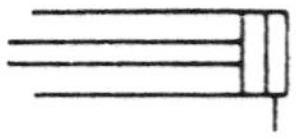

Single acting actuator
(load returns the piston)

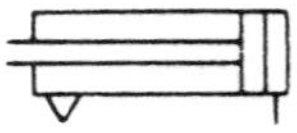

Single acting actuator
(spring returns the piston)

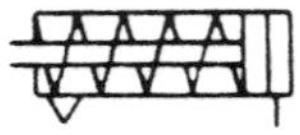

Double acting actuator

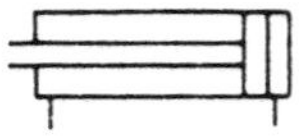

Differential actuator with oversize rod

Double acting actuator with double ended rod

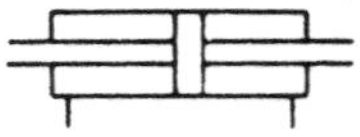

Piston with adjustable end cushioning

Piston with fixed end cushioning

Telescopic, single acting actuator

Telescopic, double acting actuator

Pressure intensifier

Rodless acting, Double acting with cushioning

Double acting with impact stroke

Rodless with magnetic coupling

Valve control mechanisms

Undefined control

Hand lever (rotary or linear)

Push button

Foot lever

Cam roller

Plunger (piston or ball)

Spring

Detent mechanism

Pressure relief

Pressure applied

Pneumatic pilot

Hydraulic pilot

Solenoid

Solenoid/hydraulic pilot

Pneumatic/hydraulic pilot

Two stage amplifier

One way trip lever

Wire sensor (whisker)

Key operator

Air spring (internal)

Spring centred

Directional control valves

Directional control valve with two discrete positions

Directional control valve with three discrete positions

Directional control valve with significant cross-over positions

Valve with two discrete and an infinite number of intermediate throttling positions

Valve with three discrete and an infinite number of intermediate throttling positions

Two position, two port valve

Two position, three port valve

Two position, four port valve

Two position, five port valve

Three position, four port valve with fully closed centre configuration

Detailed symbol of pilot operated pressure relief valve (compound relief valve)

Port labelling:
    Working lines   A, B
    Pilot lines   X, Y
    Pressure line   P
    Tank line   T

Simplified symbol of compound relief valve (Pilot flow externally drained)

(Pilot flow internally drained)

Port labelling with numbers

Brake valve

Check valve

Unloading valve (accumulator charging valve)

Spring loaded check valve

Pilot operated check valve

Counter balance valve (back pressure valve)

OR function valve

AND function valve

Sequence valve with remote control (external pilot)

Deceleration valve

Deceleration valve

Sequence valve with direct control (internal pilot)

Pressure controls

Offloading valve

Throttling orifice normally closed or normally open (*optional)

Pressure relief valve (fixed)

Pressure reducing valve (fixed)

Pressure relief valve (adjustable)

Pressure reducing valve (adjustable)

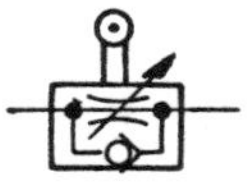

Fluid plumbing and storage

Pressure source	
Working line, return line, feed line	
Pilot control line	
Drain line	
Enclosure line	
Flexible line	
Electric line	
Pipeline connections	
Cross pipeline (not connected)	
Air vent	
Reservoir with inlet below fluid level	
Reservoir with inlet above fluid level	

Pilot operated pressure reducing valve

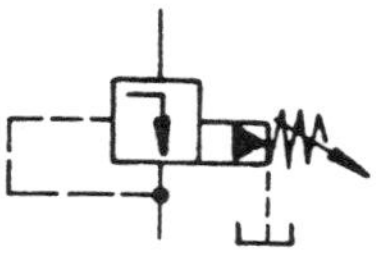

Pressure reducing valve with secondary system relief

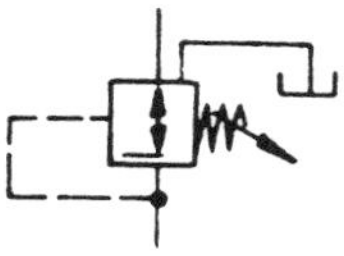

Pneumatic pressure sequence valve

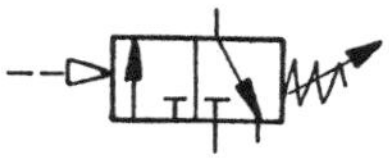

Pneumatic pressure reducing valve with secondary relief

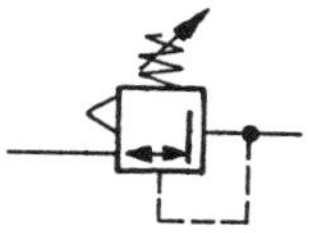

Flow controls

Throttle valve not affected by viscosity

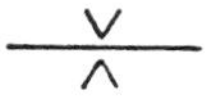

Throttle valve (fixed)

Throttle valve (adjustable)

Flow control valve, pressure and temperature compensated

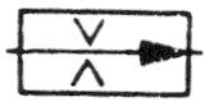

Flow control valve with reverse free flow check

By-pass flow control valve

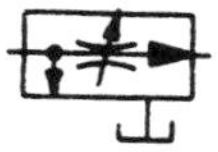

Flow divider

Pneumatic flow control valve with reverse free flow check

Quick exhaust valve

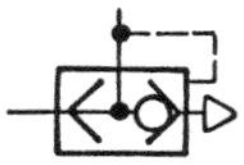

Pneumatic cam operated deceleration valve

Miscellaneous symbols

Electric motor

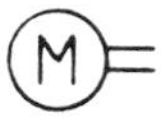

Heat engine

Electric motor with pump and drive coupling

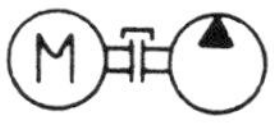

Plugged line

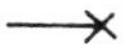

Plugged line with take-off line

Quick connect coupling

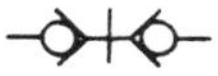

Rotary connection

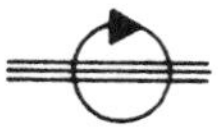

Accumulator

Filter, strainer	
Cooler with coolant lines	
Heater	
Pressure gauge, pressure indicator	
Flow meter	
Thermometer	
Pressure switch (electrical)	
Shut-off valve	
Airline lubricator	
Water separator with automatic drain	
Complete air service unit	
Air drier	
Visual indicator	
Pneumatic counter with set/reset	

Exhaust silencer	
Logic symbols for step-counter	
Classical and commercial version	
Step-counter with end plates	

Electrical symbols

Electrical contact (n/o)	
Electrical contact (n/c)	
Contactor relay	
Set/reset relay (latching relay)	
Set/reset flip-flop, single output	
Set/reset flip-flop for complementary output	
Electrical switch	
Solenoid coil	
Solenoid coil	

2 Fluid power for formulae

	Calculation parameter	Formulae with standard symbols	Coherent and/or commonly used units
Physical principles	Combined gas laws	$\dfrac{p_1 \times V_1}{T_1} = \dfrac{p_2 \times V_2}{T_2}$	$\dfrac{paA \times m^3}{K_1} = \dfrac{PaA = m^3}{K_2}$
	Kinetic energy	$E = \dfrac{(m \times v^2)}{2}$	$J = \dfrac{k \times m/s}{2}$
	Lifting force	$F_L = m \times g \times \sin \alpha$	$F_L = kg \times 9.81 \times \sin \alpha$
	Frictional force (horizontal load)	$f = m \times g \times \mu$	$N = kg \times 9.81 \times \mu$
	Frictional force (inclined load)	$F = m \times g \times \mu \times \cos \alpha$	$N = kg \times 9.81 \times \mu \times \cos \alpha$
	Inertia force	$F_m = m \times a$	$F_m = kg \times m/s^2$
	Acceleration	$\dfrac{v^2}{2 \times s}$	$m/s^2 = \dfrac{m/s}{2 \times s}$
Linear	Total piston force	$F_T = F_L + F_F + f_m$	$N_{TOT} = N_L + N_F + N_m$
Actuators	Force output including friction	$F = \dfrac{p \times A \times \eta}{100}$	$N = \dfrac{Pa \times m^2 \times \eta}{100}$
	Extension force	$F_{EXT} = \dfrac{p \times d^2 \times 0.7854 \times \eta}{100}$	$N_{EXT} = \dfrac{Pa \times m^2 \times 0.7854 \times}{100}$
	Retraction force	$F_{RET} = \dfrac{p \times (d_p^2 - d_r^2) \times 0.7854 \times \eta}{100}$	$F_{RET} = \dfrac{Pa \times (m_p^2 - m_r^2) \times 0.785}{100}$
	Compression ratio	$\epsilon = \dfrac{p\ atm + p\ gauge}{p\ atm}$	$\epsilon = \dfrac{Pa\ A + Pa\ G}{Pa\ A}$
	Cylinder consumption	$Q = \epsilon \times d^2 \times 0.7854 \times L \times n$	$m^3/min = \epsilon \times m^2 \times 0.785$
	Rotary power output	$P = 2 \pi \times n \times T$	$W = \dfrac{2 \pi \times RPM \times Nm}{60}$
Air motors	Torque	$T = F \times L_r$	$N_m = N \times m_r$
Air receivers	Receiver volume	$V_R = \dfrac{15 \times Q \times P_1}{\Delta P_2 \times C}$	$m^3 = \dfrac{15 \times m^3/min \times kPaA}{kPa \times no\ of\ starts/h}$
Compressed air distribution	System leakage rate	$Q_L = \dfrac{Q_c \times t}{T \times t}$	$m^3/min = \dfrac{m^3/min \times min}{(min + min)}$

3   Units of measurement and their symbols

Mechanical oscillations are commonly expressed in cycles per unit of time, and rotational frequency in revolutions per unit of time. Since "cycle" and "revolution" are not units, they do not have internationally recognised symbols. However, they are often expressed as abbreviations, and in English the common expressions for them are r.p.m. (revolutions per minute), and c/min (cycles per minute).

The bar and Pascal are given equal status as pressure units. The American and Australian fluid power industries still do not agree on one preferred unit. However, the Pascal is the SI unit for pressure. Thus, common current usage in industry opts for the multiples kiloPascal (kPa) and megaPascal (MPa) and for the bar, which is equivalent to 100 kPa. The European fluid power industry predominantly uses the bar as the preferred unit, and both equipment and product information from European countries has its pressure ratings specified in bars.

Symbols used in this book (base units)

Quantity	Symbol	SI unit	Other recognised units
Area	A	m^2	mm^2, km^2, cm^2
Acceleration	a	m/s^2	
Displacement	V	m^3	mL, cm^3
Flow rate	Q	m^3/s	L/min
Force	F	N	kN, MN
Frequency	f	Hz	1/s
Circle constant	π	3.1416	
Length	l	m	mm, cm, km
Mass	m	kg	
Moment	M	Nm	kNm, MNm
Power	P	W	kW, MW
Pressure	p	Pa	kPa, MPa, BAR
Radius	r	m	mm, cm
Revolutions	n	1/s	1/min
Temperature	T	K	°C
Torque	M	Nm	kNm, MNm
Velocity	v	m/s	m/min, km/h
Time	t	s	ms, min, h, d
Viscosity (DYN)	η	Pa.s	$\dfrac{N.s}{m^2}$
Viscosity (KIN)	ν	m^2/s	mm^2/s = 1cSt
Volume	V	m^3	mL, cm^3
Work	W	J	kJ, mJ

Commonly used area and volume conversions

$1m^2 = 10000\ cm^2 = 1000000\ mm^2$

$1m^3 = 1000\ dm^3 = 1000\ L = 1000000\ cm^3 = 1000000000\ mm^3$

Commonly used pressure conversions

1 bar = 100000 Pa 1 bar = 100 kPa 1 bar = 0.1 MPa	1 MPa = 10 bar 1 MPa = 1000 kPa 1 MPa = 1000000 Pa
1 kPa = 1000 Pa 1 kPa = 0.01 bar 1 kPa = 0.001 MPa	1 bar = 14.5 psi 100 kPa = 14.5 psi 1 MPa = 145 psi

Prefixes for fractions and multiples of base units

Fraction	Prefix	Symbol	
10^{-18}	atto	a	
10^{-15}	femento	f	
10^{-12}	pico	p	
10^{-9}	nano	n	
10^{-6}	micro	μ	0.000001
10^{-3}	milli	m	0.001
10^{-2}	centi	c	0.01
10^{-1}	deci	d	0.1

Multiple	Prefix	Symbol	
10^{1}	deca	da	10
10^{2}	hecto	h	100
10^{3}	kilo	k	1000
10^{6}	mega	M	1000000
10^{9}	giga	G	1000000000
10^{12}	tera	T	

Examples for using fractions and multiples of base units
*units used in Europe

Fract./Mult.	Symbol	Pa	m	L	N	W
10^{-3}	m		mm	mL		mW
10^{-2}	c		cm			
10^{-1}	d		dm*	dL*		
10^{1}	da				daN*	
10^{2}	h			hL		
10^{3}	k	kPa	km			kW
10^{6}	M	MPa			MN	MW

4 Conversion table

Dots indicate SI units.

TO CONVERT	→	TO	→	MULTIPLY BY

FORCE

	TO CONVERT	TO	MULTIPLY BY
10^5	dyne	newton (N)	10^{-5}
0.10197	kilogram-force (kgf)	● newton (N)	9.806650
7.233	poundal (pdl)	newton (N)	0.1383
0.2248	pound-force (lbf)	newton (N)	4.448
2.2046	pound-force (lbf)	kilogram-force (kgf)	0.4536
0.1004	ton-force (UK)	kilonewton (kN)	9.964
32.174	poundal (pdl)	pound-force (lbf)	0.0311
1.000	kilopond (kp)	kilogram-force (kgf)	1.000

TORQUE

	TO CONVERT	TO	MULTIPLY BY
0.7376	pound-force foot (lbf ft)	● newton metre (Nm)	1.356
0.1020	kilogram-force metre (kgf m)	newton metre (Nm)	9.807
8.851	pound inch (lb in)	newton metre (Nm)	0.1130
1.39×10^{-3}	ounce inch (oz in)	gram-force centimetre (gf cm)	72.01

PRESSURE and STRESS

	TO CONVERT	TO	MULTIPLY BY
9.869×10^{-3}	atmosphere (atm)	kilopascal (kPa)	101.30
0.1450	pound-force/in^2 (psi)	kilopascal (kPa)	6.89476
0.01020	kilogram force/cm^2 (kgf/cm^2)	kilopascal (kPa)	98.0665
0.06804	atmosphere (atm)	pound-force/in^2 (psi)	14.70
20.89	pound-force/ft^2 (lbf/ft^2)	kilopascal (kPa)	0.04788
0.01	bar	● kilopascal (kPa)	100
10.000	millibar (mbar)	kilopascal (kPa)	0.1000
33.86	millibar (mbar)	inches mercury (inHg)	0.02953
68.95	millibar (mbar)	pound-force/in^2 (psi)	0.01450
7.501	torr or mm mercury (mmHg)	kilopascal (kPa)	0.1333
0.2953	inches mercury (inHg)	kilopascal (kPa)	3.386
760.0	torr or mm mercury (mmHg)	atmosphere (atm)	1.316×10^{-3}
4.015	inches water (inH$_2$O)	kilopascal (kPa)	0.2491
13.60	inches water (inH$_2$O)	inches mercury (inHg)	0.07355
1.000	newton/metre^2 (N/m^2)	● pascal (Pa)	1.000
0.0648	ton-force (UK)/inch^2	megapascal (mPa)	15.44
6.895×10^{-3}	megapascal (mPa)	pound-force/inch^2 (psi)	145.0

DENSITY

	TO CONVERT	TO	MULTIPLY BY
0.062428	pound/foot (lb/ft^3)	kilogram/metre^3 (kg/m^3)	16.0185
0.010022	pound/gal (UK)	kilogram/metre^3 (kg/m^3)	99.776
10^{-3}	gram/centimetre^3 (g/cm^3)	● kilogram/metre^3 (kg/m^3)	1000
7.5248×10^{-4}	ton/yard^3	kilogram/metre^3 (kg/m^3)	1328.94
0.160544	pound/gal (UK)	pound/foot^3 (lb/ft^3)	6.22884

ENERGY, WORK, and HEAT

	TO CONVERT	TO	MULTIPLY BY
10^7	erg	joule (J)	10^{-7}
0.7376	foot pound-force (ft lbf)	● joule (J)	1.3558
0.2388	calorie (cal)	joule (J)	4.1868
0.1020	kilogram-force metre (kgf m)	joule (J)	9.8066

MULTIPLY BY ← **TO** ←   **TO CONVERT**

TO CONVERT →		TO →		MULTIPLY BY
9.478×10^{-4}	British thermal unit (Btu)	joule (J)		1055.1
0.3725	horsepower hour (hph)	megajoule (MJ)		2.6845
1.3410	horsepower hour (hph)	kilowatt hour (kWh)		0.7457
0.2778	kilowatt hour (kWh)	● megajoule (MJ)		3.600
3412.1	British thermal unit (Btu)	kilowatt hour (kWh)		2.931×10^{-4}
9.478×10^{-3}	therm	megajoule (MJ)		105.51

MASS

0.035274	ounce (oz)	gram (g)		28.3495
2.20462	pound (lb)	● kilogram (kg)		0.453592
1.10231	ton US (short ton)	tonne (t)		0.907185
0.984207	ton UK (long ton)	tonne (t)		1.01605
19.6841	hundred weight (cwt)	tonne (t)		0.05080
1.429×10^{-4}	pound (lb)	grain		7000
0.03108	slug	pound (lb)		32.174
0.01	quintal	kilogram (kg)		100
0.001	kip	pound (lb)		1000
5.0	carat	gram (g)		0.200

LENGTH

0.0393701	inch (in)	millimetre (mm)		25.40
3.28084	feet (ft)	metre (m)		0.3048
1.09361	yard (yd)	● metre (m)		0.914400
0.621371	mile	kilometre (km)		1.60934
0.0497	chain	metre (m)		20.1168
4.97	link	metre (m)		0.21168
0.5468	fathom	metre (m)		1.8288
39.370	thou or mil	millimetre (mm)		0.02540
0.001	millimetre (mm)	micron (μm)		1000
10^{10}	angstrom (A)	metre (m)		10^{-10}
1.644×10^{-4}	mile UK nautical	feet (ft)		6080
5.396×10^{-4}	mile UK nautical	metre (m)		1853.2
5.399×10^{-4}	mile International nautical (n mile)	metre (m)		1852.00
1.894×10^{-4}	mile	feet (ft)		5280

AREA

2.471	acre	hectare (ha)		0.4047
2.066×10^{-4}	acre	sq. yard (yd^2)		4840
2.471×10^{-4}	acre	sq. metre (m^2)		4047
1.973×10^{3}	circular mil	sq. millimetre (mm^2)		5.067×10^{-4}
1.2732	circular mil	square mil		0.7854
0.1550	sq. inch (in^2)	sq. centimetre (cm^2)		6.4516
10.764	sq. feet (ft^2)	● sq. metre (m^2)		0.09290
1.562×10^{-3}	sq. mile	acre		640
3.861×10^{-3}	sq. mile	hectare (ha)		259.0

VOLUME

0.2200	gal (UK)	● litre (l)		4.546
0.2642	gal (US)	litre (l)		3.785
0.0238	barrel (US)	gal (US)		42
0.03532	cu. feet (ft^3)	litre (l)		28.32
0.1605	cu. feet (ft^3)	gal (UK)		6.229

MULTIPLY BY ← **TO** ← **TO CONVERT**

TO CONVERT →		TO →	MULTIPLY BY
35.31	cu. feet (ft³)	● cu. metre (m³)	0.02832
1.308	cu. yard (yd³)	cu. metre (m³)	0.76456
0.061026	cu. inch (in³)	cu. centimetre (cm³) or millilitre (ml)	16.39
0.0352	fluid ounce (fl oz)	millilitres (ml)	28.41
1.76×10⁻³	pint (UK) (pt)	millilitre (ml)	568.2
1×10⁻³	litre (l)	millilitre (ml)	1000
1×10⁻³	cu. metre (m³)	litre (l)	1000

VELOCITY

0.0393701	inch/sec (in/s)	centimetre/sec (cm/s)	2.540
3.28084	feet/sec (ft/s)	● metre/sec (m/s)	0.304800
1.9685	feet/min (ft/min)	centimetre/sec (cm/s)	0.5080
0.621371	miles/hr (mph)	kilometre/hr (km/h)	1.609344
2.23694	miles/hr (mph)	metre/sec (m/s)	0.44704
0.5397	knot (UK)	kilometre/hr (km/h)	1.853
0.5400	knot International (kn)	kilometre/hr (km/h)	1.852

ACCELERATION

3.28084	feet/sec² (ft/s²)	metre/sec² (m/s²)	0.304800
0.10197	gravitational acceleration (g)	● metre/sec² (m/s²)	9.806650

POWER and HEAT FLOW RATE

3.4121	British thermal unit/hr (Btu/h)	watt (W)	0.2931
0.8598	kilocalorie/hour (kcal/h)	● watt (W)	1.163
0.7376	foot pound-force/sec (ft lbf/s)	watt (W)	1.3558
0.1020	kilogram-force metre/sec (kgf m/s)	watt (W)	9.807
1.360	metric HP	kilowatt (kW)	0.7355
1.3410	horsepower (hp)	kilowatt (kW)	0.7457
7.457×10⁻⁴	megawatts (MW)	horsepower (hp)	1341
0.2843	ton of refrigeration	kilowatt (kW)	3.517
8.333×10⁻⁵	ton of refrigeration	British thermal unit/hr (Btu/h)	12000
1.000	joule/sec (J/s)	watt (W)	1.000
1.818×10⁻³	horsepower (hp)	foot pound-force/sec (ft lbf/s)	550

ILLUMINATION

0.0929	foot candles	lux (lx)	10.764
0.0929	lumen/foot² (lm/ft²)	lux (lx)	10.764
0.0929	candela/sq. foot (cd/ft²)	● candela/sq. metre (cd/m²)	10.764

ANGULAR MEASURE

57.296	degree (...°)	radian (rad)	0.017453
9.5493	revs per min (rpm)	radian/sec (rad/s)	0.10472

ELECTROMAGNETIC

2.778×10⁻⁴	ampere hour (Ah)	coulomb (C)	3600
10⁴	gauss	weber/sq metre (Wb/m²)	10⁻⁴
2.5407	microhms/cm³	microhms/in³	0.3937
10⁵	gamma	gauss	10⁻⁵

MULTIPLY BY ← TO ← TO CONVERT

TO CONVERT	⟶	TO	⟶	MULTIPLY BY

KINEMATIC VISCOSITY

10^6	centiStoke	square metre/sec (m²/s)		10^{-6}
10.764	square feet/sec	square metre/sec		0.09290
10^6	square millimetre/sec	square metre/sec (m²/s)		10^{-6}
10^4	Stoke (St) (cm²/s)	square metre/sec		10^{-4}

DYNAMIC VISCOSITY

10^3	centipoise (cP)	pascal second (Pa s)		10^{-3}
2.419	pound/ft hr (lb/ft h)	centipoise (cP)		0.4134
0.02089	pound-force sec/foot² (lbf s/ft²)	pascal second (Pa s)		47.88
1.000	gram/metre sec (g/m s)	centipoise (cP)		1.000

HEAT TRANSFER

0.1761	Btu/ft² h°F	watts/metre² K (W/m² K)		5.678
6.933	Btu in/ft² h°F	watts/metre K (W/m K)		0.1442
2.388×10^{-4}	Btu/lb°F	joule/kilogram K (J/kg K)		4186.8
4.299×10^{-4}	Btu/lb	joule/kilogram (J/kg)		2326
2.388×10^{-4}	kilocalories/kg	joule/kilogram (J/kg)		4187

SOUND LEVEL

0.1151	neper	decibel (Db)		8.686

MULTIPLY BY	⟵	TO	⟵	TO CONVERT

PHYSICAL CONSTANTS

3.141593	π
2.718282	e
2.302585	$\log_e 10$

Index

absolute ambient pressure   58
  humidity   224
  minimal system pressure   58
  speed consistency   62
  zero pressure   5
absorption   228
  dryers   228–229
  economy of   231
  law   137
acceleration   2, 44
  constant   98
accuracy of switching   88
actuation methods   87
  "head-on"   88
actuator barrel   41, 87
  calculations   48, 49
  construction   41
  control   22
  control multiple   24
  dynamic force   53
  end caps   41
  failure   42
  force   36
  force calculations — dynamic thrust   52
    static thrust   49
  force theoretical   48, 49
  internal friction   50
  locking units   47
  mounting methods   42
  mountings   41
  movements — repeated   97
    simultaneous   99
    uniform   24
  nomogram   52–53
  position   90
  reversal   108
  rotary   6, 7, 8, 9, 36
  sizing   47
  sizing nomogram   52, 53
  undersizing   53, 54
  unloading   54
actuators double acting   38
  double ended   38, 193
  hydro-check   8
  linear   9, 68, 205, 234
  rodless   44, 45, 46, 47
  rotary   9, 68
  semi-rotary   36, 38
  single acting   36
  special   43
  swivel   8, 47, 68
  symbols   43
adiabatic compression   207
adjacency rule   142
adjustable orifice   140
adsorption   228
  dryers   228, 229
advantages of pneumatics   8
after cooler   55, 220, 222, 234, 238
aim of compressed air drying   227

air   224, 242
  ambient   205, 224
  barrier   84
  borne oil fog   201
  characteristics   68
  cleaner   220
  compressed   4
  compressibility   8
  compression process   207
  consumption calculations   55, 58
    non contact sensing   81, 82, 88
  contamination   194
  cooling   209
  distribution system   55, 232
  diverter valve   201
  dried   4
  dryer   55
  dryers   234
  expansion   207
  exhaust noise   245
  flow   194
  humidity   205
  inlet temperature   227–228
  leakage   234, 235
    chart   234
    flow rates   234
    tests   235
  line filter (separator)   29, 193, 243
    element   194
    lubricants   202
    lubrication   198, 199
    lubricator   198
  motor (see also air motors)   234, 268
  noise — exhaust   245
  presses   37
  pressure   205, 207
    barometric   205
  quality   237
  receiver   55, 220, 221, 238
    calculation   7
  service unit   193, 240
    installation   224, 202
    lubricator   245
  silencer   220
  storage   221
  supply   193
  to air heat exchanger   229
  tools   72, 193
air motor   68, 234
  differing air pressures   76
  displacement   70
  efficiency   71
  flow regulation   75
  friction losses   75
  governed   70
  maximum power output   70
    practical torque   72
    stall points   72
    torque   72
  performance   69, 77
  porting arrangements   69

  power curves   71
    output formula   70
  pressure regulation   75
  reversal of working pressure   76
  rotation   70
  sizing   75, 76, 77
  speed   69, 70
    curves   71
  table   76, 77
  torque (turning moment)   69, 70
    calculations   77
  ungoverned   70
  uses   68
  working pressure   76
air motor types axial piston   69, 71, 74
  gear   68, 69, 71
  piston   68, 69
  piston link   74
  radial piston   69, 71, 73
  turbine   68, 69, 75
  vane   68, 69, 70
allocation of i/o channels   176
alternative cascade method   125
altitude   4, 5, 205
Amontans Law   48, 52
amplifier valve   81, 85
ancillary equipment   220, 234, 240
AND function   10, 18, 20, 26, 34, 89, 91, 95, 96, 125, 163
AND gates   160
annular area   44, 58
anti locking vent   239
applications — pressure reducing valves   187
area   2
  annular   44–58
  driving surface   69
  effective piston   49
  formula   2
assigning input output channels   175, 178
associative law   137
asynchronous circuit   93
Atlas Copco radial piston motor   73
atmosphere   4
atmospheric conditions   224
  pressure   4, 5
authorities statutory   231
auto cycling   166
automatic drain   194, 221, 222, 231, 238, 239
  sequencing   134
availability of air   8
axial flow compressors   216
axial piston motors   69, 71, 74, 75

back pressure   51, 52, 62
  sensor   78
barometric air pressure   205
barrel — cylinder   40
basic cycle selection mode   166
  emergency stop module   105, 168

bellows 37
binary counting 155
   movement 25
   movement expression 91
   reduction control 155
   valve 140
bistable 22
Boolean algebra 96, 137, 144, 153
   identities 157
   postulates 137
   switching equations 142
   theorems 137
boring machine 68
bottom out 88
   dead centre 1, 209
bottle sorter 149, 150
bowl filter 194
   guards 195
   lubricant 200, 201
   plastic 195
Boyles law 6, 7, 8, 39, 48, 52
brakes pneumatic piston rod 108
brand name 240
breakaway pressure 51
   torque 69
bronze sintered silencers 245
buckling length of piston rod 55
bus bar distribution system 90

calculation air consumption 55
   motor torque 77
   compression ratio 58
   cylinder 48, 49
   compressor 211
   condensate drop out 226
   dynamic thrust 52
   dynamic force 52
   external forces 48
   flow rate 58
   free buckling length 55
   hardware cost 130
   intensified pressure 65
   main pipe system 236
   minimum piston diameter 51
   near static force 50
   retraction force 49
   static force 50
cam angles 88
capacity control 215, 221, 227
carriage 45
   guidance 46
cartridge separator 213
cascade advantages 130
   alternative method 125
   circuit 240
   circuit emergency stop 129
   circuit construction 127
   control problem 130
   method alternative 125
   method Festo 125
   method Martonair 134
   method Rohner 125
   method Salek 125
   valves 127
cassette 174, 185, 190
C.E.T.O.P. 10
cental processing unit (C.P.U.) 158, 163
centrifugal force 70, 195
   compressors 216
centigrade 7
Charles law 7, 8
check valve 2, 61, 201, 213, 215
checking program 184, 185

chemical industry 212
   process 227
circling 93
circular area 2
circuit array 90
   cascade 240
   combinational 91, 138
   construction 89, 127, 242
   construction costs 130
   dead end 197
   design 93, 137
   designer 89, 104
   designers 57, 115
   diagram 241
   documentation & construction 16
   draining 240
   installation 91
   interlock 83
   interlocking 90
   labelling 25, 89, 91
   logic control 240
   pneumatic step-counter 93
   power 91, 240
   sequential 91
   step-counter 240
clamping 37
   applications 52
   cylinder 37
clamp & drill machine 163
cleanliness 193
clear carry instruction 177
clock 177
coalesce 201
code gray 142
   mirror 142
coil bound 88
   number 177
cold countries 227
collective equation 148
cooler oil 214
cooling air 209, 227
   compressor 210
   compressed air 237
   mechanism 207
   process 229
   spaces 213
combinational circuit 91, 138
   circuit design 137
   machinery 8
   pattern circuits 153
comparison of design concepts 133
   cascade & step counter 133
compare instruction 177
compatible P.L.C. controllers 174
complex circuits 89
   logic control 205
compressibility 8
commercial step counter modules 113, 115
commissioning 242
compressed air
   cooler 220
   cooling 237
   drying 224
   drying aim 227
   servicing 193
compression adiabatic 207
   chamber 210
   curves 212
   element 207
   isothermal 207
   polytropic 207
   process 207
   ratio 206
   spaces 213

compressor 240
   ancillary equipment 220
   calculation 211
   capacity control 215, 221
   capacity controller 193
   control 217
   cooling 210
   design principles 209
   designers 207
   diaphragm 212
   double acting 210
   location 219
   oil flooded 214
   oil free piston 212
   rotary vane 213–214
   savings 198
   single acting 209
   staging 210
compressors 9
   air 205
   air cooling 210, 215
   axial flow 216
   centrifugal 216
   continuous flow 209
   diaphragm 212
   double acting 210
   dynamic 209, 215
   helical lobe 214
   mammoth 209
   mobile 209, 213, 215
   multi stage 209
   oil free 212, 215
      flooded 213, 214, 215
   piston 209, 210, 217
   positive displacement 200, 212, 215
   reciprocating piston 209
   rotary 209
   rotary screw 214
   screw 209
   single acting 209, 210
   single stage 209
   two stages 210–211
   vane 204, 212, 217
   water cooling 210
component labelling 243
components — fluid power 240
   control 243
   mounting 243
concrete vibrators 72
condensate 194, 210, 222, 234, 238
   collection 221
   drain 222
   droplets 195
   drop out 224
   drop out calculations 226
   level indicator 195
condensation 237, 238
confirmation signal 93, 163
constant speed 60
   speed control 217
   speed phase 53
contact sensing 79
construction — circuit 89, 240, 242
contamination 8, 38
continuation mode 110, 171, 172
console 158, 178
   display 181
consumption free air 234
   tools/processes 234
continous cycling 127
   flow compressors 209
   sequencing 134
control electrical 242
   cabinets 242, 243

capacity 216
circuits 193
combinational 9
compressor capacity 217
constant speed 217
element 243
engineers 159
exhaust regulation 219
flow 9
grip arm 217, 219
inlet valve unloading 217
intake throttling 218
problem 130, 140, 144, 145, 147, 149
    analysis 89, 91
    review 153
    sequential 9
    start, stop 217
    time & pressure 34
conversion units 206
corrosion 224, 231
    free 238
counters 176
cross over condition 20
Crouzet step counter 115
cushioning 39, 40, 42, 54, 60, 63, 65
cut in pressure 193
cut in/ cut out pressure differential 217, 234
cycle selection 91, 106, 166, 168
    extended 107, 166
    module 106
    module for P.L.C. 166
    combined with emergency stop 107
    combined with auto manual selection 108
cycling auto/manual 166
cylinder first stage 210
    high pressure 211
    second stage 211
    barrel 41, 42
cylinders
    aluminium 41
    brass 41
    double acting 38
    impact 44
    side loading 42
    single acting 36
    speed maximum 44
    stainless steel 41

damping, pulsation 209, 221
dead end circuits 197
debugging 185
deceleration 39, 44
deep bore drill rigs 73
definition, power 8
deliquescent desiccant 230, 231
deleting instructions 184
demarcation problems 158
De Morgans theorem 137, 147
dentist drill 68
design, concepts 132, 134
    tools 137
    method, logic 93
    method, pneumatic step count 93, 95
    principles, compressor 209
    sequential circuits 91
    system 232
developed force 36
devices, compressed air 244
dew point 225
    atmospheric (A.D.P.) 225
    curves 226
    free air 226

graph 225
    pressure (P.D.P.) 225, 227, 228
    specified 230
diagram, circuit 241
    traverse time 91, 93, 171
    step motion 93
diaphragm, compressor 212
diffuser 215
directional vanes 195
    control valves 10
dirt 193, 220, 222, 243
distribution, air 232
    air signals 243
    system 222, 227, 235, 237
distributive law 137
documentation 240
"don't care" combinations 143, 148, 149, 152, 153
double acting, actuator sizing 52
    acting linear actuator 4, 38, 42
    ended actuators 38, 42
    pilot control 22
drain automatic condensate 194, 238, 239
drain points 238
drills 68
drilling, head 104
    machine 117
drip rates 242
drive pressure 52
driving surface area 69
drop legs 238
drop out, condensate 224
dryer 240
    concepts & terminology
    performance 226
    pressure drop 229
    selection criteria 227, 228
dryers, 234
    absorption 228, 229, 230
    adsorption 228, 229, 230
    refrigeration, (low temperature) 228, 229
drying, 224
    agents 230, 231
    energy costs 227, 229
    methods 229
dual coding 140
dynamic compressors 209, 215
    thrust calculations 52

editing, the program 185, 186
efficiency, exhaust 62
effective, area 2
    force 51
    piston area 49, 52
effect of actuator undersizing 54
electric, motor 9, 209, 215, 217, 220
    switch gear 140
electrical control boxes 242
    filament regeneration 229
    limit switches 38
    power supply 231
    pressure switch 212
    systems 158
electrically operated valves 8
electricians 159
electronic controller 160
    controller programming 159
    programmable controllers 93, 158
    programmable memory 160
    step counter module 163
    switching peculiarities 159

element, filter, 194
    compression 207, 209
emergency sequence 240
    signals 105
    situation 104, 171
    stop 91, 104, 105, 168
    stop control 168
    stop modification module 111, 171
    stop module 105, 168
    stop program 129
    stop with manual & auto input 106
    stop with multiple input 106
end instruction 163
    position, cushioning 39, 63, 65
    position, sensing 78
energy, heat 207
    costs, drying 227
    kinetic 39, 40, 44, 52, 207
    potential 8, 71, 73
    stored 36
engineers maintenance 79
entering the program 182
environmental conditions 40
eprom 176
equipment ancillary 140
    compressor ancillary 220
erase existing memory 182
error, fatal/non fatal 187
    testing 187
errors programming 181
Euler formulae 55
European standard 205
example program 185
exhaust, air noise 245
    efficiency 61
    ports 199
    port speed control 65
    regulation 219
    silencer 61, 244
expansion chamber 245
explosion risk 8
extended cycle selection module 166
external by pass control 219
extraction of minimised equations 143

failure power 159, 176, 177
false output form 147
fast signal processing 243
fault finding 89, 91, 125, 240, 243
feed back signal 93
felt pad separator 213, 214
FESTO actuator end cavity volumes 58
FESTO cascade method 125
FESTO P.L.C. 59
FESTO rodless actuators 44
filter (separator) 193
    airline 243
    bowl 194
    element 194
    intake silencer 220
    screen 201
    servicing 194
filtration 243
final pipe diameter 237
first stage cylinder 210
fittings airline 230, 235, 240, 243, 244
flat slide valves 16
flexible tubing 244
flow 3
    axial 215
    capacity of dryer 277
    chart 192
    centrifugal 215
    control 60

control methods   60, 62
rate calculation   58
rate   205, 232, 234
rate of air motor   69, 70
through orifice   4
fluid   1
connection lines   12
flow direction   9
power circuit   240
power circuit structure   92
power circuitry   158
power components   240
power system   1
fog   192, 224
food processing   212
force   18
calculated   51
developed   36
dynamic thrust calculations   52
effective   51
formula   2
gravitation   2
horizontal   48
imbalance   52
impact   44
incline   48
inertia   48
lifting   48
multiplication   36
output   47
static thrust   49
theoretical actuator   49
total   48
transmission   1
forced relay setting   188
forces external   48
forging machine   117
formula Euler   55
free air   4, 58, 205
air chart   234
air consumption   234
air delivery F.A.D.   194
air dew point   226
air flow total   227
falling   60, 62
reverse flow   60
friction   3
co-efficient   48
force   48
free valve   18
loss   50, 52
fringe condition circuits   92
condition modules   91, 240
function key   180, 187
functions and   158
arithmetic computation   159
counter   159
inhibition   159
memory   158
micro processor   176
quick search editing   181
search   187

galvanized steel pipe   237
gap sensor   83
gas   1
laws   6
ideal   7
general law   8
volume   7
gauge pressure   5
Gay Lussacs law   8
gear air motor   69, 71, 72
applications   73

general gas laws   8
German step counters   113
graph — dew point   225
air consumption for proximity sensors
82
graphical aids   93
gravitational force   2
gravity   48
gray code   142
grip arm   217, 219
group memory   125
grouping   127

hardware calculations   130
cost   129
hazardous situations   202
"head on" actuation   88
heat   38
exchange   221
exchanger "air to air"   229
energy   207
energy exchanger   210
exchange medium   213
heated fan flow regeneration   229
helical lobe compressor   214
high pressures   211
holding relays   166, 176
hot surfaces   243
horizontal direction force   48
hospitals   212
hostile environment   38
hydrochech actuators   8, 62
hydraulic speed control   62
oil   62
hysterisis   87
humidity   220, 205
absolute   224
relative   224, 227

identical AND function   101, 129
idle speed   219
impact control   43, 54
cylinder   44
force   44
impeller   216
imperial standard   205
impulse valve   30, 31, 34, 120
indexing table   120
industrial applications   194
areas   4
automation   158
manufacturing plant   227
shop air   212
inertia   8
inhibition function valve   20, 96, 159
initial position   10, 90
inlet temperature   227, 231
valve unloading   217, 219, 220
input-output monitor   189
program addresses   179
input relays   176
inserting instructions   184
installation — air servic unit   202
of compressor   219
instruction clear-carry   177
compare   177
deleting   184
end   163
inserting   184
no end   186
OR   183
set clear (STC)   187
sequential address   163
intake filter/silencer   220

throttling   218
intensification   65
intercooler   209, 210, 222, 238
interlocking circuits   90
machine program   79
internal auxiliary relays   163, 176
combustion engine   9, 215
friction   50
valve mechanism   10
international standards   89
i/o assigning   178
channel allocation of   176
I.S.O. class V.G.10 as per 3448   202
standards   10
isothermal   207
item number   240

Karnaugh Veitch maps   96, 137, 142,
144, 150, 151, 152
keep relays   166
Kelvin   7
key board   179, 180
kinetic energy   187

labelling circuits   89, 91
labelling components   243
ladder diagram   163, 174, 181
ladder logic   159
laws absorption   137
Amontans   48, 52
associative   137
Boyles   7, 8, 48, 52
Charles   7, 8
De Morgans   137
distributive   137
gas   6, 8
Gay Lussacs   8
Pascals   48, 70, 197
leakage, air   234
chart   234
external   40, 232
flow rates   234
internal   69
losses   72
prevention   234
tests   234
levers and linkages   36
lifting forces   48
light weight object sensing   81
light mineral oil   202
limit switches   163
valve control   25
valves   25, 198, 240
linear actuators   8, 9, 36, 198, 205, 234
linear actuator sizing charts and
nomograms   55
lintra rodless actuator   46
liquid   1
level sensing   80
list parts   240, 242
load deceleration   60
supported   62
loading and verifying program   19
loading point   218
location of compressors   8
locking units   47
logging industry   37
logic control circuit   240
design method   93
functions   8, 97, 141, 142
steps   144, 145
switching equations   96
looping   142, 143, 152
low temperature dryers   229

lubricant bowl 200, 201
    manufacturers 202
    transfer 199
lubrication air 198, 199
lubricator 193
    air 245
    drip rates 242
    macro (oil fog) 200
    micro (oil mist) 201
machine
    clamp & drill 163
    commissioning 242
    complex manufacturing 158
    control 78
    designer 79, 93
    filling 150
    operators 202
    packaging 144, 150
    safety 79, 105
    sorting 144
    tool 104
machinery combinational 8
    manually activated 243
    pneumatically controlled 238
    sequential 8
macro lubricator 200
magnetic coupling 44
main pipe system 235
    calculation 235
maintenance and trouble shooting 187
    engineers 79
    personnel 89
manual cycling 166
manually drained filters 194
manual signal input devices 243
manufacturers specifications 194
martonair cascade method 125, 130, 134
    rodless actuator 45
mass 2, 48
materials which attack bowls 195
maximum cylinder speed 44
    force output 47
    speed of actuator 60
    system pressure 198, 227
        temperature 227
    torque 72
memory clearing operation 182
    electronic 160
    power 132
memory relay temporary 176
    restrictor 132
    shunting 132
    valves 90, 92, 95, 124, 150
mercury barometer 5
metal guard 195
    worker unions 158
micro lubricator 201
    processor 158
minimal system pressure 49, 50, 58
minimised switch equations 137, 143, 153
minimisation process 148
minimum operating pressure 234
mirror code 142
mist 224
    generator 201
    oil 245
Mitsubishi Melsec F series P.L.C. 159
mobile compressors 214, 215
mode Program 181
    Monitor 181
    Run 181
Module
    basic cycle selection 166

basic cycle selection with emergency stop 105, 168
commercial step counter 113
cycle selection combined with emergency stop 107
cycle selection with auto manual selection 108
emergency stop modification 111
emergency stop modification for P.L.C. 171
emergency stop with manual and automatic input 106
emergency stop with multiple input 106
extended cycle selection 107, 166
    shunt selector 164
    start stop 147
    step counter 168
Modules
    fringe condition 240
    step counter 198
moisture 230
    drop out 232
    separator 220, 222
molecules 207
momentary signal 150
monitor input/output 189
    key 187
    mode 180, 181
monitored pressure 217
mono stable 22
motorising 215
motors 193
    air 68, 69, 70, 71, 75
    electric 209, 215
mounting
    components 243
    methods 42
    valves 88, 243
multi staging 215
near static force calculation 50
neutral position 10
Newton 2
no end instruction 186
noise 243
    exhaust 245
nomogram 52, 53, 235
non contact type sensing 78
    air barrier 84
    back pressure 79
    equipment 194
    gap sensor 83
    liquid level 80
    proximity sensor 81
    reed sensor 87
non oil flooded compressor 222
    return valve (see check valve)
normally closed contacts 160, 163
    valves 160
    open contacts 163
    timers 33
    valves 160
NOT function valve 18, 20, 96
number — item/part 240
numeric keys 179
nut and olive 244

oil 222
    airborne 201
    carry over 214
    cooler 214
    feed rates 202
    flooded type 213
    fog 200

fragmentation 201
free 212, 213, 214, 215
    air delivery 215
    piston compressor 212
    injected 202
    light mineral 202
    lubrication 212
    mist 201, 245
    mist (fog) 198
    particle size 201
    separator 214, 215
    viscosity 202
oils 240
Omron C20 159, 163, 164, 172, 176, 178
one way trip valves 78, 79
on-off valve 18
operator safety 79
opposing signal 30, 94, 129
    spring force 37
optimal mechanical advantage 43
    speed control 65
    optional combinations 148
    OR functions 10, 12, 18, 20, 26, 27, 34, 61, 89, 91, 95, 96, 97, 113, 115, 116, 129, 133, 137, 143, 153, 158, 171
orifice 4
    adjustable 140, 196
    flow control valve 60
Origa rodless actuator 46, 37
outlet port 209
    valves 210
output relay 163, 164, 176
    torque 72, 234
over heating 219
overload protection 8
over lubrication 202
oversize piston rod 42
oversizing piston rod 58

part description/number 240
parts list 240, 242
Pascal 3
Pascal's Law 1, 4, 48, 70, 197
performance charts 37
periodic servicing frequency 240
personal computers 158
pharmaceutical industry 212
physical process 227
pilot signals 12, 22
pipes and fittings 232, 235
pipe copper 238
    diameter 237
    galvanized steel 238
    grading 236
    length 232, 235
        actual 235
        calculated 235
        equivalent 235
    take offs 238
piping 244
    system 3
piston 36, 40, 41, 209
    air motors 69
    area 2
        calculation 49
        effective 49
    compressors 209, 217
    diameter calculation 51
        smallest possible 51
    rod 40, 41
piston rod buckling 55
    coupling (self aligning) 42
    flexing 42
    guidance 38

oversize   42
oversizing   58
side stress   243
stability
speed   47
plastic bowl   195
fittings   244
tubing   243, 244
P.L.C.   93, 129, 158, 163
console   174
display   191
compatible   174
Festo F.P.C. 201   159, 201
Mitsubishi, Melsec F series   159
Omron C20   159, 163, 164, 172, 178
step counter control   172
sysmac C series   176, 178
pneumatic actuator   4
load   48
air barrier sensor   149
control   160
machinery   238
systems   43
cushioning   39, 54, 65
fittings   244
logic circuitry   158
mechanism for industry   124
pilot signal   160
piston rod brakes   108
power transmission   8
sequential controls   159
step-counter circuit design   93
system   55, 234
complex   202
flow control operating   60
systems   6, 8, 9, 194
pneumatically operated machinery   243
clutch   147
points of use   227, 232, 237
polycarbonate plastic bowl   195
polytropic compression   207
poppet ball/disc   10
valves   16, 17, 18
port labelling/numbering   12
position initial   90
sensed control   25
sensing   22, 78, 93
accuracy   80
positive displacement compressors   209, 215
potential energy   8, 9, 71, 73
power   241
circuit   240
definition   8
failure   159, 176, 177
formulae   80
loss   234
memory   132
output air motor   70
requirement   205
savings   212
source   210
to weight ratio   72
transmission   8
valve   10, 22, 65, 163, 198
preparation signal   171
pressure   278
absolute   4, 5, 207
zero   5
adjusted   195
ambient   205
atmospheric   4
back   51, 52
breakaway   51

calculation   65
cut in   193
dew point (P.D.P.)   225, 228
diaphragm   196
differential   3, 6, 9, 216, 219
across orifice   4
cut in/cut out   212, 234
drive   52
drop   3, 194, 232, 234, 238
silencers   245
table   236
through driers   227
exposed driving surface area   69
fluctuations   195
formulae   2
gauge   5, 61, 81, 193
in liquids   6
intake   205
intensification   65
lines   90
loss   232
low   80, 207
minimum operating   234
monitored   217, 218
ratio   244
pressure reducing valve   50, 51, 195
application   197
with relief   197
without relief   195
regulator   193, 195
safety margin   234
sensing   34
point   217
sequence valve   34
system   193, 234
vacuum   5, 81, 207
problem analysis   89
processes chemical   227
physical   227
production lines   72
program checking   184, 185
data   163
debugging   190
entering   182
loading   192
mode   180, 181
saving   190
selector   153
selector circuits   122
programmable controllers   93, 158, 172
programming & control applications   172
console   178
difficulties   158
electronic controller   159, 172
errors   182
instructions   168
step counter   163

proximity sensing   81
air consumption   83
pulsation   215
damping   209, 221
pulse tank   209
purging process   230
push button selection   134
push in/push on fittings   244

quick exhaust speed control   61
valve(s)   61, 243
quiet zone   195

rack and pinion drive   47, 68
radical piston motor(s)   69, 71, 73, 74
ram   176

rapid check of counter timer values   190
relay status   190
rate-flow   232
rates — oil feed   202
flow   25
ratio pressure   244
receiver air   221, 228, 238
volume   221
unit   84
reciprocating piston rod   38
piston compressors   209
reducing valve   195
applications   197
reed sensor   87
reflex senor   81
refrigeration dryer   229
periodic maintenance   230
regeneration dryer desicant   229
regulation exhaust   219
regulation pressures   242
regulator pressure   193
relative humidity   224, 227
relief valve   219
relay
contact searching   186
numbers   176
set/reset forced   188
relays elctronic   159, 163, 176, 177
remote control   104
repeated actuator movements   197
steps   172
reservoir   30, 31
replaceable cartridge oil separator   213, 215
restrictor   60, 132
retentive relays   176
review of control problems   153
RMIT   93
robot   117
rod side stress   243
rodless actuators   44, 46
rod wiper ring   40
Rohner   93
cascade method   125, 130
rotary actuators   9, 36, 68
movement   68
torque output   68
continuous flow compressors   209
screw compressors   214, 215
vane advantages   21
compressors   213, 214
ratings   213
rotor   213
run mode   180, 181
running program   185
torque   69

safety   8
air relief valve   221
guards   104
machine and operator   79
margin pressure   234
overtravel   88
Salek Cascade method   125
sampling press   147
saturation   224
quantity   224
savings compressor   198
savings, power   212
work   212
scan time   177
screw compressors   214
drivers   72
male and female   214

seals 40, 41
search facilities 174, 186, 187
second stage cylinder 211
selection criteria dryers 227
selector valves 10
self aligning piston rod coupling 42
semi rotary actuators 36, 68
sender unit 84
sensing accuracy 78, 79, 80, 81, 82, 83
  light weight object 81
  position 22
  proximity 81, 158
sensor back pressure 79, 80
  flow 201
  gap 83
  heat 158
  level 158
  limit 158
  liquid level 80
  pressure 158
  reed 87
  reflex 81
  tube 80
separator cartridge/felt pad 213
  moisture 220, 222
  oil 213
sequence emergency 240
  program 127
  step 93
  valve 34
sequential address instruction 163
  control 9, 91, 93
  control circuits 90, 198
  machinery 8
series arrangements 34
  connection 99, 100, 129
  function 26, 128, 135, 160
  function on P.L.C. 160
servicing 240
  compressed air 193
  information 240
servo valve 218
set carry instructions 177
  commands 128
settings initial 242
shake proof fasteners 243
shock absorption 37
  waves 245
short strokes 37
shunt module (selection) 165
shunting memory 132
shunt valve 10
side loading 42
shuttle valve 26
side stress 243
signal confirmation 93, 163
  distributing 243
  disturbing pilot 94
  feedback 93
  input devices 243
  inversion 3, 8, 149
  labelling 93
  lines 243
  opposing 94, 95, 124
  pilot 198
  preparation 128, 171
  processing fast 243
  substitute preparation 113
signal commands 12
silencers 62, 220, 244, 245
  cleaning 245
silk screen printing 133
simultaneous actuator movements 99
single acting actuators 36, 37

sizing system 65
skipping motion 117
sliding spool 10
  vanes 70
solenoid signal commands 163
solid particles 4
sonic airflow 4, 52
spring machine 144
special actuators 43
  auxiliary relays 176
  centre condition 20
specifications manufacturers 194
specific contact search 187
  instruction search 187
speed consistency 62
  constant 60
  control 4, 43, 54
    "meter in" 63 66
    "meter out" 51, 52, 63, 66
    summary of installation methods 66
  erratic 54
  hydraulic check 62
  hydrocheck 8, 62
  maximum 44, 60
  quick exhaust 61
  summary types 65
  torque characteristics 70
  uniform 62
spool 10, 16
  and valve body 160
  spool-poppet valves 18
  spring actuated valves 13
  spurious signals 153
  stall point 72
  standards European 150, 202, 205
  Imperial 205
  International 89
  S.I. 206
start command 96
start not permitted 145, 147
start permitted 145, 147
start stop control 217
  provision 240
start up 242
starting torque 70
static force calculation 50
static thrust calculation 49
statutory authorities 231
steam turbines 215
step commands 93
step-counter method 93
step-counter circuit 240
step-counter circuit design 93, 95, 113,
  163
  advantages 129
  module 165
  modules 97
  programming 163
step-counters, crouzet 115
  French 115
  German 113
  for selected parallel programs 120
  for simultaneous parallel programs
    121
  with defined skipping 118
  with interim start 116
  with repeated steps 120
  with undefined skipping 118
  motion diagram 93
step sequence 95
stop instantaneously 108, 168
stopping modes 104, 110, 168
storage volume 237
stored air 8

stroke 36, 40
subsonic air flow 4
suction 199
suggested lubricants 202
sump pumps 72
supersonic 245
supported load (suspended) 62
surface area driving 69
switch positions 10
switching accuracy 88
  equations 96, 97, 100, 102, 103, 128,
    137, 138, 143, 157, 162
  positions 10
swivel actuator 8, 47, 68
symbols 43, 90
sysbus 176
Sysmac C Series PLC 176, 178
system demand 218
  economy 237
  pressure 193
  pressure maximum 198
  minimal 49, 50
  sizing 65
    demand 238

take off pipes 238
tape recorded program 159
technical work 71
tee connection 26, 90, 97
teflon 212, 237
temperature, air 227
  centigrade 7
  compressed air 207
  inlet 227, 231
  Kelvin 7
temporary memory relays 176, 177
testing for errors 187
theoretical actuator force 49
  performance curves 207
thread sealing 237
three port valves 12
  position valves 10, 20
throttling intake 218
throttling function 60
time and pressure control valve 240
  setting valves 240
  sensed control 34
time delay 26, 137
  sequencing 102
  valves 29, 30, 33, 34
time lines 93
timer function 159, 164
  forced set 188
  placement 102
timers 176
timer valve changing 189
toggle disc valves 18
top dead centre 210
torque breakaway 60
  calculations 7
  maximum 72
  practical 72
  output 68, 234
  running 69, 73
  starting 70, 73
torricelli 6
total free air flow 227
  system pressure 234
trapped pilot signal 24
traverse-time diagram 91, 93, 134, 171,
  240
true poppet valves 18
truth tables 137, 138, 140, 142, 145

tubing   240, 243, 244, 293
   colour coding   293
turbine air motor   69, 75
turbulence   237, 238, 245
twin pressure valve   20
   poppet valve   26
two position valves   90
two stage amplifier valve   85

ungoverned air motor   70
uniform actuator movement   24
unit, sender and receiver   84
unloaded maximum speed   70
unloading — actuator   217, 218
   grip arm   217, 219
   inlet valve   217
   valve   218

vacuum generator   133
   partial   209
   pressure   5, 81
   range   207
valve actuation methods   10, 13
   amplifier   81, 52
   body   12
   classification   12
   cam roller   61
   conversion   20
   cross over condition   20
   diverter   201
   electrical   13
   fluidic back pressure   78
   hydraulic   13
   manual   13
   mechanical   13, 78
   mechanism   12
   memory   90, 92, 95, 124, 150
   mounting   88, 243
   one way trip   78, 95
   plunger   78
   pneumatic   13

port labelling   12
ports   12
pressure regulator (reducer)   193, 195
relief   219
roller   22, 78
roller lever   61
safety air relief   221
servo   218
solenoid   13, 168
symbols   10
switching positions   10
unloading   218
whisker   78
valves   193, 240
   check   61, 201
   cycle selection   134
   directional control   10, 12
   flat slide valve   16
   flow control   9, 60
   friction free (toggle disc)   18
   impulse   30, 34, 120
   inlet   210
   limit   240
   normally closed   160
      open   160
   on-off   10
   outlet   210
   poppet   16
   position   78
   position sensing   22–25
   power   10, 27, 163, 198
   pressure sequence   34, 242
   quick exhaust   61, 243
   relief   219
   selector   10
   shunt   10
   shuttle   26
   solenoid   13, 168
   spool   16
   spool/poppet   18
   three position   10

time delay   29, 242
time & pressure control   240
twin pressure   20
two position   10, 90
   pressure   26
   twin pressure   20
vane air motor   69, 70, 71, 72
vane areas   70
   number and position   72
vanes   70
   directional   195
   movable   213
variable restrictor   60
   speed control   217, 219
venturi action   199
   nozzle   201
   principle   83
   tube   199
   vacuum generator   133
verifying the program   192
vessel drying   230
vibration   243
volume   7
volume end cavity   58
   reduction   209, 210
   storage   237
volute   216

waste liquid discharge   231
water   4, 193, 242
   condensate   234, 238
   cooled   210
   level   194
   vapour   4, 224, 230
ways   10
weighing machines   81
weight detector station   150
woodpecker motion control   103
work savings   212

yes function valve   18, 20, 96, 149